TRANSACTIONS

OF THE

STATE AGRICULTURAL SOCIETY,

WITH REPORTS OF

COUNTY AGRICULTURAL SOCIETIES,

FOR 1851.

PUBLISHED BY ORDER OF THE LEGISLATURE.

VOL. III.

J. C. HOLMES,

SECRETARY OF THE MICHIGAN STATE AGRICULTURAL SOCIETY.

LANSING:
INGALS, HEDGES & CO., PRINTERS TO THE STATE.

......

1852.

CONTENTS.

ILLUSTRATIONS.

OFFICERS FOR 1852.

PRESIDENT.

JAMES B. HUNT, Pontiac.

EXECUTIVE COMMITTEE.

GEORGE C. MUNROE, Jonesville,
GROVE SPENCER, Ypsilanti,
WALTER WRIGHT, Adrian,
IRA PHILLIPS, Armada,
JEREMIAH BROWN, Battle Creek,
ANDREW Y. MOORE, Schoolcraft,
W. H. MONTGOMERY, Monroe,
TITUS DORT, Dearborn,
M. SHOEMAKER, Jackson,
A. N. HART, Lapeer.

TREASURER.

H. H. BROWN, Detroit.

SECRETARY.

J. C. HOLMES, Detroit.

ANNUAL REPORT

Of the Michigan State Agricultural Society for 1851.

To the President of the Senate:

I herewith present the third annual report of the Michigan State Agricultural Society, together with the reports of its auxiliaries, the county agricultural societies, as far as I have been able to procure them. Several county societies have been organized during the past year; but as they have neglected to make returns of their proceedings, I can make no definite statement respecting them. We are happy in being able to state that, since the organization of the State Society, the attention of our farmers has been called more particularly, than formerly, to the necessity of procuring the best breeds of horses, neat cattle, sheep and swine.

Thinking it to be of vast importance to the prosperity of our State, that the raising of stock should occupy a prominent place in our agricultural pursuits, and that we should fully understand what are the best breeds to introduce into the State, I have procured statements from several breeders of stock, respecting the valuable points attributed to the particular variety raised by them; these statements will be found in the following pages.

The contrast so very apparent between the stock exhibited at our first and third fairs, in favor of the latter, speaks volumes in praise of the spirit manifested by our farmers in their endeavors to procure the very best breeds of the different kinds of stock.

We find that from the fickleness of the seasons, or imperfect tillage, or perhaps from the two combined, together with the fluctuations of the market, that it will not answer for us to place so much dependence as we have heretofore done, upon our wheat crop, therefore the raising of sheep, pork and neat cattle is attracting the attention of agriculturists throughout the State.

We hope to see the dairy and the orchard receive a greater share of attention than they now receive from our enterprising farmers, for we think that either of these branches may, if well attended to, become as great, if not greater sources of profit, than the raising of wheat. We would by no means recommend the abandonment of the wheat crop, but recommend to the farmer not to trust entirely, as is now too often the case, to his wheat crop; for if it should prove a failure, the years work is lost; whereas, if he has his flock of sheep, his herds of cattle, or his dairy or orchard to turn to, he is not left entirely desolate.

Wool growing has become a very successful and prominent feature in the products of our State, and several of our enterprising farmers have at great expense, sought for, and we think, obtained some of the very best animals for the improvement of our sheep, that could be found. They will undoubtedly receive their reward for their praiseworthy endeavors to raise the flocks of Michigan sheep to the highest standard of excellence.

With these brief remarks, I direct your attention to the following pages.

J. C. HOLMES,
Sec'y Mich. State Agricultural Society.

Detroit, March, 1852.

OFFICERS

Of the Michigan State Agricultural Society, Elected September 27th, 1850.

The annual meeting of the Michigan State Agricultural Society was held at the committee room, on the show ground at Ann Arbor, September 27th, 1850—the President, Gov. E. Ransom in the chair.

The following officers for the ensuing year, were elected:

For President—Hon. James B. Hunt, of Pontiac, Oakland Co.

" Recording Secretary—J. C. Holmes, of Detroit.

" Treasurer—Hon. John J. Adam, of Clinton, Lenawee Co.

EXECUTIVE COMMITTEE.

Name.	*County.*	*Town.*
Titus Dort,	Wayne,	Dearborn.
Saml. M. Bartlett,	Monroe,	Lasalle.
Michael Shoemaker,	Jackson,	Jackson,
Payne K. Leech,	Macomb,	Utica.
Jereh. Brown,	Calhoun,	Battle Creek.
Walter Wright,	Lenawee,	Adrian.
John Hamilton,	Genesee,	Flint.
Grove Spencer,	Washtenaw,	Ypsilanti.
F. V. Smith,	Branch,	Coldwater.
Wm. H. Edgar,	Kalamazoo,	Kalamazoo.

VICE PRESIDENTS.

Name.	*County.*	*Town.*
Oka Town,	Allegan,	Otsego.
John Bowne,	Barry,	Hickory Corners.
L. H. Merrick,	Berrien,	Niles.
Harvey Haines,	Branch,	Coldwater.
Zenos Tillotson,	Calhoun,	Marshall.

Name.	*County.*	*Town.*
E. S. Smith,	Cass,	Cassopolis.
Saml. Ashman,	Chippewa,	Saut St. Mary.
Morris S. Allen,	Clinton,	Dewitt.
Wm. Hammond,	Eaton,	Charlotte.
Caleb S. Thompson,	Genesee,	Grand Blanc.
Lewis T. Miller,	Hillsdale,	Moscow.
E. B. Danforth,	Ingham,	Lansing.
George B. Kellogg,	Ionia,	Matthewton.
Edward Belknap,	Jackson,	Portage Lake.
Andrew Y. Moore,	Kalamazoo,	Schoolcraft.
George J. Barker,	Kent,	Grand Rapids.
Harvey Gray,	Lapeer,	Lapeer.
Amos Hoag,	Lenawee,	Adrian.
George Galloway,	v[illegible]ngston,	Hamburg.
Charles O'Malley,	Mackinaw,	Mackinaw.
Ira Phillips,	Macomb,	Romeo.
Jackson Kinne,	Monroe,	Erie.
Linus Cone,	Oakland,	Troy.
Timothy G. Eastman,	Ottawa,	Polkton.
Norman Little,	Saginaw,	Saginaw.
John Clark,	St. Clair,	St. Clair.
Joseph R. Williams,	St. Joseph,	Constantine.
Isaac Gale,	Shiawassee,	Hartwellville.
F. S. Finley,	Washtenaw,	Ann Arbor.
Jona. Shearer,	Wayne,	Plymouth.
John McKinney,	Van Buren,	Paw Paw.

CORRESPONDING SECRETARIES.

Name.	*County.*	*Town.*
O. C. Comstock, Jr.,	Calhoun,	Marshall.
Justus Gage,	Cass,	Dowagiac.
Shel. McKnight,	Chippewa,	Saut St. Mary.
J. Foot Turner,	Clinton,	Dewitt.
John D. Burns,	Eaton,	Charlotte.
E. P. Gale,	Genesee,	Davidsonville.
Wm. T. Howell,	Hillsdale,	Hillsdale.

Name.	*County.*	*Town.*
Whitney Jones,	Ingham,	Lansing.
Cyrus Lovell,	Ionia,	Ionia.
James Depue,	Jackson,	Spring Arbor.
N. A. Balch,	Kalamazoo,	Kalamazoo.
Henry Seymour,	Kent,	Grand Rapids.
Lorenzo Morse,	Lapeer,	Farmers Creek.
A. G. Eastman,	Lenawee,	Adrian.
N. J. Isbell,	Livingston,	Howell.
Chas. E. Avery,	Mackinaw,	Mackinaw.
J. H. Butterfield,	Macomb,	Utica.
E. G. Morton,	Monroe,	Monroe.
Andrew C. Walker,	Oakland,	Farmington.
Silas G. Harris,	Ottawa,	Grand Haven.
Hiram L. Miller,	Saginaw,	Saginaw.
Chas. A. Loomis,	St. Clair,	St. Clair.
Lewis A. Leland,	St. Joseph,	Colon.
John B. Bloss,	Shiawassee,	Shiawassee.
Fitz H. Stevens,	Van Buren,	Paw Paw.
Mark Norris,	Washtenaw,	Ypsilanti.
Phineas Homan,	Wayne,	Detroit.

EXECUTIVE MEETING.

The Executive Committee held its third annual meeting at Detroit, December 11th, 1850.

Present, the President, Hon. James B. Hunt, of Oakland county;

Titus Dort, of Wayne county;

Samuel M. Bartlett, of Monroe county;

Michael Shoemaker, of Jackson county;

Payne K. Leech, of Macomb county;

Jereh. Brown, of Calhoun county;

Walter Wright, of Lenawee county,

John Hamilton, of Genesee county;

Grove Spencer, of Washtenaw county;

F. V. Smith, of Branch county;

Also, F. S. Finley, Esq., Vice President of this society for the county of Washtenaw;

Prof. J. Holmes Agnew, Secretary of the Washtenaw Co. Agricultural Society; Linus Cone, Esq., delegate from the Oakland Co. Agricultural Society.

The meeting was called to order by the President, at 2 o'clock P. M. The record of the proceedings of the annual meeting of the society held at the village of Ann Arbor, Sept. 25th, 26th and 27th, were read and approved. The report of the Secretary and the Treasurer were read and referred to appropriate committees.

On motion of J. Holmes Agnew,

Resolved, That a committee be appointed to report a time and place for holding the society's third annual fair.

Messrs. Agnew, Bartlett and Brown were appointed said committee.

On motion of Mr. Bartlett,

Resolved, That an appropriation of two thousand dollars be made for the payment of premiums to be awarded at the society's third annual fair.

On motion of Mr. Spencer,

Resolved, That a committee of seven be appointed to report rules and regulations to be observed at the third annual fair.

Messrs. Spencer, Holmes, Finley, Cone, Hamilton, Brown and Hunt were appointed said committee.

On motion of Mr. Brown,

Resolved, That a committee of five be appointed to report a list of judges to serve at the third annual fair.

Messrs. Brown, Dort, Wright, Hamilton and Leech were appointed said committee.

On motion of Mr. Agnew,

Resolved, That a committee of three be appointed to draw up a petition to be presented to the Legislature for an appropriation of the sum of six hundred dollars annually for the next three years for the use of the society. Also to print for the use of the society, at least two thousand copies of the society's transactions for 1850.

On motion of Mr. Bartlett,

Resolved, That a committee of five be appointed to draft a premium list for the ensuing year.

Messrs. Bartlett, Spencer, Cone, Shoemaker and Smith were appointed said committee.

Mr. Agnew, from the committee on time and place for holding the third annual fair, made the following report:

The committee on the time and place for holding the third annual fair, recommend that said fair be held on the third Wednesday, Thursday and Friday, 17th, 18th and 19th days of September, 1851, and that Detroit be the place, on condition that its citizens raise the sum of fifteen hundred dollars within four months from the date hereof, to defray the local expenses of obtaining and fitting up the grounds under the direction of the committee of management. In case Detroit declines, then the village of Jackson shall next have the privilege on condition of raising the sum of $1,500 as above; and in case of Jackson's failure, then either of the towns west on the line of the Central railroad shall have the same privilege on the same terms.

The above report was unanimously adopted.

[NOTE.—In consequence of the State agricultural societies of New York and Ohio selecting the 17th, 18th and 19th days of September for holding their fairs, the time for holding our fair was subsequently postponed to the 24th, 25th and 26th September.]

The committee appointed to examine the Treasurer's report, reported that they had examined said report and vouchers and found them correct.

On motion of Mr. Shoemaker,

Resolved, That Gen. Lewis Cass be invited to deliver the annual address before the society at its third annual fair.

The committees appointed to report rules and regulations, a premium list, and list of Judges for the fair of 1851, made the following reports, which were adopted:

RULES AND REGULATIONS,

To be observed at the Third Annnal Fair of the Michigan State Agricultural Society, to be held at Detroit, on Wednesday, Thursday and Friday, September 24th, 25th and 26th, 1851.

The committee on the reception of strangers and members of committees, have their head-quarters at the Michigan Exchange, at Detroit, Tuesday evening the 23d, and until 10 o'clock A. M. of Wednesday, of the 24th September. After that hour, and during the exhibition, at the office, on the ground.

A Register will be opened at the room of the committee, where strangers and members of commitees will please register their names immediately upon their arrival.

Members of the execntive and viewing committees are requested to report to the secretary, at the committce's room, at the Michigan Exchange, on Tuesday, the 23d, at 7 o'clock P. M., in order that all vacancies may be filled at a meeting of the Executive Committee, to be holden on the 24th, at 10 o'clock A. M., at the society's tent, on the show ground.

It is of the utmost importance that committees should be at their post at the hour appointed for their meeting.

Any person may become a member of the Society, for one year, by paying one dollar into the treasury, and then may continue a member by paying 50 cents per annum.

Members of the Society, and all who may become such at the time of the Fair, will be furnished with tickets, which will admit the person and his wife, and children under eighteen years of age, to the exhibition, at all times during the continuance of the show.

Single tickets 12½ cents, admitting one person, will be for sale at the besiness office, on Thursday morning, 25th.

Exhibitors, who intend to compete for premiums, must become members of the Society.

Persons wishing to enter their farms for premium, will give notice to either member of the committee on farms for their county.

In order to save time and confusion, exhibitors are requested to enter and arrange their articles, as far as possible on Tuesday, the 23d, so that all may be in readiness for examination by the Judges on Wednesday, the 24th, at 2 o'clock P. M.

Persons employed for the purpose, and wearing appropriate badges, will be in readiness on the ground to receive the articles intended for exhibition in their several departments.

Exhibitors will be careful to have their animals and articles arranged in their appropriate places, and in season, otherwise they will be overlooked by the viewing committees.

All articles intended for exhibition must be entered at the business office, at the entrance of the show ground, before entering the enclosure. Exhibitors of stock should be very careful to enter their animals in their appropriate class: any grade animal entered as a full blood, will be excluded from competition in the class to which it belongs.

All stock must be accompanied with a concise written statement of the pedigree, feeding, and other circumstances in relation to the character and condition of the animals. Horses will be entered under the head of Stallions. Matched horses, single horses, or breed mares and colts under four years of age, according to the fact. A short pedigree in writing will be required. For fat cattle, a short statement of the kind of food, manner and length of time of feeding, &c., will be required.

Cards will be furnished with the number as entered at the office. Exhibitors will be careful to place these cards upon or near the articles or animals, so that the judges will have no difficulty in finding them. No animals or articles entered for exhibition can be taken from the ground before the close of the fair, except by permission of a member of the Executive Committee; and no premium will be paid on animals or articles removed in violation of this rule.

—

INSTRUCTIONS TO JUDGES.

The judges on animals will have regard to the symmetry, early maturity, size, and general characteristics of the breeds which they judge. They will make proper allowances for age, feeding, and other circumstances on the character and condition of the animals. They are expressly required not to give encouragement to over fed animals. No premiums to be awarded to bulls, cows or heifers, which shall appear to have been fattened for the butcher; the object being to have superior animals of this description for breeding.

No persons whatever will be allowed to interfere with the judges during their adjudications. The judges on stock, if not satisfied as to regularity of the entries in their respective classes, will apply to the Secretary for information; and should there be any doubt, after examination of their coming within the regulations, or if any animal is of such a character as not to be entitled to exhibition in competition, they will report the facts to the Executive Committee, that such course may be adopted as the case may require.

FAT CATTLE.

The judges on fat cattle will give particular attention to the animals submitted to them for examination. It is believed that, all other things being equal, those are the best cattle that have the greatest weight in the smallest superfices. The cattle exhibited in this class will be weighed, and the judges will take measures to give the superfices of each, and publish the result with their reports.

GENERAL RULES.

When there is but one exhibitor, although he may show several animals in a class, only one premium will be awarded; that to be first or otherwise, as the merits of the animal or article may be adjudged. And a premium will not be awarded when the animal or article is not worthy, though there be no competition.

No viewing committee, with the exception of the committee on miscellaneous premiums, shall award any discretionary premiums, without the previous permission of the executive committee. Whenever, however, articles of merit, superior in their character are presented, and which are entitled to special commendation, the judges are desired to notice them particularly, and refer them to the consideration of the executive committee. No animal or article can take more than one premium, except as specially provided.

As one great object of the Society is to collect valuable information upon subjects connected with agriculture, the several viewing committees are requested to gather all the information possible from exhibitors in their several departments, and embody in their reports all valuable information thus elicited, and make their reports as full as time and circumstances will permit.

Judges are requested to make their reports to the Secretary, at the business office, by 1 o'clock P. M., Thursday, 25th.

Stock, to compete for premiums, must be exhibited by the owner or his agent.

Domestic manufactures, needle, shell-work, &c., must have been manufactured in the State, and within the year, except such articles as have not been heretofore exhibited, and such articles, together with fruits, flowers, vegetables, &c., must be the production of the exhibitor, in order to entitle the competitor to a premium.

All animals and all articles brought out of the State for exhibition, shall be entered and marked as foreign, and shall not come in competition with animals or articles owned in the State; but animals or articles of the State may come in competition with said foreign animals or articles, although they may be entered for competition in the State. When there is no foreign stock to compete with, no premium will be awarded in that class.

Prize animals, articles and implements at the previous exhibitions, will be allowed to compete for the prizes; but they must receive a higher prize, or in a different class, to entitle them to a premium. Should the same premium heretofore given them be awarded, they will receive a certificate to that effect, instead of the prize.

Persons exhibiting several articles, will, in making their entries, have written lists of their articles, with the name of the exhibitor, and place of residence attached, to hand to the book-keepers at the business office.

Exhibitors of stock, will, in making out their lists, give all the information possible concerning their animals. They will also be careful to enter their stock in the class in which they belong.

In the horticultural department, it is particularly desired that complete and correct memorandums should be rendered.

—

ORDER OF PROCEEDINGS.

On Wednesday, the 24th of September, the Executive Committee will hold a meeting at the Society's tent, on the show ground, at 10 o'clock A. M., for the purpose of filling any vacancies that may occur in the viewing committees.

All members of viewing committees, who may be on the ground, will be particular to attend this meeting.

The viewing committees will receive their committee books from the Secretary at 2 o'clock P. M. of the 24th, and commence their examinations immediately thereafter.

PLOWING MATCH.

The plowing match will take place at half-past 9 o'clock A. M., on Thursday 25th.

ANNUAL ADDRESS.

The annual address will be delivered by the Hon. Lewis Cass, on the show ground, at 2 o'clock P. M. of the 25th.

REPORTS.

Immediately after the address, the reports of the judges will be read. The judges will be expected to give the reasons for their decisions, embracing the valuable and superior qualities of the animals or articles to which premiums are awarded.

PREMIUMS.

Premiums will be forwarded to the persons to whom they are awarded, at as early a day after the fair as practicable.

ANNUAL MEETING.

The annual meeting of the society for the choice of officers, and for the transaction of such other business as may come before the meeting, will take place at the society's tent, at 10 o'clock A. M. of the 26th.

SALE OF STOCK AND IMPLEMENTS.

A public sale of stock and implements will commence at 10 o'clock A. M. of the 26th.

RULES FOR THE PLOWING MATCH.

Persons entering for the plowing match will observe the following rules and regulations:

1st. Each team will be required to plow one quarter of an acre. In order to insure good work, one hour and ten minutes will be allowed in which to perform it.

2d. The team may consist of one or two pair of horses or oxen, provided they are managed by the plowman.

3d. The plow may be held by the competitor, or such person as he may select; but the name of the plowman in all cases to be given.

4th. Each plowman will be required to mark out his land by plowing at least four back furrows, from a strip left for that purpose, before commencing on time.

5. The furrow must be at least seven inches deep, and not over ten inches wide.

The excellence of the work to consist, 1st. In leaving the furrow slice light and friable. 2d. In so disposing the sod and vegetable matter as to insure their ready decomposition.

All jointed and sub-soil plows to be excluded from competition with other plows, but it is desired that they should be entered, that their merits may be tested and premiums awarded them if merited.

LIST OF PREMIUMS

For the Annual Fair, to be held at Detroit, on Wednesday, Thursday and Friday, September 24th, 25th and 26th, 1851.

CATTLE.

CLASS I—SHORT HORNS.

Judges—B. Pierson, Flint, Genesee County; C. M. Giddings, Cleveland, Ohio; Epaph. Ransom, Kalamazoo.

Best bull, 5 years old or over, medal and	$8 00
2d do do do	7 00
3d do do do	5 00
Best bull 3 years old, and under 5, medal and	8 00
2d do do do	7 00
3d do do do	5 00
Best bull 2 years old,	8 00
2d do do	5 00
3d do do	3 00
Best bull, 1 year old,	5 00
2d do do	3 00
3d do do	2 00
Best bull calf,	3 00
2d do	2 00
3d do Transactions for 1850.	
Best cow, 5 years old or over, medal and	8 00
2d do do do	7 00
3d do do do	5 00
Best cow, 3 years old, and under 5,	8 00
2d do do do	7 00
3d do do do	5 00

Best 2 years old heifer, ... $5 00
2d do do 3 00
3d do do 2 00
Best 1 year old heifer, ... 3 00
2d do do 2 00
3d do do Allen on Domestic Animals.
Best heifer calf, ... 3 00
2d do 2 00
3d do Transactions.

CLASS II—DEVONS.

Judges—James B. Wells, Franklin, Oakland county; Wm. B. Canfield, Mt. Clemens, Macomb county; Peter Latshaw, Erie, Monroe county.

Best bull, 5 years old or over, medal and $8 00
2d do do do 7 00
3d do do do 5 00
Best bull 3 years old and under 5, medal and 8 00
2d do do do 7 00
3d do do do 5 00
Best 2 years old bull, ... 8 00
2d do do 5 00
3d do do 3 00
Best 1 year old bull, ... 5 00
2d do do 3 00
3d do do 2 00
Best bull calf, ... 3 00
2d do 2 00
3d do Transactions for 1850.
Best cow, 5 years old or over, medal and 8 00
2d do do do 7 00
3d do do do 5 00
Best cow, 3 years old, and under 5, ... 8 00
2d do do do 7 00
3d do do do 5 00
Best heifer, 2 years old, ... 5 00
2d do do 3 00
do do 2 00

Best heifer, 1 year old,				$3 00
2d	do	do		2 00
3d	do	do Allen on Domestic Animals.		
Best heifer calf,				3 00
2d	do			2 00
3d	do Transactions for 1850.			

CLASS III—HEREFORDS.

Judges—Same as short horns.

Best bull, 5 years old or over, medal and				8 00
2d	do	do	do	7 00
3d	do	do	do	5 00
Best bull, 3 years old, and under 5,				8 00
2d	do	do	do	7 00
3d	do	do	do	5 00
Best bull 2 years old,				8 00
2d	do			5 00
3d	do			3 00
Best bull, 1 year old,				5 00
2d	do			3 00
3d	do			2 00
Best bull calf,				3 00
2d	do			2 00
3d	do Transactions.			
Best cow, 5 years old or over, medal and				8 00
2d	do	do	do	7 00
3d	do	do	do	5 00
Best cow, 3 years old, and under 5,				8 00
2d	do	do	do	7 00
3d	do	do	do	5 00
Best heifer, 2 years old,				5 00
2d	do	do		3 00
3d	do	do		2 00
Best heifer, 1 year old,				3 00
2d	do	do		2 00
3d	do	do Allen on Domestic Animals.		

Best heifer calf, ... $3 00
2d do 2 00
3d do Transactions.

CLASS IV—AYRSHIRES.

Judges—Same as for Devons.

Best bull, 5 years old or over, medal and 8 00
2d do do do 7 00
3d do do do 5 00
Best bull, 3 years old, and under 5, ... 8 00
2d do do do 7 00
3d do do do 5 00
Best bull 2 ys. old, ... 8 00
2d do 5 00
3d do 3 00
Best bull 1 yr. old, ... 5 00
2d do 3 00
3d do 2 00
Best bull calf, ... 3 00
2d do 2 00
3d do Transactions.
Best cow, 5 years old or over, medal and 8 00
2d do do do 7 00
3d do do do 5 00
Best cow, 3 years old, and under 5, ... 8 00
2d do do do 7 00
3d do do do 5 00
Best heifer 2 ys. old, ... 5 00
2d do 3 00
3d do 2 00
Best heifer, 1 yr. old, ... 3 00
2d do 2 00
3d do Allen on Domestic Animals.
Best heifer calf, ... 3 00
2d do 2 00
3d do Transactions for 1850.

CLASS V—ALL CATTLE NOT INCLUDBD IN THE PRECEDING CLASSES.

Judges—H. Lothrop, Schoolcraft; Chester Yost, Ypsilanti; John Montgomery, Eaton Rapids.

Best bull, 5 years old or over, medal and				$8 00
2d	do	do	do	7 00
3d	do	do	do	5 00
Best bull 3 years old and under 5,				8 00
2d	do	do	do	7 00
3d	do	do	do	5 00
Best bull 2 years old,				8 00
2d	do	do		5 00
3d	do	do		3 00
Best bull 1 year old,				5 00
2d	do	do		3 00
3d	do	do		2 00
Best bull calf,				3 00
2d	do			2 00
3d	do Transactions for 1850.			
Best cow, 5 years old or over, medal and				8 00
2d	do	do	do	7 00
3d	do	do	do	5 00
Best cow 3 years old and under 5,				8 00
2d	do	do	do	7 00
3d	do	do	do	5 00
Best heifer 2 years old,				5 00
2d	do	do		3 00
3d	do	do		2 00
Best heifer 1 year old,				5 00
2d	do	do		3 00
3d	do	do Allen on Domestic Animals.		
Best heifer calf,				3 00
2d	do			2 00
3d	do Transactions for 1850.			

CLASS VI—WORKING OXEN.

Judges on Steers and Working Oxen—Ashael Brown, Coldwater; Isaac I. Voorhies, Pontiac; H. S. Holcomb, Jackson.

Best yoke working oxen, 4 years old or over, medal and				$10 00
2d	do	do	do	10 00
3d	do	do	do	5 00

Best 10 yoke working oxen, from one county,					$20 00
2d	do	do	do	do	10 00

CLASS VII.—STEERS.

Judges—Same as Working Oxen.

Best yoke 3 years old,			$10 00
2d	do	do Buel's Farmer's Companion, and	5 00
3d	do	do Transactions, and	2 00
Best yoke 2 years old,			6 00
2d	do	do	4 00
3d	do	do Transactions, and	2 00
Best yoke 1 year old,			5 00
2d	do	do	3 00
3d	do	do Transactions.	

CLASS VIII—FAT CATTLE.

Judges—Same as on Fat Cattle and Milch Cows.

Best pair fat oxen, medal and			$10 00
2d	do	do	10 00
3d	do	do	5 00
Best fat ox,			5 00
2d	do		3 00
3d	do Transactions.		
Best fat cow,			5 00
2d	do		3 00
3d	do Transactions.		
Best fat steer, 3 years old,			5 00
2d	do	do	3 00
3d	do	do Transactions.	
Best fat heifer, 3 years old,			5 00
2d	do	do	3 00
3d	do	do Transactions.	

Applicants for premiums on fat cattle must furnish particular statements of the manner of feeding, kind, quantity and cost of food, and all the expenses connected with the fattening; otherwise they will be excluded. Animals exhibited in pairs cannot compete for single premiums.

CLASS IX—FAT CATTLE FED ON HAY AND GRASS ALONE AFTER ONE YEAR OLD.

Judges on Fat Cattle and Milch Cows—Charles Fitzhugh, Saginaw; George Clark, Lapeer; Asahel Beach, Battle Creek.

Best pair fat oxen, medal and		$10 00
2d do		10 00
3d do		5 00
Best single fat ox,		5 00
2d do		3 00
3d do	Transactions for 1850.	
Best fat cow,		5 00
2d do		3 00
3d do	Transactions.	
Best fat steer, 3 years old,		5 00
2d do do		3 00
3d do do	Transaction.	
Best fat heifer, 3 years old,		5 00
2d do do		3 00
3d do do	Transactions.	

Exhibitors of fat cattle offered as grass fed, must have with them the affidavit of the breeder that they have been fed on grass and hay alone, since one year old; otherwise they will be excluded.

CLASS X—MILCH COWS.

The cow to be kept on grass only, during the experiment, and for fifteen days previous to each period of trial.

The time of trial, from tenth to twentieth of June, and from twentieth to thirtieth of August.

Statement to be furnished, containing: 1st, the age and breed of cow, and time of calving; 2d, the quantity of milk, in weight, also of butter, during each period of ten days; 3d, the butter made to be exhibited with cow, at the fair, in Detroit, and the statement to be verified by the affidavit of the competitor and one other person conversant with the facts.

Judges—Same as on Fat Cattle.

Best milch cow, medal and	$8 00
2d do Allen on Domestic Animals, and	8 00
3d do	5 00

CLASS XI—FOREIGN CATTLE.

Judges—Edward Belknap, Henrietta, Jackson county; Charles E. Stuart, Kalamazoo; A. S. Brooks, Novi, Oakland county.

Best short horn bull, 2 years old and over, diploma and....$8 00
do do heifer or cow, do diploma and............8 00
do Devon bull, 2 years old and over, diploma and..........8 00
do " heifer or cow, do " 8 00
do Hereford bull, diploma and..........................8 00
do " heifer or cow, 2 years old and over, diploma and..8 00
do Ayrshire bull, 2 years old and over, diploma and.......8 00

HORSES.

CLASS I.—FOR ALL WORK.

Judges—George A. Monroe, Jonesville; Alfred J. Boss, Pontiac; Daniel B. Hibbard, Jackson.

Best stallion, 4 years old or over, medal and...............$8 00
2d " do 8 00
3d " do Youatt on horse, and........3 00
Best brood mare, do with foal at her foot, medal and 8 00
2d do do do 8 00
3d do do do Youatt on Horse, and 3 00
Best stallion 3 years old,.................................7 00
2d do do 5 00
3d do do 3 00
Best mare 4 years old,.....................................5 00
2d do 3 00
3d do Youatt on the Horse, and Transactions.
Best stallion 2 years old,.................................3 00
2d do Youatt on the Horse.
3d do American Veterinarian.
Best mare 2 years old,.....................................3 00
2d do Youatt on the Horse.
3d do American Veterinarian.
Best stallion 1 year old,..................................3 00
2d do Transactions and American Veterinarian.
3d do Transactions.

Best mare 1 year old, ...$3 00
2d do Transactions and American Veterinarian.
3d do Transactions.

CLASS II.—DRAUGHT HORSES.

Judges—On draught, matched and single horses; Benjamin F. Fifield, Monroe; William H. Colman, Battle Creek; Harvey Seely, Pontiac.

Best stallion, 4 years old or over, medal and $8 00
2d " do 8 00
3d " do Youatt on the Horse, and...3 00
Best brood mare, 4 years old or over, with foal at her foot, medal and 8 00
2d brood mare, 4 years old or over, ...8 00
3d do do Youatt on the Horse and...3 00
Best stallion, 3 years old, ...7 00
2d do 5 00
3d do 3 00
Best mare 3 years old, ...5 00
2d do 3 00
3d do Youatt on the Horse, and Transactions.
Best stallion 2 years old, ...3 00
2d do Youatt on the Horse.
3d do Cole's Am. Veterinarian and Transactions.
Best mare 2 years old 3 00
2d do Youatt on the Horse.
3d do Diseases of Animals and Transactions.
Best stallion 1 year old, ...3 00
2d do Transactions and American Veterinarian.
3d do Transactions.
Best mare 1 year old, ...3 00
2d do Transactions and American Veterinarian.
3d do Transactions.

CLASS III.—BLOOD HORSES.

Judges—Isaac Wixom, Argentine, Genesee County; Eber Adams, Adrian, Lenawee County; George W. Howe, Detroit.

Best stallion 4 years old or over, medal and $8 00
2d do 8 00
3d do Youatt on the Horse and 3 00
Best brood mare, 4 years old or over, with foal at her foot, medal and 8 00
2d do with foal at her foot, ... 8 00
3d do with foal at her foot, Youatt on Horse and 3 00
Best stallion 3 years old, ... 7 00
2d do 5 00
3d do 3 00
Best mare 3 years old, ... 5 00
2d do 3 00
3d do Youatt on the Horse and Transactions.
Best stallion 2 years old, ... 3 00
2d do Youatt on the Horse.
3d do American Veterinarian and Transactions.
Best mare 2 years old, ... 3 00
2d do Youatt on the Horse.
3d do American Veterinarian and Transactions.
Best stallion 1 year old, ... 3 00
2d do American Veterinarian and Transactions.
3d do Transactions.
Best mare 1 year old, ... 3 00
2d do American Veterinarian and Transactions.
3d do Transactions.

CLASS IV—MATCHED AND SINGLE HORSES.

Judges—Same as class 2, Draught Horses.

Best pair matched horses, ... $10 00
2d do do 8 00
3d do do 5 00
4th do do 3 00
Best single horse, ... 5 00
2d do Youatt on the Horse, and 3 00
3d do 3 00
4th do 2 00

SHEEP.

CLASS I—LONG WOOLED.

Judges—On long wooled, middle wooled, and fat sheep; Andrew Y. Moore, Schoolcraft, Kalamazoo county; A. J. Kinney, Erie, Monroe county; C. A. Chipman, Rochester, Oakland county.

Best buck over 2 years old, medal and				$5 00
2d	do	do	do	5 00
3d	do	do	do	3 00
Best buck 2 y's. old, or under,				5 00
2d	do	do		3 00
3d	do	do		2 00
Best pen of 5 ewes, medal and				5 00
2d	do	do		5 00
3d	do	do	Am. Shepherd, and	3 00
Best pen of 5 buck lambs,				5 09
2d	do	do		3 00
3d	do	do	Transactions.	
Best pen of 5 ewe lambs,				5 00
2d	do	do		3 00
3d	do	do	Transactions.	

CLASS II—MIDDLE WOOLED.

Judges—Same as class 1.

Best buck over 2 years old, medal and				5 00
2d	do	do		5 00
3d	do	do		3 00
Best buck 2 years old or under,				5 00
2d	do	do		3 00
3d	do	do		2 00
Best pen of 5 ewes, medal and				5 00
2d	do	do		5 00
3d	do	do	Am. Shepherd, and	2 00
Best pen of 5 buck lambs,				5 00
2d	do	do		3 00
3d	do	do	Transactions.	
Best pen of 5 ewe lambs,				5 00
2d	do	do		3 00

3d best pen 5 ewe lambs, Transactions.

[This class includes Southdown, Norfolk, Dorset, &c.]

CLASS III—MERINOS.

Judges—On Merinos and Saxons, William O'Harra, Toledo, Ohio; Charles Dickey, Marshall, Calhoun county; Amos Hoag, Adrian; Edward Sawyer, Grand Blanc; George Redfield, Cassopolis.

Best buck over 2 years old, medal and $5 00
2d do do do 5 00
3d do do do 3 00
Best buck 2 y's old or under, ... 5 00
2d do do 3 00
3d do do 2 00
Best pen 5 ewes, medal and 5 00
2d do do 5 00
3d do do Am. Shepherd, and 2 00
Best pen 5 buck lambs, ... 5 00
2d do 2 00
3d do Transactions.
Best pen 5 ewe lambs, ... 5 00
2d do 3 00
3d do Transactions.

CLASS IV—SAXONS.

Judges—Same as Class 3.

Best buck over 2 years old, medal and $5 00
2d do do 5 00
3d do do 3 00
Best buck 2 years old or under, ... 5 00
2d do do 3 00
3d do do 2 00
Best pen 5 ewes, medal and 5 00
2d do 5 00
3d do American Shepherd and 2 00
Pest pen 5 buck lambs, ... 5 00
2d do do 3 00
3d do do Transactions.
Best pen 5 ewe lambs, ... 5 00
2d do do 3 00
3d do do Transactions.

CLASS V—NATIVES AND GRADES.

Judges—On Natives and Grades, and Shepherd's Dogs; John Starkweather, Ypsilanti; George Macomber, Mt. Clemens; H. K Farrand, St. Joseph.

Best buck over 2 years old, medal and			$5 00
2d	do	do	5 00
3d	do	do	3 00
Best buck 2 years old or under,			5 00
2d	do	do	3 00
3d	do	do	2 00
Best pen 5 ewes, medal and			5 00
2d	do		5 00
3d	do	Am. Shepherd and	2 00
Best pen 5 buck lambs,			5 00
2d	do	do	3 00
3d	do	do Transactions.	
Best pen 5 ewe lambs,			5 00
2d	do	do	3 00
3d	do	do Transactions.	

FAT SHEEP.

Judges—Same as Class 1.

Best fat sheep,			$3 00
2d	do		2 00
3d	do	American Shepherd.	

CLASS IV—FOREIGN SHEEP.

Judges—Jonathan Shearer, Plymouth; John B. Bloss, Shiawassee, Shiawassee county; George Willard, Ida, Monroe county.

LONG-WOOLED.

Best buck,	$5 00
do pen of 5 ewes,	5 00
do pen 5 buck lambs,	3 00
do do ewe lambs,	3 00

MIDDLE-WOOLED.

Best buck,	$5 00
do pen 5 buck lambs,	3 00
do do ewes,	5 00
do do ewe lambs,	3 00

MERINOES.

Best buck,..$5 00
do pen 5 buck lambs,..............................3 00
do do ewes,.......................................5 00
do do ewe lambs,..................................3 00

SAXONS.

Best buck,..$5 00
do pen 5 buck lambs,..............................3 00
do do ewes,.......................................5 00
do do ewe lambs,..................................3 00

SHEPHERD'S DOG.

Best shepherd's dog,..............................$5 00
2d do do American Shephard.

Evidence to be furnished of the thorough training of the dog; otherwise no premium can be awarded.

SWINE.

Judges—Jacob Summers, Utica, Macomb county; Wm. Ten Eyck, Dearborn; Alvarado Brown, Quincy, Branch county.

Best boar over two years old,.....................$5 00
2d do do 3 00
3d do do Transactions.
Best boar over 1 year old,........................5 00
2d do do 3 00
3d do do Transactions.
Best boar over 6 months and under one year,.......3 00
2d do do do do Transactions.
Best breeding sow over 2 years old,...............5 00
2d do do do 3 00
3d do do do Transactions.
Best breeding sow over 1 year old,................5 00
2d do do do 3 00
3d do do do Transactions.
Best sow 6 months old and under one year,.........3 00
2d do do do do Transactions.
Best lot of pigs not less than 5, and under 10 months old,....5 00
2d do do do do do 3 00
3d do do do Transactions.

POULTRY.

Judges—Melancthon Freeman, Kalamazoo; S. Gillett, Detroit; J. G. Cornell, Spring Arbor.

Best lot of Dorkings, not less than three, 1 cock and 2 hens,				$3 00
do	Polands,	do	do	3 09
do	large fowls,	do	do	3 00
do	turkeys,	do	do	3 00
do	ducks,	do	do	3 00
do	Guinea fowls,	do	do	3 00
do	geese,	do Cock's Amer. Poultry Book, and		1 00
do	poultry owned by exhibitor, statements to be furnished and verified, Bement's Amer. Poulterer's Comp'n and			4 00
Best exhibition of pigeons,				2 00

FARM IMPLEMENTS.

CLASS I.

Judges—On Farm Implements, Classes 1 and 2, R. T. Merrill, Birmingham, Oakland county; Edward Smith, Clinton; Hiram B. Mather, Niles.

Best farm wagon,	$5 00
do harrow,	3 00
do corn cultivator,	3 00
do fanning mill,	5 00
do corn stalk cutter,	5 00
do straw cutter,	3 00
do corn and cob crusher, by horse power,	5 00
do clover machine,	5 00
do flax and hemp dresser,	5 00
do horse cart for farm,	3 00
do ox cart,	3 00
do horse rake,	2 00
do ox yoke,	2 00
do roller for general use,	5 00
do clod crusher and roller combined,	5 00

CLASS II.

Best plow harness,	$2 00
do wagon do	2 00

Best carriage harness, ... $2 00
do harness for general purposes, ... 2 00
do riding saddle, ... 1 00
do dozen axes, ... 2 00
do churn 2 00
do cheese press, ... 2 00
do six milk pans, ... 2 00
do potato washer, ... 2 00
do grain cradle, ... 2 00
do six hand rakes, ... 2 00
do do hay forks, ... 2 00
do do manure forks, ... 2 00
do do grain or cradle scythes, ... 2 00
do do grass scythes, ... 2 00
do hay rigging, ... 2 00
do lot of grain measures, ... 2 00
Best dozen wire tied brooms, ... 2 00
2d do do do 1 00
Best do twine do 2 00
2d do do do 1 00

CLASS III.

Judges—Class 3 and 4: P. R. Adams, Tecumseh; Isaac Lewis, Monroe; A. A. Wilder, Detroit; Asa Parish, Coldwater; J. E. Beebe, Jackson.

Best horse power for general purposes, on the sweep level principle, diploma and $5 00
do horse power for general purposes, on the railroad or endless chain principle, diploma and 5 00
do one horse power, ... 5 00
do thresher to be used with steam or horse power, ... 5 00
do seed planter, for hand or horse power, for hills and drills, diploma and 3 00
do wheat drill, not less than six drills, ... 3 00
do grain drill, with apparatus for depositing manure, ... 3 00
do cultivator and drill combined, ... 3 00
do broad cast sower, diploma.
do wheat cultivator, diploma and 2 00

Best portable saw-mill for wood, fences, and for farm use, diploma.
do corn sheller, horse power, diploma and 2 00
do corn sheller, hand power, diploma and 1 00
do vegetable cutter, diploma.

Best and most numerous collection of Agricultural Implements manufactured in this State, by, or under the supervision of the exhibitor; materials, workmanship, utility, durability and prices to be considered.

CLASS IV.—MACHINERY AND IMPLEMENTS.

For the best and most useful machine or implement for the farmer, either newly invented or an improvement on any now in use, medal and $10 00

Medals or diplomas will be awarded for articles of mechanical ingenuity and machinery deemed useful.

PLOWS.

Judges—On plows and plowing, Linus Cone, Troy, Oakland Co.; Jonathan Dayton, Grand Blanc, Genesee Co.; Wm. Dougherty, Berrien Springs, Berrien Co.

Best sod plow for stiff soils, furrow not less than 7 inches in depth, nor over 10 inches in width, diploma and $5 00
2d best do do 5 00
Best sod plow for light soils, furrow 6 & 12 inch., diploma and 5 00
2d do do 5 00
Best do for fallows or old land, diploma and 5 00
2d do do 5 00
Best subsoil plow, diploma and 5 00

PLOWING MATCH—WITH HORSES.

First premium, ... $10 00
2d do Gardner's Farmer's Dictionary and 7 00
3d do 5 00

WITH OXEN—SINGLE TEAMS.

First premium, ... $10 00
2d do Gardner's Farmer's Dictionary and 7 00
3d do 5 00

BOYS UNDER 18 YEARS OF AGE, WITH HORSES OR OXEN.

First premium, medal and 3 00
2d do 3 00
3d do 2 00

BUTTER.

Judges—On butter, cheese, sugar, honey and bee hives, Austin Wales, Detroit; Mrs. Isaac I. Voorhies, Pontiac; Mrs. H. E. De-Garmo, Ann Arbor; Mrs. J. Shearer, Plymouth; Mrs. K. S. Bingham, Green Oak, Livingston Co.

Best lot of butter (quality as well as quantity considered) made from five cows, in thirty consecutive days, 15 lbs. of the butter to be exhibited,............................$7 00
2d best do do 5 00
3d do do do Webster's Encyclopedia of Domestic Economy.
4th do do do 2 00
Best ten lbs. of butter made in June,......................3 00
2d do do Transactions for 1850 and 1 00
3d do do 2 00
Best fifteen lbs. of butter made any time,..................3 00
2d do do Transactions and....1 00
3d do do 2 00
4th do do 8th vol. Mich. Farmer.

The exhibitors must state in writing, the time when the butter was made, the number of cows kept on the farm, the mode of keeping, the treatment of the cream and milk before churning, winter and summer, the method of freeing the butter from the milk, the quantity and kind of salt used, and whether saltpetre or other substance has been employed.

CHEESE.

Best cheese one year old and over, not less than 25 lbs.,....$5 00
2d do do do Webster's Encyclopedia of Domestic Economy.
3d best cheese one year old and over, not less than 25 lbs.....2 00

A statement of the manner of making the cheese must accompany each sample.

SUGAR.

Best ten lbs maple sugar,..............................$5 00
2d do do 3 00
3d do do 8th vol. Michigan Farmer.

HONEY AND BEE-HIVES.

Best 10 lbs. honey,....................................$3 00
2d do 2 00
3d do 8th volume Michigan Farmer.
Best bee hive with description,..........................3 00

The honey must be taken up without destroying the bees; the kind of hive to be specified.

DOMESTIC MANUFACTTRES.

CLASS I.

Judges—J. R. Kellogg, Allegan; H. N. Munson, St. Clair; Jos. Rhodes, Adrian.

Best pair woolen blankets, diploma and..................$4 00
2d do do Transactions and..................2 00
Best 10 yards flannel, diploma and......................4 00
2d do do 3 00
Best 10 yards woolen cloth, diploma......................4 00
2d do do Transactions and..................2 00
Best 10 yards woolen carpet, diploma and..................5 00
2d do do 3 00
3d do do 2 00
Best hearth rug,......................................3 00
2d do Transactions.
3d do 8th vol. Michigan Farmer.
Best 10 yds rag carpet, diploma and......................3 00
2d do 2 00
3d do Transactions.
Best pair of woolen knit stockings, Transactions and..........1 00
2d do do 1 00
Best do do socks,..........................2 00
3d do do 1 00
Best do do mittens,..........................1 00
Best woolen coverlet, diploma and........................1 00
2d do Transactions.
Best piece of broadcloth, diploma.
2d do Transactions.
Best piece sattinett, diploma.
2d do Transactions.

Best woolen shawl,................................$3 00
2d do 2 00
Best sample of woolen yarn, not less than one pound,........1 00
" do worsted do 1 00
" pair of " stockings,..........................2 00

CLASS II.

Judges—J. Penny, Grand Rapids; S. C. Hammond, Detroit; John Palmer, Detroit.

Best 10 yds linnen,................................$5 00
2d do 3 00
3d do Transactions.
Best 10 yds tow cloth,..............................5 00
2d do 3 00
3d do Transactions.
Best 10 yds linnen diaper,..........................5 00
2d do 3 00
Best pair knit cotton stockings,.....................2 00
do wove do 2 00
do knit linnen stockings,......................2 00
Best pound of linen thread,.........................2 00

Discretionary premiums will be awarded to articles of merit not included in the above list.

CLASS III.

Judges—H. P. Baldwin, Detroit; B. F. Eggleston, Jackson; F. Buhl, Detroit.

Best pair cowhide boots,............................$3 00
2d do 2 00
3d do Transactions.
Best pair calf boots,................................3 00
2d do 2 00
3d do Transactions.
Best pair men's cowhide shoes,.......................2 00
2d do do Transactions.
Best pair ladies slippers,...........................2 00
2d do Transactions.
Best pair ladies calf bootees,.......................2 00
2d do do Transactions.

Best pair of lasts, and not less than four pair, ... $2 00
Best overcoat, diploma and 4 00
2d do 3 00
Best dress coat, diploma and 3 00
2d 3 00
Best pair pants, diploma and 2 00
2d do Transactions.
Best vest, diploma and 2 00
2d do . Transactions.
Best silk or fur hat, diploma and 2 00
2d do Transactions.
Best straw hat, ... 3 00
2d do 2 00
3d do 1 00

CLASS IV.

Judges—Nathaniel Phillips, Ypsilanti; W. W. Calkins, Michigan Centre; A. Tuttle, Detroit.

Best two horse carriage, diploma and $5 09
2d do Transactions and 2 00
Best one do diploma and 5 00
2d " do 3 00
Best bedstead, diploma and 2 00
" sofa, ... 3 00
" bureau, diploma and 2 00
" six chairs, " 2 00
" table, " 2 00
" rocking chair, " 2 00
" sett of horse shoes, diploma and 1 00
" lot of horseshoe nails, not less than one pound, ... 1 00
" " chisels, diploma and 3 00
2d " " 2 00
Best lot of edge tools manufact'd at one establishm't, diploma & 5 00
" " coopers tools, diploma.
" flour barrel, ... 1 00
" pork barrel, ... 1 00
" wash tub, ... 1 00
" water pail, ... 1 00

Best pannel door,..$2 00
" lot of window sash,..2 00
" cooking stove, diploma.
" parlor stove, "
" pump,..3 00

PAINTINGS, DRAWINGS AND DAGUERREOTYPES.

Judges—F. E. Cohen, Detroit; Henry Ledyard, Detroit; Edward Lawrence, Ann Arbor.

Best specimen of animal painting in oil, by Michigan Artist,..$3 00
" do do water colors, "
Downing's Cottage Residences.
" specimen of cattle drawing, by Mich. Artist,............2 00
" " daguerreotype,........................2 00
2d " " 1 00
Best oil or water color painting, by Mich. Artist,............3 00
" specimen of statuary,........................3 00

NEEDLE, SHELL, AND WAX-WORK.

Judges—Rev. J. A. Baughman, Detroit; Miss Caroline K. Sawyer, Grand Blanc, Genesee county; Mrs. A. C. Hubbard, Detroit.

Best ornamental needlework,........................$3 00
" ottoman cover, Downing's Cottage Residences.
" table cover, ...2 00
Best group of flowers,........................2 00
" fancy chair-work with needle, Downing's Cottage Residences.
" variety of worsted work,........................2 00
" worked collar,........................2 00
" worked quilt other than silk,........................3 00
" white quilt,........................2 00
" silk quilt,........................3 00
" portfolio, marked,........................2 00
" silk bonnet,........................2 00
" straw bonnet,........................2 00
" lace cape,........................2 00
" two lamp mats,........................1 00
" ornamental shell work,........................2 00
2d do do 1 00
Best specimen wax flowers,........................2 00
2d do do 1 00

Best specimen artificial flowers, other than wax,............$2 00
2d do do do 1 00

FLOWERS.

Judges—D. C. Walker, Romeo, Macomb Co.; Mrs. Mark Norris, Ypsilanti; Mrs. D. M. Brown, Battle Creek.

Best and greatest variety and quantity of cut flowers,........$3 00
2d do do do 1 00

DAHLIAS.

Best and greatest variety, Beck's Botany of the United States.
2d do do 1 00
Best 12 dissimilar blooms, one volume of Hovey's Magazine of Horticulture.
2d do do 1 00
Best single Dahlia,..1 00

ROSES.

Best and greatest variety,..............................2 00
" 10 dissimilar blooms,..................................2 00

GENERAL LIST.

Best 6 varieties of Phlox,................................2 00
" and greatest variety verbenas,........................2 00
" do do indigenous flowers,........................2 00
" collection greenhouse plants owned by one person,........3 00
" floral design, 1 vol. Downing's Horticulturist.
2d do 2 00
Best hand bouquet, flat,..................................2 00
2d do do 1 00
Best do round, Beck's Botany of the U. S.
2d do do 1 00
Best basket bouquet with handle,..........................2 00
For the most beautifully arranged basket of flowers,..........2 00

FRUIT.

APPLES.

Judges—Wm. T. Howell, Hillsdale; J. W. Scott, Adrian.

For the best and greatest variety of good table apples, three of each variety, named and labelled, grown by exhibitor, one vol. Hovey's fruits and fruit trees, with colored plates.

2d best do do ..$5 00

3d best do do

Thomas' Fruit Book and 2 00

Best ten varieties table apples, ... 5 00

2d best do do 3 00

Best six winter varieties, 1 vol. Hovey's Mag. of Horticulture.

2d best do Downing's Fruits and Fruit Trees.

For best fall seedling apple for all purposes, with description, the history of its origin, &c., ten specimens to be exhibited, .. 4 00

2d best do do 8th vol. Michigan Farmer.

Best seedling winter apple, ten specimens, with description as above, Colman's Agricultural Tour.

PEARS.

Best and greatest number of varieties of good pears, named and labelled, grown by exhibitors, 1 vol. Hovey's Fruits and Fruit Trees, with colored plates.

2d best do do 1 vol. Downing's Horticulturist.

3d best do do 2 00

Best collection of Autumn pears, first rate, named and labelled, Landscape Gardening and Architecture.

2d do do 8th vol. Michigan Farmer.

Best fall seedling pear, not less than ten specimens, with description, history of its origin, &c., ... 4 00

PEACHES.

Best ten varieties, labelled, 1 vol. Hovey's colored fruits.

2d do do 1 vol. Downing's Horticulturist.

Best ten specimens, ... 3 00

2d do Downing's Fruits and Fruit Trees.

Best seedling variety, six specimens, with description, history of its origin, &c., ... 2 00

2d do do 8th vol. Michigan Farmer.

PLUMS.

Best collection of plums: six specimens each, ... 3 00

" four varieties, six specimens each, ... 2 00

Best twelve plums, choice variety, Downing's Fruits and Fruit Trees.

2d do do Thomas' Fruit Book.

Best seedling plum with description, 8th vol. Michigan Farmer.

NECTARINES AND APRICOTS.

Best and greatest number of good varieties of each fruit, six specimens of each variety,..........................$3 00

Best six specimens of any good variety, Downing's Fruits and Fruit Trees.

QUINCES.

Best twelve quinces of any good variety,..................$5 00

2d do do do 1 vol. Hovey's Magazine.

3d do do do 8th vol. Michigan Farmer.

GRAPES.

Best and most extensive collection of good native grapes, grown in open air,....................................$5 00

2d do do Allen on the Grape, and....2 00

Best dish of native grapes,.............................2 00

2d do Allen on the Grape.

WATERMELONS.

Best four specimens of any variety,......................$2 00

2d do do 8th vol. Michigan Farmer.

MUSKMELONS.

Best four specimens of any variety,......................$2 00

2d do do 8th vol. Michigan Farmer.

All fruit offered for premiums must be offered by exhibitor.

The fruit exhibited, for which premiums are awarded, to be at the disposal of the Executive Committee.

Discretionary premiums will be awarded for choice fruits not enumerated.

VEGETABLES.

Judges—John Chamberlain, Pontiac; George Henting, Marshall; Cyrus Lovell, Ionia.

Six best stalks celery, Transactions for 1850.

Three best heads cauliflower, Leibig's Agricultural Chemistry.

do do brocolis,................................$1 00

Twelve best white table turnips, ... $1 00
do carrots, ... 1 00
do table beets, ... 1 00
do parsnips, ... 1 00
do onions 1 00
Three best heads cabbage, ... 2 00
2d do do 8th vol. Michigan Farmer.
Twelve best tomatoes, ... 1 00
do sweet potatoes, ... 1 00
Two best vegetable eggs, ... 1 00
Best half peck Lima beans, ... 1 00
do Windsor beans, ... 1 00
Best bunch double parsley, ... 1 00
Three best crookneck squashes, Johnson's Agric'l Chemistry.
do of any other variety for cooking, ... 1 00
Best and largest pumpkin, Gardner's Farmer's Dictionary.
do do 1 00
Twelve best ears seed corn, ... 2 00
Best half-peck table potatoes, ... 2 00
2d do do Gaylord & Tucker's Am. Husbandry.
Best and greatest, distinct variety of vegetables raised by exhibitor, ... 5 00

Discretionary premiums will be awarded for choice garden products exhibited, not above mentioned.

MISCELLANEOUS ARTICLES.

Judges—Charles G. Hammond, Detroit; Mark Norris, Ypsilanti; E. G. Morton, Monroe.

GRAIN, FLOUR AND SEEDS.

Best crop of wheat not less than 5 acres, medal and $7 00
2d do do do Transactions and 4 00
Best crop of spring wheat not less than 5 acres, ... 8 00
2d do do do Colman's Tour and .. 2 00
Best crop of Indian corn not less than 5 acres, ... 8 00
2d do do Downing's Landscape Gardening & 2 00
Best crop of Barley not less than 2 acres, ... 8 00
2d do do Leibig's Agricultural Chemistry & 3 00

Best crop of rye not less than two acres, Colman's Tour.
do oats, do do do
do potatoes, not less than one acre, ... $4 00
do carrots, not less than a quarter of an acre, ... 4 00
Best acre of broom corn, ... 3 00
do clover seed, ... 4 00
Best sample of winter wheat, not less than one bushel, ... 3 00
do spring wheat, do Johnston's Ag. Chemistry.
do flour from the least quantity of wheat, not less than one barrel, ... 5 00
do do (without regard to quantity of wheat used) not less than one barrel, ... 5 00
2d best do do 3 00
3d do do 8th vol. Michigan Farmer.
Best sample of Indian corn, not less than one bushel, Transactions.
do oats, do do do

Awards on field crops will be made by the Executive Committee, at its annual meeting in December.

Persons making applications for premiums on crops, must forward to the Secretary by the 1st of December, 1851, full statements of the variety, number of bushels, and mode of cultivation of the articles for which they are competitors. The affidavit of the competitor should accompany his statements.

ESSAYS.

Judges—Warner Wing, Monroe; Charles A. Loomis, St. Clair; A. N. Hart, Lapeer, Lapeer county.

Best essay on the cultivation of wheat, ... $15 00
do do Indian corn, ... 15 00
do do potatoes, ... 15 00
do raising sheep, ... 15 00
do any other subject connected with agriculture, ... 15 00

All essays for which premiums are awarded will be considered the property of the Society.

MANAGEMENT OF FARMS.

Premiums are to be awarded by the Executive Committee, at its annual meeting, in December, 1851.

6

For the best cultivated farm, regard being had to the quantity and quality of the produce, the manner and expense of cultivation, and the actual products.

The persons making application for the premiums, must answer the following questions:

To all who furnish full answers to the questions, premiums will be given, consisting of the Society's Diploma, and one or two volumes of the Society's Transactions, according to the value of such reports.

SOILS, &C.

1. Of how much land does your farm consist? and how much wood, waste and improved land respectively?

2. What is the nature of your soil and subsoil? Is there limestone in it?

3. What do you consider the best mode of improving the different kinds of soil on your farm? of clay soil if you have it; of sandy soil, and of gravelly soil? Answer respectively.

4. What depth do you plow? What effect has deep plowing had on various soils?

5. Have you made any experiments to test the difference in a succeeding crop, between shallow, common or deep plowing?

6. Have you used the subsoil plow? and what have been its effects on different soils and crops? Have you drained any of your lands? If so, what soils, and with what results?

7. What trees and plants were indigenous to your soil? Give the names of each.

MANURES.

8. How many loads of manure (30 bushels per load) do you usually apply per acre? How do you manage your manure? Is it kept under cover? or are there cellars under your barns or stables for receiving it?

9. What are your means and what your manner of making and collecting manure? How many loads of manure do you manufacture annually? How many do you apply?

10. How is your manure applied; whether in its long or green state, or in compost? For what crops, or under what circumstances do you prefer using it, either in a fresh or a rotten state?

11. Could you not cheaply, essentially increase your supply of manure by a little extra labor?

12. Have you used lime, plaster, guano, salt, or any substance not in common use as manure? In what manner were they used, and with what results?

TILLAGE CROPS.

13. How many acres of land do you till? and with what crops are they occupied, and how much of each crop?

14. What is the amount of seed planted or sown for each crop? the time of sowing—the mode of cultivating, and of harvesting, and the product per acre? Have any insects been found injurious to your crops? If so, describe them, and the remedies adopted. Have you made, or can you give an estimate of the value of fertalizing matter taken from the soil by an acre of wheat, estimating 20 bushels per acre?

15. What kind and quantity of manure do you prefer for each, and at what times, and in what manner do you apply it?

16. How deep do you have manure covered in the earth, for different crops and different soils?

17. Have your potatoes been affected with any peculiar defect or disease, and have you been able to discover any clearly proved cause for it, or found any remedy?

GRASS LANDS, &C.

18. What kind of grasses do you use? How much seed of clover or the various kinds of grass do you sow to the acre? At what season of the year do you sow? and what is the manner of seeding? What kinds of grass are best adapted to lands used for dairy purposes?

19. How many acres do you mow for hay, and what is the average product? At what stage do you cut grass, and what is your mode of making hay?

20. Is any of your mowing land unsuitable for the plow, and what is your mode of managing such land?

21. Have you practiced irrigating or watering meadows or other lands, and with what effect? What is your particular mode of irrigation, and how is it performed?

22. Have you reclaimed any low, bog or peat lands? What was the mode pursued, the crops raised, and what the success?

23. Have you succeeded in eradicating the weeds from your farm, if so, by what methods, and what weeds are the most troublesome?

DOMESTIC ANIMALS.

24. How many oxen, cows, young cattle and horses do you keep, and of what breeds are they?

25. Have you made any experiments to show the relative value of different breeds of cattle or other animals for particular purposes, and with what results?

26. What do you consider the best and cheapest manner of wintering your cattle; as to feed watering and shelter?

27. How much butter and cheese do you make annually, from what number of cows, and what is your mode of manufacture.

28. How many sheep do you keep? Of what breed or breeds are they? How much do they yield per fleece, and what price does the wool bring? How many of your sheep usually produce lambs, and what number of lambs are usually reared? How much will your sheep or lambs sell for per head to the butcher?

29. What do you consider the best and cheapest manner of wintering your sheep as to food, watering and shelter? How many in proportion to your flock (if any) do you lose during the winter? What difference, if any, between the fine and coarse wooled sheep, in these respects?

30. How many swine do you keep; of what breed are they; how do you feed them; at what age do you kill them, and what do they weigh when dressed?

31. What experiments have you made to show the relative value of potatoes, turnips and other root crops, compared with Indian corn, or other grain, for feeding animals, either for fattening or for milk?

FRUIT.

32. What is the number of your apple trees? Are they of natural or grafted fruit? and chiefly of what varieties?

33. What number and kind of fruit trees, exclusive of apples, have you? and what are among the best of each kind?

34. What insects have attacked your trees, and what method do you use to prevent their attacks?

35. What is your general management of fruit trees?

36. What other experiments or farm operations have produced interesting or valuable results?

FENCES, BUILDINGS, &C.

37. What is the number, size, and general mode of construction of your farm buildings, and their uses?

38. What kind of fences do you construct? What is the heighth and length of each kind? And their cost and condition? Have you constructed any wire fences? If so, what has been its cost, and what its advantages, and how made?

39. To what extent are your various farming operations guided by accurate weighing and measuring? And to what degree of minuteness are they registered by daily accounts?

40. Do you keep regular farm accounts? Can you state the annual expense in improving your farm, and the income from it, with such precision that you can at the end of the year strike an accurate balance of the debt and credit? Would not this practice conduce very much to close observation, careful farming, and in the end much improve your system, as well as better your fortune?

41. Give the annual receipts and expenditures on your farm, specifying each.

The persons making applications for these premiums, must submit written answers to questions which will be furnished by the Secretary, on application. The statement to be returned by the Corresponding Secretary of each county to the Recording Secretary at Detroit, on or before the 10th day of December, 1851.

The Judges of Farms, consist of the Vice President and Corresponding Secretary for each county, who, together with one other appointed by the Executive Committee, will examine and report upon all farms that may be offered for premium in their respective counties.

Persons wishing to enter their farms for premium will notify the examining committee for their county, who will at the proper time make their examinations in accordance with the above regulations.

JUDGES ON FARMS.

To act in concert with the Vice President and Corresponding Secretary of their respective counties:

Name.	*P. O. Address.*	*County.*
Calvin C. White,	Gun Plains,	Allegan.
Hiram Lewis,	Prairieville,	Barry.
Harvey Haines,	Coldwater,	Branch.
Andrew L. Burke,	Berrien Springs,	Berrien.
H. Cook,	Homer,	Calhoun.
Cyrus Bacon,	Edwardsburgh,	Cass.
H. M. Dodge,	Sault Ste Marie,	Chippewa.
Wm. H. Faxon,	Duplain,	Clinton.
M. S. Brackett,	Bellevue,	Eaton.
Charles Gibson,	Grand Blanc,	Genesee.
Elias G. Dilla,	Jonesville,	Hillsdale.
John Thomas,	Lansing,	Ingham.
Richard Dey,	Ionia,	Ionia.
Simeon J. Strong,	Parma,	Jackson.
John Milham,	Kalamazoo,	Kalamazoo.
James Davis,	Grand Rapids,	Kent.
Wm. Hemingway,	Hadley,	Lapeer.
George E. Pomroy,	Clinton,	Lenawee.
Richard P. Bush,	Howell,	Livingston.
Samuel Abbot,	Mackinaw,	Mackinaw.
Calvin Pierce,	Utica,	Macomb.
Thomas Clark,	Monroe,	Monroe.
Harrison Voorhies,	White Lake,	Oakland.
Henry Pennoyer,	Grand Haven,	Ottawa.
Noah Beach,	Bridgeport,	Saginaw.
D. B. Harrington,	Port Huron,	St. Clair.
Wm. Johnson,	Nottawa,	St. Joseph.
J. M. Hartwell,	Hartwellville,	Shiawassee.
S. H. Merrick,	Niles,	Van Buren.
John Lake,	Ypsilanti,	Washtenaw.
A. L. Stevens,	Nankin,	Wayne.

On motion of Mr. Spencer,

Resolved, That Titus Dort, of Wayne county, act as chairman *pro tem.* of the executive committee for the ensuing year.

On motion of Mr. Bartlet,

Resolved, That the chairman *pro tem.*, the secretary, and Mr. Shoemaker, be a committee, to be denominated the "business committee," with power to transact all busines for the ensuing year, that the executive committee could if in session.

On motion of Mr. Bartlet,

Resolved, That no indebtedness shall be created without the consent of a majority of the business committee, and no account shall be paid until it shall have been audited, and certified as correct by two of said committee.

On motion of Mr. Spencer,

Resolved, That the compensation of the business committee, except the Secretary, be two dollars per day, and necessary traveling expenses while actively employed on business for the society, to be paid, aside from the traveling expenses, by the individual members of the executive committee, if not allowed by the society at its annual meeting in September, 1851.

On motion of Mr. Smith,

Resolved, That the thanks of the executive committee of the Michigan State Agricultural Society, be tendered to Gen. J. E. Schwarz, for his liberality in giving them the free use of his office for the transaction of business during their present session.

Adjourned.

J. C, HOLMES,
Sec. Mich. State. Ag. Society..

EXECUTIVE MEETING.

A meeting of the Executive Committee was held at the Minhigan Exchange, at Detroit, September 23d, at 8 o'clock P. M., the Presidont, Hon. J. B. Hunt, in the Chair.

The following members answered to their names:

Titus Dort, Wayne,

S. M. Bartlett, Monroe,

M. Shoemaker, Jackson,

J. Brown, Calhoun,

Walter Wright, Lenawee,

Grove Spencer, Washtenaw.

On motion,

Resolved, That during the Fair the members of the Executive Committee be stationed at and have the general superintendence of the following departments, viz:

Hall of Fine Arts—Grove Spencer.

Floral Hall—J. Brown.

Domestic Manufactures—M. Shoemaker.

Gates and Fences—Titus Dort.

Sheep, Swine and Poultry—W. Wright.

Horses—S. M. Bartlett.

Cattle—P. K. Leech.

Farm Implements—A. C. Walker.

Machinery Ground, and general direction of articles—C. G. Hammond.

To forward articles from Central Railroad—F. V. Smith.

Plows and Plowing—Jonathan Shearer.

After some general conversation upon plows, plowing, and the affairs of the Society, the committee adjourned.

THIRD ANNUAL EXHIBITION.

The Third Annnal Exhibition of the Society was held at Detroit on Wednesday, Thursday and Friday, the 24th, 25th and 26th days of September, 1851.

The ground selected for the occasion was upon a part of each, the Cass and DeGarmo Jones farms, lying between Third and Fourth streets, and north of the Chicago turnpike.

The first day of the Fair was cloudy, but not unpleasant; the time was principally taken up in entering stock and articles for exhibition.

The executive committee held a meeting at the committee room, on the show ground, at 10 o'clock, A. M., for the purpose of filling vacancies in the viewing committees.

The 25th was clear, wind south, warm and very pleasant. The grounds were so crowded with people that the viewing committees found great difficulty in performing the duties assigned them. At 2 o'clock P. M., a very eloquent address was delivered by Gen. Lewis Cass.

The morning of the 26th was ushered in by heavy showers, which continued at intervals until 10 o'clock, A. M.

The reading of reports of committees commeneed at 10 A. M.

The election of officers for the ensuing year took place at the committee room at 12, M., which resulted as follows:

President.—Hon. James B. Hunt, of Pontiac, Oakland Co.

Secretary.—J. C. Holmes, Detroit.

Treasurer.—H. H. Brown, Detroit.

On motion of P. R. Adams,

Resolved, That the thanks of the Michigan State Agricultural Society, and the citizens generally, be tendered to Gen. Lewis Cass, for his able, eloquent and instructive address, delivered at the Society's

Third Annual Fair; and that he be invited by the executive committee to furnish them with a copy for publication.

On motion of Titus Dort, of Wayne,

Resolved, That the second article of the constitution be amended by striking out in the tenth line as recorded, all after the word "until" to the word "provided," and insert the following: "The annual meeting of the executive committee."

On motion of Grove Spencer, of Ypsilanti,

Resolved, That the compensation of the business committee, except the Secretary, be two dollars per day and necessary traveling expenses while actually employed on business of the Society.

On motion,

Resolved, That the thanks of the Society be tendered to Gen. Lewis Cass and Mrs. De Garmo Jones, for the free use of the grounds occupied by the Society for its third annual fair; also to the ladies and gentlemen who kindly volunteered their services in tastefully and appropriately decorating Floral Hall. Also to Miss Duffield and associate ladies for the arduous duties performed by them as Superintendents of the Hall of Fine Arts.

Philo Parsons, of Detroit; Thomas B. Skinner, of Battle Creek; and A. N. Hart, of Lapeer, were appointed delegates to represent the Society at the Annual Exposition of the American Institute of New York, to be held in October next.

The executive committee elected for the ensuing year, voted to hold their annual meeting at the village of Jackson, on the second Tuesday of December next, at 12 o'clock, M.

At half-past two, P. M., the meeting adjourned *sine die.* During the afternoon of the 26th, a sale of stock took place, at which much of the stock on the ground changed owners.

EXHIBITORS,

And Articles Exhibited, September 24th, 25th and 26th, 1851.

BULLS.

Lewis Nash, Plank Road, Durham, "Jim Splendor," 5 years old.
Silas Sly, Plymouth, 2 Durhams, each 1 year old.
David Brooks, Avon, Livingston co., N. Y., Durham, 3 years.
Edward Belknap, Henrietta, Durham, 4 years.
F. Gardner, South Lyons, Durham, grade, 4 years.
J. B. Ward, Farmington, do 3 years.
F. V. Smith, Coldwater, Devon, "Charley," 6 years.
do do do "Duke of Devon," 2 years.
E. Butterworth, do do 1 year.
B. W. Philips, Lagrange, Durham, 5 years.
David Vickery, Gidley's Station, Durham, 4 years.
Rev. Chas. Fox, Grosse Isle, cross of Durham and Devon.
A. Spaulding, Grass Lake, Grade, 3 years.
L. Potts, Bridgewater, Durham, 1 year.
J. Freeman, Manchester, do grade, 1½ year.
John Thomas, Oxford, Devon, 4 years.
D. M. Uhl, Ypsilanti, Durham, 1 year.
S. M. Bartlett, Lasalle, Devon, 3 years.
J. D. Yerkes, Northville, Grade, 1 year.
H. King, Detroit, Ayrshire, 3 years.
George E. Pomeroy, Clinton, Devon.
Andrew Y. Moore, Schoolcraft, Grade.
O. B. Blackmar, Moscow, Devon, 4 years.
G. P. Bennet, Jackson, do 2 years.
Ira Phillips, Romeo, Durham, 4 years.
S. G. Pattison, Marengo, Durham.

H. E. DeGarmo, Ann Arbor, Durham, 1 year.
John Amer, Gosfield. C. W., do 6 years.
N. D. Bingham, Clarkston, Native, 2 years.
Elijah Leland, Quincy, Devon, 5 years.
do do do 4 years.
Robert Ferguson, Almont, Durham, 3 years.
Luther Shaw, Chesterfield, Grade, 2 years.
Isaac Askew, Amherstburg, C. W., Durham, "Guelph," 4 years.
do do cross of Durham and Ayrshire, 1 year.
Gideon Stodard, Litchfield, Grade, 2 years.
Wm. Canfield, Mt. Clemens, Grade, 1 year.
J. G. Warren, Coldwater, Devon grade.

COWS.

F. S. Finley, Ann Arbor, Durham grade, 8 years.
do do do do 3 years.
Silas Sly, Plymouth, do 3 years.
do do 2 do 2 years.
do do do 1 year.
Edward Belknap, Henrietta, Durham, 10 years.
do do 2 do 2 years.
J. Milham, Kalamazoo, do 3 do
J. B. Ward, Farmington, 2 do 3 do
F. V. Smith, Coldwater, Devon, 8 years.
do do do 6 do
do do do 4 do
do do do 2 do
E. Butterworth, do do 7 do
B. W. Phillips, Lagrange, Durham grade, 4 years.
Thomas Clark, Lapeer, Grade.
Myron Gates, Plymouth, Durham, 6 years.
G. W. Howe, Detroit, Grade, 1½ years.
David Thompson, Detroit, Native, 14 years.
Charles E. Stuart, Kalamazoo, Red Durham, 2 years.
David C. Vickery, Parma, 2 Grade, 3 years.
John Starkweather, Ypsilanti, Durham, 3 years.
James Depue, Spring Arbor, Grade.

J. Freeman, Manchester, Durham grade, 2 years.
D. M. Uhl, Ypsilanti, do do 6 do
do do do do 7 do
H. King, Detroit, 20 Native.
Samuel Blackwood, Novi, Grade, 4 years.
George E. Pomeroy, Clinton, Durham.
do do do grade.
A. R. Chapman, Novi, do do 5 years.
John Renwick, do do do 3 do
Andrew Y. Moore, Schoolcraft, do 6 do
do do 2 do 4 do
do do do 2 do
O. B. Blackmar, Moscow, Devon, 4 years.
do do do 3 do
do do Native, 8 do
S. G. Pattison, Marengo, 2 Durham, 2 years.
David Williams, Royal Oak, do 7 do
H. E. DeGarmo, Ann Arbor, do 1 do
J. Mulholland, Monroe, do 9 do
J. C. White, Detroit, do
Isaac Askew, Amherstburgh, C. W., Ayshire grade, 4 years.
do do do do 1 do
John Paton, do Durham, 2 do
do do do 5 do
Thomas Bigley, Detroit, Grade, 2 years.
John Hamilton, Amherstburgh, C. W., Durham, 4 years.
do do do 1 do
Reuben Towne, Detroit, Native.
David Williams, Royal Oak, Durham, 6 years.
Ambrose Burr, Plymouth, Native. 5 years.

CALVES.

F. S. Finley, Ann Arbor, Durham grade bull, 3 months old.
Silas Sly, Plymouth, Durham bull, 4 months old.
" " " heifer, 6 "
David Brooks, Avon, N. Y., Durham heifer 6 months old.
" " " " 3 "
" " " bull, 3 "

J. B. Ward, Farmington, " " 6 years old.
" " " heifer, 5 weeks old.
F. V. Smith, Coldwater, Devon heifer, 4 months old.
" " " 3 "
" " " bull, 3 "
B. W. Phillips, Lagrange, Durham grade, 10 weeks old.
Myron Gates, Plymouth, "
David Thompson, Detroit, 2 Native,—twins—4 months old.
David C. Vickery, Parma, grade bull.
" " " heifer.
J. D. Baldwin, Ann Arbor, grade bull, 11 months old.
James DePue, Spring Arbor, grade.
J. Freeman, Manchester, grade bull, 5 months old.
John Thomas, Oxford, 2 Devon heifers, 2 months old.
E. M. Crippen, Coldwater, 2 Devon heifers, 3 months old.
Sam'l Blackwood, Novi, grade, 5 months old.
Geo. E. Pomeroy, Clinton, Durham grade.
Geo. Dunlap, Northville, Durham grade, 3 months old.
John Renwick, Novi, " " 5 "
Andrew Y. Moore, Schoolcraft, grade bull, 5 "
" " 3 grade heifers, 5 "
O. B. Blackmar, Moscow, grade bull, 6 months old.
" " Devon bull, 4 "
" " " heifer, 2 "
S. G. Pattison, Marengo, Durham, 4 months old.
David Williams, Royal Oak, Durham, 4 "
J. Mulholland, Monroe, Durham grade heifer, 5 months old.
C. H. Williams, Coldwater, Devon grade, 4 months old.
" " " 4½ "
Isaac Askew, Amherstburgh, C. W., Durham, 7 "
" " " grade, 7 months old
" " Ayrshire grade, 7 "
" " Devon, 4 months old.
Thomas Bigley, Detroit, grade.
John Hamilton, Amherstburgh, C. W., Durham, 14 months old.
Reuben Towne, Detroit, Native.
David Williams, Royal Oak, Durham bull, 3 months old.

WORKING OXEN AND STEERS.

S. A. Randall, Norvell, 1 yoke steers, 3 years.

John Starkweather, Ypsilanti, 1 yoke working oxen, Grade, 7 years.

"	"	1	"	"	"	4	"
"	"	1	"	steers,	"	2	"
John Renwick, Novi,		1	"	"	"	3	"

STALLIONS.

D. McComber, Baltimore, 3 years.

Isaac Schram, Grand Blanc, bay, "Michigan Cock Fighter," 8 y'rs.

B. W. Phillips, Lagrange, blood, "Ivanhoe," 12 years.

John Summers, Shelby, " 7 years.

B. D. Brown, Detroit, "Highlander," 8 years.

Chency Hill, Richfield, for all work. "Victor," 12 years.

Major Ward, Redford, " "Young Zack," 5 years.

F. E. Eldred, Detroit, blood, "Black Eagle," 8 "

E. Stephenson, Chatham, C. W., for all work, 4 "

James Clisby, Quincy, blood, 7 years.

G. Knapp, Albion, for all work, 3 years.

J. L. Graham, Three Rivers, Messenger, 3 years.

Horace Smith, Tekonsha, draught, 4 "

John Renwick, Novi, blood, 3 "

Andrew Y. Moore, Schoolcraft, blood, "Bucephalus," 8 years.

George Chamberlain, Redford, blood, 3 years.

J. Hammond, Spring Arbor, for all work, "Sir Henry Duroc," 9 y'rs.

John Amer, Gosfield, C. W., " 6 years.

John Eckless, Plymouth, " 3 "

F. McHardy, Almont, draught, "Black Sampson," 10 years.

Thomas Jackson, London, C. W., draught, 4 "

Luman Fuller, Milford, blood, "Cock Fighter," 6 "

John Parker, Kalamazoo, for all work, "Green Mountain Morgan."

John B. Swan, Birmingham, blood, 3 years.

J. S. Smith, Plymouth, draught, "Plow Boy," 8 years.

Z. S. Flanders, Nottawa, blood, "Proud American," 5 years.

A. D. Morton, Marengo, " "Michigan Grey," 3 "

John Hamilton, Flint, " "Telegraph," 6 "

C. A. Green, Troy, " "Sir Archy Lightfoot," 8 years.

Moses Rumsey, Trenton, " "Mahomet," 4 "

S. M. Lathrop, Plymouth, draught, "Plow Boy," 8 years old.

L. Green, Farmington, for all work, 4 years.

Jacob Van Antwerp, Dryden, for all work, 11 years.

James Dougall, Windsor, C. W., poney, 3 "

B. Mosier, Farmington, for all work, 3 "

J. W. Sage, Memphis, " 7 "

P. L. Carter, Jackson, blood, "Glenco," 6 "

" " " "Jo. Miller," 5 "

MARES.

Edward Belknap, Henrietta, bay, 5 years old.

Geo. Clark, Lapeer, brood mare and colt.

Jas. O'Neil, Brownstown, for all work, 4 years old.

Joseph Tireman, Greenfield, 1 mare, for all work, with foal at foot, 9 years old.

" " do do 2 yr. old.

Walter Brewster, Battle Creek, brood mare, 3 years old.

B. W. Phillips, Lagrange, brood mare with foal at foot.

Thos. Clark, Lapeer, 1 " " 10 yrs old.

" " " " 7 "

J. T. Willson, Jackson, 1 mare for all work, 4 years old.

George Culver, Farmington, 1 brood mare with foal at ft, 10 yrs old.

Frank Lombard, Lapeer, 1 mare for all work, 4 years old.

Willard White, Southfield, 1 mare for all work, with foal at foot, 8 years old.

F. E. Eldred, Detroit, 1 brood mare, 8 years old.

John Starkweather, Ypsilanti, 1 brood mare for all work, 6 yrs old.

D. M. Uhl, Ypsilanti, brood mare for all work, with foal at foot, 13 years old.

H. B. Brevoort, Detroit, 1 brood mare with foal at foot, 7 years old.

Orson Ingalls, Almont, 1 " " 8 "

Andrew Y. Moore, Schoolcraft, 1 blood mare, "Jane McClure," 6 years old.

Geo. Dunlap, Northville, 1 brood mare with foal at foot.

O. B. Blackmar, Moscow, 1 brood mare, 15 years old.

B. L. Skiff, Borodino, 1 brood mare for all work, 6 years old.

A. Laplant, Detroit, 1 French brood mare for all work, with foal at foot, 12 years old.

L. Green, Farmington, 1 brood mare for all work, with foal at foot, 4 years old.

C. L. Crouse, Hartland, 1 mare for all work, 4 years old.

Barney Mosier, Farmington, 1 brood mare, 5 years old.

" " " 12 "

Wm. Maiden, Plank Road, 1 " with foal at foot.

C. W. Green, Farmington, 1 " 10 years old.

" " " 7 "

B. F. Spaulding, Dowagiac, 1 blood mare, 7 "

Z. Bird, Plymouth, 1 brood mare for all work, 6 "

E. Adams, Davidsonville, 1 mare for buggy or saddle, 6 years old.

Chas. Sly, Farmington, 1 blood mare. 5 years old.

MATCHED AND SINGLE HORSES.

H. O. Bronson, Jackson, 1 pr. matched horses.

H. R. Johnson, Detroit, 1 gelding, 4 years old.

A. Fairfield, Livonia, 1 span for plowing.

R. Lee, Novi, 1 pair of matched, 4 years old.

G. W. Hudson, Detroit, 1 pair ponies.

F. W. Backus, " 1 black gelding, 6 years old.

J. S. Brown, Windsor, 1 pr. matched, 4 "

J. Simmons, Farmington, 1 sorrel gelding, 6 "

J. W. Durham, Southfield, 1 pr. matched, 6 "

J. A. Austin, Plymouth, 1 sorrel gelding, 6 "

John Thomas, Oxford, 1 pr. matched, 6 & 7 "

Phineas White, Lapeer, " 6 "

A. Knapp, Northville, 1 pr. draught mares.

S. J. Freeman, Borodino, 1 pr. matched horses, 3 years old.

Judge Beecher, Genesee. 1 gelding, 4 "

J. J. Joyce, Plymouth, 1 pr. horses for plowing, 5 "

Robert Ferguson, Almont. " "

H. N. Howard, Pontiac, 1 gelding, 6 "

Henry Bogart, Novi, 1 pair matched.

Bronson Howard Detroit, 1 Shetland poney.

Austin Wales, Roseville, 1 pr. matched.

Geo. W. Rood, Lapeer, 1 gelding, 4 "

C. L. Crouse, Hartland, 1 " 5 "

R. W. P. Ingersoll, London, 1 gelding 3 years old.

E. O Parker, " 1 " 3 "

David Ellis. 1 " 6 "

David Sacket, Redford, 1 pr. matched mares, 2 "

A. Knapp, Northville, 1 pr. horses for plowing.

W. Lake, Flint, 1 pr. matched.

M. Rudd, Cassopolis, 1 gelding, 7 "

E. D. Minton, Marengo, 1 pr. horses for plowing.

A. Wattles, Troy, 1 pr. matched.

John S. Bagg, Detroit, 1 gelding, 6 "

Samuel Rogers, Almont, 1 pr. white Messenger 5 & 6 "

" " 1 bay saddle horse, 5 "

Luther Green, Farmington, 1 pr. matched geld. 6 "

George Clark, Lapeer, 1 pair horses for plowing.

T. Clark, " " "

COLTS.

Martin L. Cole, Climax, black stud, 2 years old.

George Clark, Lapeer, colt with the dam.

" " mare, 1 year old.

Joseph Tireman, Greenfield, stallion, 1 year old.

" " colt with the dam.

Isaac Schram, Grand Blanc, 1 span bay colts, 2 years old.

B. W. Phillips, Lagrange, colt with the dam.

Thomas Clark, Lapeer, 2 colts with the dams.

Myron Gates, Plymouth, stallion, 2 years old.

Frank Lombard, Lapeer, stallion, 2 years old.

Willard White, Southfield, stallion, 1 year old.

Wm. S. Carr, Manchester, stallion, 3 years old.

James Clisby, Quincy, stallion, 1 year old.

D. M. Uhl, Ypsilanti, colt with the dam.

" " 1 colt, 2 years old.

H. B. Brevoort, Detroit, 1 colt with the dam.

Orson Ingalls, Almont, 1 "

Sam'l Harring, Calamo, 1 stallion, 2 years old.

George Dunlap, Northville, 1 colt with the dam.

Wm. Anderson, Utica, stallion, 2 years old.

Peter Synder, Detroit, gelding, 2 years old.

A. Laplant, Detroit, poney, 3 years old.
" " " 2 "

Wm. Maiden, Plank Road, 1 blood colt.

George Culver, Farmington, 1 colt with the dam.

SHEEP.

A. L. Bingham, W. Cornwall, Vt., 20 French Merino ewes.
" " 2 " " bucks.

F. S. Gale, Bridgeport, Vt., 20 Spanish " ewes.
" " 10 " " bucks.
" " 3 French " "

E. Arnold, Dexter, 5 Spanish " ewes.
" " 2 " " bucks, 2 & 3 ys.

R. R. Wright, Waybridge, Vt. 1 " " buck, 1 year.
" " 1 Rambulet " 4 "
" " 15 ewes.

G. W. Gale, Ypsilanti, 5 Spanish Merino ewes.
" " 1 " " buck.

Edward Belknap, Henrietta, 1 Leicester " 2 years.

J. Milham, Kalamazoo, 5 Merino bucks, 1 year old.
" " 5 " ewes, 2 "
" " 5 " buck lambs.

E. T. Lovell, Climax, 4 " bucks, 4 years old.
" " 1 $\frac{3}{4}$ Merino & $\frac{1}{4}$ Saxon buck, 4 ys
" " 1 Merino buck, 1 year old.

J. B. Ward, Farmington, 3 Spanish Merino bucks.

C. A. Green, Troy, 1 pen of French and Sp. ewe lambs, 5 mo. old.
" " 1 " " buck " "
" " 1 grade buck, 3 years old.
" " 1 fat sheep, Native.

A. Beach, Dearborn, 1 Merino buck, 3 years old.
" " 1 Southdown " 5 "
" " 1 grade Merino lamb, 5 months old.

Wm. Maiden, Plank Road, 5 Southdown ewes.
" " 5 " " lambs.

Joab Polhemas, Marshall, 1 Merino buck, 2 years old.

Phillip Minns, Detroit, 1 Leicester " 2 "

Freeman Webb, Pinkney, 6 Merino " 1 "

Freeman Webb, Pinkney, 6 Merino buck lambs.
" " 3 " "
Silas Sly, Plymouth, 1 saxon buck, 3 years old.
" " 6 " 1 "
G. M. Simonson, Royal Oak 1 saxon buck 3 years old.
D. D. Gillett, Sharon, 1 saxon buck, 2 years old.
" " 1 " 1 "
" " 5 saxon ewes.
" " 5 ewe lambs.
" " 5 buck lambs.
J. P. Gillett, " 1 buck 2 years old.
" " 1 buck 1 year old.
" " 5 ewes.
" " 5 ewe lambs.
Washington Heath, Plymouth, 5 Leicester ewes, 3 years old.
Alonzo Henry, Canton, 1 merino buck, 2 years old.
" " 5 " ewe lambs.
Washington Heath, Plymouth, 1 Leicester buck, 3 years old.
Cyrus Stone, York, 10 Leicester ewe lambs.
" " 2 " buck lambs.
" " 1 " " 2 years old.
" " 3 " bucks, 2 "
J. A. Austin, Plymouth, 10 Spanish merino bucks, 1 year old.
Chas. Rich, Lapeer, 5 merino ewe lambs.
" " 19 " bucks, 1 year old.
James H. Fellows, Manchester, 1 French and Spanish merino buck, 2 years old.
Wait Peck, Manchester, 1 saxon buck, 2 years old.
Hiram Taylor, Romeo, 5 merino ewes, 1 and 2 years old.
" " 5 " buck lambs.
" " 3 " bucks, 1 year old.
J. D. Yerkes, Northville, 1 Southdown buck, 2 years old.
J. H. Murray, Farmington, 15 merino bucks.
Wm. Ten Eyck, Dearborn, 5 grade ewes.
" " 1 grade buck, 3 years old.
" " 1 merino buck, 1 year old.

Hiram Goodrich, Oceola, 2 pauler merino bucks, 2 and 4 years old.
" " 2 " ewes, 1 and 4 "
Wm. Whitfield, Waterford, 2 Southdown bucks, 6 and 9 "
" " 3 " " 1 "
" " 5 " buck lambs.
" " 5 " ewe lambs.
" " 5 " ewes.
P. K. Leach, Utica, 5 merino ewes, 1 year old.
J. H. Butterfield, Utica, 5 merino ewes.
N. Dickinson, Romeo, 5 merino ewe lambs.
" " 5 " "
" " 5 " buck lambs.
" " 1 " " 1 year old.
" " 1 " " 2 "
John Kirk, Dearborn, 5 grade ewes.
Wm. J. Bingham, Shoram, Vt., 15 merino ewes.
" " 2 " bucks.
" " 1 French merino buck.
Harris Newton, Avon, 1 French and Spanish merino buck, 2 yr. old.
" " 5 " " " ewe lambs.
" " 1 " " " buck.
Wm. Whitfield, Pontiac, 5 merino grade buck lambs.
Alonzo Henry, Borodino, 5 French and Spanish merino ewes.
J. Marvin, Monroe, 6 merino bucks, 1 to 3 years old.

SWINE.

Edward Belknap, Henrietta, 1 Suffolk boar, 9 months old.
R. B. Merrit, Battle Creek, 1 boar, 2 years old.
J. Brown, " 1 " 8 "
G. Knapp, Albion, 1 brood sow, 2 "
G. P. Bennet, Jackson, 2 pigs, 5 mo. old.
" " 1 pig, 5 weeks old.
J. C. White, Detroit, 1 Byfield boar.
" " 1 " sow.
E. Quirk, Dearborn, 1 boar, 6 mo. old.

POULTRY.

Wm. Maiden, Plank Road, 4 turkies.
" " 6 Poland fowls.

John Griffith, Farmington, 2 pr. goslings.

T. Wiley, Detroit, 8 Bantam chickens.

M. Freeman, Kalamazoo, Royal Cochin China fowls.

" " Shanghae fowls.

" " Dorking "

" " English game fowls.

" " Black Poland "

" " White " "

" " Game & Dorking, (crossed,) fowls.

" " Kent Co. fowls.

" " Game and Malay (crossed) fowls.

" " Kent Co. & Dorking, " "

" " Sebright Bantam fowls.

Amos Mead, Mead's Mills, lot wild turkeys.

J. S. Miller, Detroit, 4 white turkeys.

" " 5 Poland fowls.

R. Windfield, " lot carrier and tumbler pigeons.

J. Winder, " 1 pair Cochin China fowls.

James Dougall, Windsor, C. W., lot of Dorking fowls.

" " lot of fancy pigeons.

J. Perkins, York, lot of turkies and geese.

MISCELLANEOUS ANIMALS.

D. D. Gillet, Sharon, 1 Shepherd's dog.

J. P. Gillet, " 1 "

Isaac Cozzens, Detroit, 1 Spanish Jack, 3 ys. old.

J. C. White, " 1 Jack.

Loren Flint, Novi, 1 deer.

F. C. Chambers, Detroit, 1 deer.

FARM IMPLEMENTS AND MACHINERY.

Messrs. Nixon & Voorhees, Adrian, Moore's improved seed planter or grain drill.

Messrs. King & Walker, Lancaster, Pa., Moore's imp. seed planter.

Messrs. Johnston, Wayne & Co., Detroit, 1 eighteen horse high pressure engine.

Messrs. John Paton & Co., Detroit, 1 two horse close carriage.

" " " 1 four seated Rockaway.

John Daines, Birmingham, 1 drain tile machine.
Michigan Central Railroad, Detroit, 1 passenger car.
L. H. Hubbard, Mt. Clemens, cast iron beam plow.
O. C. Mosier, Jackson, sod plow.
" " plow for old lands.
Paul Coffin, Byron, stump extractor.
Messrs. Hallock & Raymond, Detroit, sewing machine.
D. D. Gillet, Sharon, straw cutter.
R. Simmons, Farmington, grain cradle.
C. H. Bennet, Plymouth, cider press and grinder.
Messrs. Beckford & Hoffman, Macedon, N. Y., grain drill.
James H. Fellows, Manchester, subsoil plow.
James Stell, Tecumseh, cutting machine for straw and corn stalks.
J. B. Tillinghast, Graham's Station, Ohio, churn.
E. H. Davis, Kalamazoo, a two horse wagon.
Gates & McKnight, Chicago, Illinois, screw cutting machine.
Augustus Day, Detroit, hydraulic churn.
Messrs. D. O. & W. S. Penfield, Detroit, 2 Emery & Co. improved railroad horse power.
" " " 2 Emery & Co., overshot thresher and separator.
" " " 2 feed mills.
" " " 8 straw cutters.
" " " 1 fanning mill.
" " " 1 vegetable cutter.
" " " 2 water rams.
" " " 1 sausage meat cutter.
" " " 1 " stuffer.
" " " 3 thermometer churns.
" " " 3 root pullers.
" " " 3 sett grind stone hangings.
" " " 2 corn shellers.
" " " 1 dozen cast steel hoes.
" " 1 1 digging fork.
" " " manure forks, four sizes.
" " " 1 hand rake.

D. O. & W. S. Penfield, Detroit, brush hooks.

" " " pruning saws and chiseles.

" " " 1 pair pruning shears.

" " " transplanting trowels.

" " " bull rings.

" " " cow ties.

" " " apple parers.

" " " hay knives.

" " " road scraper.

" " " 30 plows—assorted.

" " " 1 cook stove.

" " " 1 Smith's ventilating smut machine.

" " " 1 set horticultural tools and chest.

" " " scuffle hoes.

" " " pruning saws.

" " " folding harrows.

" " " ladies' weeding rakes.

" " " seed planter and drill.

" " " 1 dozen ox bow fastenings.

B. B. Morris & Co., Pontiac, 2 cultivators, with spring steel teeth.

A. Walcot, Detroit, grain drill and 4 plows.

J. S. Smith, Plymouth, 2 sod plows.

D. E. Rice, Detroit, 3 platform scales.

" " 1 letter press, dentist lathe and stamps.

Darius Hinkston, Clarkson, N. Y., Roger's & Hinkston's cast iron wheel cultivator, with improved steel teeth.

James Dawson, South Nankin, 2 scrapers.

Isaac McNeil, Newark, N. Y., Bartlett's self-acting regulator for adjusting and maintaining at a uniform height, the water in steam boilers.

Isaac McNeil, Newark, N. Y., 1 steam boiler and engine.

Edward Shepherd, Detroit, agricultural fences.

" " dairy maids.

G. Lupton, Detroit, 1 patent crown head plane.

" " 1 double carriage, self-regulating stone rubber and polisher.

Aaron Palmer, Brockport, N. Y., 1 cylinder grain drill.

A. Chope, Detroit, 1 two horse wagon with iron axle.
" " " " wooden axle.

F. F. Parker & Bro., Detroit, collection of farm implements and machines.

John W. Harrison, Niles, improvement in detaching horses from carriages.

Augustus Day, Detroit, buggy or skeleton wagon.

T. M. Clark, Lancaster, Pa., Clark's portable merchant mill.

H. Pate, Nankin, 1 two horse wagon.

John Brewer, Ypsilanti, 1 model board fence.

Messrs. DeGraff & Kendrick, Detroit, 1 hydraulic engine and pump.
" " " 1 stationary engine, 17 inch bore, 4 feet stroke.
" " " 1 stationary engine, 14 inch bore, 3 feet stroke.
" " " 1 stationary engine, 13 inch bore, 2½ feet stroke.
" " " 1 stationary engine, 11 inch bore, 2½ feet stroke.
" " " 1 stationary engine, 8 inch bore, 2½ feet stroke.

Moses Rogers, Ann Arbor, 2 straw cutters.

G. C. Van Mater, Detroit, self-propelling machine.

A. A. Holmes, Tecumseh, Crawford's clover huller.

J. M. Holbrook, Detroit, 1 miller's and inspector's brand.

Messrs. DeGraff & Kendrick, Detroit, 1 low pressure engine, on steamboat Ruby.
" " " 1 low pressure engine, on steamboat Pearl.

Rob't J. King, Adrian, seed planter or grain drill.

Messrs. Ellithorpe & Co., Detroit, hay, straw and manure forks.

E. DeForest, Avon, N. Y., 1 model horse rake.

Albert Wilber, Rives, 1 iron beam two horse plow.
" " 1 double mold board one horse plow.

C. H. McCormick, Chicago, Ill., Virginia grain reaper.

J. T. Willson, Jackson, 1 corn and cob mill.

Alexander Wattles, Troy, 1 plow.

Samuel Stanton, Plymouth, 1 Mulley harvest rake.

John Phelan, Niles, 1 two horse wagon.

Moses Stanfield, Jackson, 1 one horse buggy wagon.

Peabody & Munroe, Albion, 1 separator and horse power.

A. Smith & Son, Birmingham, 4 plows and 1 grain roller.

BUTTER AND CHEESE.

W. Lapham, Farmington, 1 new cheese.

John Griffith, " 1 cheese 1 year old.

" " 2 " less than 1 year old.

Mrs. J. J. Voorhies, Waterford, 10 lbs. butter made in June.

J. B. Springer, Livonia, 15 lbs. butter.

W. Gillett, Rome, 2 cheeses.

Luther Lappin, Farmington, 1 cheese over 1 year old.

" " 2 " under 1 year old.

Rob't Walker, Hamtramck, 12 lbs butter.

T. H. Murry, Farmington, 3 new cheeses.

George Chamberlain, Redford, 2 crocks butter.

Mrs. Titus Dort, Dearborn, 1 crock butter.

Mrs. A. L. Stevens, Nankin, 1 "

W. Dennison, Troy, 15 lbs butter.

A. C. Hubbard, Detroit, 15 lbs butter.

James Bailey, Troy, 1 crock butter.

Loren Moore, York, 1 cheese 1 year old.

" " 1 cheese less than 1 year old.

E. Cross, Redford, 15 lbs butter.

Clark Beardsley, Troy, 15 lbs butter.

James Smith, Detroit, 1 tub butter.

Mrs. Titus Dort, Dearborn, 2 new cheeses made without pressing.

SUGAR, HONEY AND BEE HIVES.

Allen Eames, Kalamazoo, 10 lbs. honey.

Amos Mead, Plymouth, 2 boxes honey.

D. White, Northville, 15 lbs. maple sugar.

A. Stockwell, Redford, 1 box honey.

S. Godfrey, Paw Paw, 1 " "

" " Godfrey's Crystal Palace bee hive.

Loren Moore, York, 1 lot maple sugar.

Mrs. Jacob Hendrickson, Pontiac, 1 piece woolen carpet.

" " 1 straw bonnet.

Mrs. Ann Jones, Dearborn, 2 pair woolen blankets.

" " 2 white woolen hose.

Mrs. H. Arnold, Ann Arbor, 1 bed quilt.

John Gray, Dearborn, 1 pair woolen blankets.

" " 1 pair white woolen hose.

Henry Blake, Detroit, Doily's, 1 knit quilt.

Joab Polhemas, Marshall, 1 riding saddle, 1 side do.

Baxter & Gallagher, Detroit, cordage.

Linus Cone, Troy. 1 bed quilt.

C. A. Hedges, Lansing, 1 specimen of book binding.

G. W. Collins, Farmington, 1 pair flannel blankets.

" " 1 pair lamp mats.

" " 1 pair crochet shoes.

" " 1 bunch sewing silk.

" " 1 pair silk hose.

" " 1 " mitts.

" " 1 pair cotton socks.

" " 1 " woolen do

" " 1 pair woolen hose, double thread.

" " 1 " " single do

" " 2 pair linen hose.

Mrs. Isaac I. Voorhies, Waterford, 1 white cotton quilt.

Mrs. Amos Brown, Detroit, 1 cotton quilt.

Mrs. D. D. Gillet, Sharon, 1 do

A. C. Walker, Farmington, 1 piece rag carpet.

James H. Fellows, Manchester, 1 pair wool socks.

" " 3 " gloves.

Edward Sawyer, Grand Blanc, 1 white cotton quilt.

" " 1 pair cotton hose.

" " 1 pair woolen mittens.

" " 1 pair linen hose.

Mrs. Maria Warner, Plymouth, 1 pair worsted hose.

" " 1 " gloves.

Mrs. Maria Warner, Plymouth, 1 pair tripple thread worsted socks.
" " 6 skeins woolen yarn.
John Thomas, Oxford, 1 double carpeting coverlet.
Mrs. Louisa C. Nash, Livonia, 1 rag carpet.
C. TenBrook, Adrian, 2 pr. woolen blankets.
A. Taylor, Kalamazoo, 7 ps. colored dressed cloth.
" " 2 ps. flannel.
Messrs. Hibbard & Davis, Milford, 1 piece doe skin cassimere.
" " " " satinett.
George Chamberlain, Redford, 2 cotton and wool coverlets.
" " " 2 woolen blankets.
Mrs. Mary Christie, Detroit, 1 bed quilt.
Mrs. B. F. Eggleston, Jackson, 1 cotton bed quilt.
Mrs. A. W. Comstock, Port Huron, 1 colored knit bed spread.
Mrs. E. Force, Manchester, 1 cotton quilt.
Messrs. Raymond & Nall, Detroit, shawls and carpeting.
Louis V. Baker, Buchanan, 1 pc. full cloth.
Edward Chase, Rose, " "
Mrs. John Palmer, Detroit, 1 pr. woolen socks.
Mrs. M. Beebe, Southfield, 1 bed quilt.
Mr. Julia Ann White, Northville, 1 rag carpet.
D. D. Gillett, Sharon, 12 yds woolen cloth.
Eagle & Elliot, Detroit, 1 overcoat, 1 dress coat, 1 pr. pants.
J. H. Murray, Farmington, 10 yds carpeting.
Francis Leslie, Dearborn, 2 pcs plaid flannel.
" " 1 net shawl, net jacket.
" " 1 white comforter.
" " 1 lb. stocking yarn.
Mrs. Alice W. Butterfield, Utica, 1 dress coat.
N. Flattery, Detroit, 1 spring lounge.
Peter Cheld, " thread, reed, ball, &c.
Valentine & Cruwell, Detroit, 1 case jewelry.
David Thomas, 1 pc. carpeting.
Mrs. A. A. Fish, Detroit, 1 chair.
H. Schanacker, " 1 upright piano.
Mrs. A. C. Hubbard, " 1 pr. woolen hose.
Wm. Snow, Detroit, specimens of wire cloth.

Mrs. J. H. C. Garvin, Washington, 1 shawl, 2 pr. hose, 1 pr. socks.
Adam Planter, Detroit, 1 steel plow.
J. P. Booth, Fentonville, 1 double harness.
John Galloway, Detroit, books.
C. Cornwall, Ann Arbor, 2 pcs. cassimere.
" " 16 pcs. full cloth.
" " 2 pcs. flannel.
Mrs. Titus Dort, Dearborn, 1 pce. rag carpet.
Mrs. Wm. Lowe, Farmington, 1 cotton quilt.
J. B. Bloss, Shiawassee, 1 pr. cotton hose, 1 pr. linen hose.
Mrs. A. L. Stevens, Nankin, 10 yds. flannel.
Mrs. F. Perrin, " 1 pr. woolen blankets.
Doct. Rose, Detroit, 1 pr. Banning's body braces.
A. Bour, " 1 pr. fancy boots.
D. Farrier, Ypsilanti, 1 rug.
Mrs. A. Bradner, Dearborn, 13 yds. rag carpeting.
Messrs. Hyde & Satchell, Detroit, 1 embroidered vest.
" " several vest patterns.
James Bailey, Troy, 10 yds. full cloth.
" " 2 coverlets, 2 blankets.
" " 10 yds. flannel.
" " 2 pr. woolen hose.
" " 1 pr. worsted hose.
Geo. Kirby & Co., Detroit, 1 lot sole leather.
" " 12 sides upper leather.
Orrin Derby, Niles, 1 Spanish saddle.
O. Starr, Royal Oak, 7 cow bells.
Chas. Piquette, Detroit, 1 case gold pens.
John Clay, Detroit, 10 articles lambs wool goods.
W. Dennison, Troy, 1 pc. rag carpeting.
" " 1 pc. gray full cloth.
" " 1 pr. worsted hose, three threads.
Benj. Lee, Detroit, 1 jar candy.
Mrs. F. Gaines, Dearborn, 1 pair blankets, 1 rug.
do do 1 pair woolen hose.
Doct. L. C. Whiting, Detroit, 1 set artificial teeth.

George Heron, Detroit, 1 carriage.
J. A. Bailey, do telegraph instruments.
C. Crosman, do electro magnetic instruments.
J. R. Grout, do ingot copper.
Wm. Tate, do hair work.
L. S. Noble, do lot of whips.
J. B. Brumfield, Plymouth, horse shoes.
Thomas Hall, Detroit, 2 set horse shoes.
A. Smith & Son, Birmingham, specimens of ingenious whittling.
Mrs. E. W. Withington, Lasalle, 1 pair woolen hose.
John Ladue, Detroit, 10 dozen fancy colored morocco linings.
George Common, Detroit, 1 overcoat.
Friend Palmer, do 1 case blank books.
Mrs. Jas. B. Hunt, Pontiac, 1 pair socks.
J. Boutwell, Ann Arbor, 1 cotton quilt.
Chas. H. Wood, Detroit, lot pressed brick.
Judson Rockwell, do 1 case electro plated ware.
Mrs. O. P. Stout, Troy, lot woolen stocking yarn.
Smith & Tyler, Detroit, 2 pair ladies' shoes.
Elisha Tyler, Detroit, Yankee knife sharpener.
Henry J. Hopkins, Detroit, copper tea kettle made out of a cent.
Howard, Watkins & Co., Detroit, lot of household furniture.
Miss J. Tooker, Superior, 1 hearth rug.
George Winter, Detroit, 1 hat.
Clark Besley, Troy, 1 pair woolen hose.
do do 1 " worsted "
do do 1 " woolen gloves.
George Doty, Detroit, 1 case silver ware.
Mrs. Moreland, Manchester, 1 bed quilt.
Vreeland & Sperry, Ann Arbor, 1 buggy.
Stevens & Zug, Detroit, 1 Jenny Lind cottage chair.
" " 1 rose wood reception "
" " 1 Voltaire arm chair of carved rose wood.
" " 1 wash stand.
" " 1 Chinese lamp stand.
" " 1 boqet table of native wood.

Stevens & Zug, Detroit, 1 trio medallion sofa frame, carved rosew'd.
" " 1 tete-a-tete, of carved rosewood.
" " 1 parlor chair of "

J. Perkins, York, farm baskets.

Messrs. F. & C. H. Buhl, Detroit, case of furs, hats, &c.

James Smith, Greenfield, 1 hearth rug.

John Sutherland, Ann Arbor, 1 patent riflle.
" " 1 double barrel rifle.
" " 1 shot gun.

J. W. Tillman, Detroit, 1 rose wood sofa in Brocatelle.
" " 1 " ladies' chair, in Plush.
" " 1 mahogony arm chair, in hair cloth.
" " 1 initiation rose cane seat arm chair.
" " 1 " rush "
" " 1 " straw "
" " 1 rose wood double fall-leaf table.
" " 1 " chair, in velvet plush.
" " 1 mahogony centre table, marble top.
" " 1 rose wood sofa.
" " 1 " corner stand.
" " 1 mahogany easy chair, in hair.
" " 4 door mats.

Mrs. Isaac J. Voorhies, Pontiac, 1 woolen blanket shawl.

J. M. Butts, Toledo, Ohio, 3 top buggies, 1 two seat carriage.

Proff. Barnes, Port Huron, 1 mammoth rustic chair.

James Smith, Detroit, 1 cotton coverlet.

J. Bloynk, Detroit, 1 seven octave square piano.

Mrs. Mark Norris, Ypsilanti, 2 pr. white woolen stockings.

Mrs. B. B. Kercheval, Detroit, 1 doll bedstead.

Alexander Wattles, Troy, 1 doz. wire-tied corn brooms.
" " 1 " twine-tied "

John Hutchins Southfield, ½ " " "

PAINTINGS, DRAWINGS AND DAGUERREOTYPES.

Miss E. G. Allen, Adrian, 3 paintings in water colors.

Miss A. F. Jenison, Jackson, monochromatic painting of the State Prison.

George H. Shearer, Detroit, 1 isomerical perspective drawing.
" " 2 water color paintings of rose and strawberry.
George H. Shearer, Detroit, 1 lead sketch.
Miss Sarah E. Smith, " 1 winter scene in water colors.
" " 1 India ink painting.
Miss Mary N. Smith, " 1 crayon picture of a child.
L. T. Ives, Detroit, 1 oil painting, cattle scene.
" " 2 portraits.
G. W. Clark, " 2 oil paintings.
J. J. Merritt, Battle Creek, 4 frames with daguerreotypes.
O. H. Moore, Schoolcraft, animal painting, in oil.
George Watson, Detroit, 2 pictures.
Mrs. M. Movius, " 2 landscape drawings.
C. C. Burt, Homer, specimens of daguerreotype.
Moses Sutton, Detroit, " "
John Atkinson, " oil painting and gilt frame.
Messrs. Lum & Ford, Detroit, design for Hotel.
" " " design for Church.
John Ladue, Detroit, 5 oil paintings.
C. Morse & Son, " specimen of Lambert's Anatomical plates.
O. D. Moore, " daguerreotypes.
R. E. Roberts, " 1 crayon picture, fac simile of peaches grown by R. E. Roberts.
F. E. Cohen, Detroit, oil painting, "Mizers."
John Atkinson, " 1 fruit piece, oil painting.
L. H. St. John, Brooklyn, lot of mezzotint, pencillings, &c.

NEEDLE, SHELL AND WAX WORK.

Mrs. Clara Davis, Ypsilanti, embroidered muslin mantilla.
Mrs. E. Fowler, Detroit, 1 crotchet quilt.
Susanna Hale, do needle work in frame.
Mrs. Griffin, do needle work wall basket.
Miss A. Lum, do 2 embroidered pictures.
Mrs. C. W. Williams, Greenfield, 1 piece needle work, scripture scene.
Mrs. Gillman, Detroit, 1 tidy, 1 piano stool cover.
do do 5 mats, 2 fine collars.
do do 5 pair candle shades.

Miss D. Brown, Ypsilanti, 1 satin emdroidered cushion.
Miss C. Davis, do 1 embroidered velvet smoking cap.
Miss E. M. Moiles, Detroit, 1 knit bed spread.
Miss A. H. Moiles, do 1 do
Messrs. S. Freedman & Bro., Detroit. 2 silk bonnets.
Miss Mary S. Palmer, do 1 piece worsted work.
Mrs. P. E. Demill, do 2 pieces fancy work.
Jona. Cudney, Flint, 1 bed quilt.
Mrs. H. E. De Garmo, Ann Arbor, 1 silk table spread.
Mrs. Peter Denoyers, Detroit, 1 tapestry work table cover.
Miss A. Clark, do lamp mats.
M. Aspinall, Detroit, 1 tidy.
Mrs. Ellwood, do 1 shell work chair.
Mrs. C. H. Spencer, Ypsilanti, 1 embroidered smoking cap.
Mrs. A. C. Hubbard, Detroit, 1 embroidered hdkf.
" " 1 collar.
Miss E. Ferris, " 1 piece fancy needle work.
Miss W. W. Ballard, Ypsilanti, 1 comforter.
" " 1 purse.
" " 1 horse net.
Mrs. J. N. Elbert, Detroit, 1 fire screen.
Mrs. Catton, " 1 tidy.
" " 1 bureau cover.
" " 1 sample fancy work.
Misses Baldwin, " 1 embroidered chair.
" " 4 do ottomans.
Mrs. G. T. Sheldon, " 1 worsted work rocking chair.
Mrs. A. A. Fish, Detroit, 1 tidy.
J. B. Bloss, Shiawassee, 1 pr. worsted slippers.
Mrs. E. Thomas, Detroit, 3 cases shirts.
Mrs. R. W. Van Fossen, Detroit, 1 sofa cushion.
" " 1 pr. ottomans.
" " 1 pr. suspenders.
" " 1 ship.
Mrs. Charles Piquette, " worsted work, "Boltan Abbey."

Mrs. H. B. Withington, Lasalle, 1 shell work gothic church.
" " worsted embroidery, girl with hat of flowers.

Mrs. W. W. Withington, Lasalle, 1 shell cottage.
" " 1 " basket.
" " 2 muslin caps.

Miss Susan Hale, Detroit, card basket, by a Miss, 3 years old.

Mrs. B. B. Kercheval, " 2 boxes arranged shells.

Mrs. Frank Woodbridge, Detroit, 1 box "

Mrs. H. DeGraff, " 1 worked worsted cushion.

Miss Culver, " 2 embroidered ottomans.

Miss F. Finley, " 1 crochet hat.

Mrs. J. B. Hunt, Pontiac, 1 needle-work Berthe cape.

Mrs. Lemcke, Detroit, 1 case ornamental hair work set in jewelry.

Mrs. F. Buhl, " 1 pr. ottomans.

Miss Cornelia Ayres, Detroit, 2 hour-glass stands.

Miss A. P. Davidson, Highland, 1 worsted ottoman cover.

Mrs. Riddett, Detroit, 1 table cover.

Miss Isabella Duffield, Detroit, 1 worsted work chair.

Mrs. J. Starkweather, Ypsilanti, 11 pcs. ornamental French needle work.

Mrs. S. Lewis, Detroit, 1 pr. embroidered suspenders.

Miss Mary N. Smith, " 2 pine burr baskets.
" " 1 port folio.

Mrs. B. G. Stimson, Detroit, 1 fancy feather brush.

Mrs. Babe, Detroit, 1 glass case, containing 2 lamp mats, 1 slumber cushion for a sewing chair, 1 toilet cushion, 1 wall basket, 1 round pin cushion, 1 pr. muffetees, 5 pr. infants socks, 2 scarfs, 1 comforter, 1 crochet work bag, 1 crochet bag for muslins 1 cover for curtain lapel, 2 crochet purses.

Mrs. William Hale, Detroit, 1 table spread, 1 work bag.

FRUIT.

N. Lapham, Farmington, six varieties apples.

Wm. Maiden, Plank Road, apples.

Chas. Fox, Gross Isle, 10 specimens peaches.

Chas. Fox, Gross Isle, 12 specimens quinces.

" " 37 varieties apples.

" " a dish of red Siberian crab apples.

Linus Cone, Troy, apples—Esopus Spitzenburg, Swaar, Roxburry Russet, R. I. greening, Famense, Norton's Melon, Fall Pippin, Westfield seek no further.

Linus Cone, Troy, 12 quinces.

Brayton Flint, Novi, a dish of peaches.

Francis Raymond, Detroit, two varieties foreign grapes.

Cyrus A. Chipman, Rochester, apples—Roxbury Russet, Swaar, R. I. greening, Snow, Golden Russet, Twenty ounce apple, Newton pippin, Calville, Sweeting, Russet, seek no further, Cabashea, Summer Queen, Garden Sweet, Holland pippin.

Cyrus A. Chipman, Rochester, pears—Stevens Genesee, Pound, Autumn Bergamont Summer Bonne Chretien, Seckel.

Isaac W. Ruggles, Pontiac, apples.

F. E. Eldred, Detroit, 10 varieties grapes.

J. Simmons, Farmington, quinces.

Ambrose Burr, Plymouth, 1 doz. quinces.

D. D. Gillett, Sharon, 6 seedling peaches.

James H. Fellows, Manchester, 18 varieties apples.

John Thomas, Oxford, 13 varieties apples.

Hubbard & Davis, Detroit, apples, Westfield seek no further, R. I. Greening, Snow, Ribston Pippin, Pownals Spitzenburgh, Esopus Spitzenburgh, Summer Queen, Yellow Bellefleur, Steele's Red Winter, Egg Top, Summer Pearmain, Summer Pippin, Fall Pippin, Holland Pippin, Baltimore, Newtown Pippin, Calville, Roxbury Russet, Jonathan, Yellow Siberian Crab, Red Siberian Crab, Cherry Crab, 20 ounce Pippin, Bough Sweet, Gate Apple, Chapman's Orange, Sweet Pearmain, Pound Sweeting, Early Russet, Red Gilliflower, Black do., Blue Pearmain, Swaar, Baldwin, Quinces, Portugal and Orange.

Albert Terry, Avon, 16 varieties apples.

do do 2 do pears.

do do 1 do quince.

do do 1 do peach.

Wm. Ten Brook, Adrian, 2 bbls. winter apples, 13 varieties.

do do 5 varieties peaches.

do do 1 variety grapes.

T. Butterfield, Brooklyn, 1 jar peaches.

A. Streeter, Bruce, 31 varieties apples.

J. B. Springer, Livonia, 1 variety, 12 specimens fall seedling apple.

A. Eames, Kalamazoo, 1 variety seedling peach.

J. C. Holmes, Detroit, pears—Bartlet, Royal Winter, New Egg, Figue d'Autonne, Countess d'Lincey, Passe Colmar, Louise Bonne d'Jersey, Epine Dumas, Duchess d'Orleans, Bleekers Meadow, Bozi d'Montigny, Doyenne d'Ore, Verte Longue, panache, Chaumontelle, Bell Lucrative, Napoleon, Swans Orange, Spanish Bonne Chretien, Vicar of Winkfield.

Grapes—Isabella, Catawba, Clinton, Mountain Sprout Water Melon.

N. Allen, Jackson, 3 specimens apples.

Ira Phillips, Romeo, 1 dish Stevens' Genesee pears.

" " 1 " quinces.

P. K. Leech, Utica, 6 varieties winter apples.

" " 10 " autumn "

H. Walker, Detroit, 15 " apples.

" " 3 " pears.

A. Stockwell, Redford, apples.

Thos. Read, Ann Arbor, quinces.

Wait Peck, Manchester, 26 varieties apples.

Titus Dort, Dearborn, 3 " "

B. G. Stimson, Detroit, 5 " grapes.

" " 7 " pears.

A. L. Stevens, Nankin, 1 doz. quinces.

C. C. Trowbridge, Detroit, 1 doz. quinces.

" " 1 dish grapes.

F. Gaines, Dearborn, 1 dish quinces.

Orson Star, Royal Oak, 1 dish greening apples.

A. Gray, Dexter, 1 doz. maiden blush apples.

Wm. Dennison, Troy, 14 varieties apples.

George Clark, Detroit, 1 dish native grapes.

" " 1 dish peaches.

" " 1 dish quinces.

John S. Bagg, Detroit, Catawba grapes.
J. B. Comstock, Franklin, 5 varieties apples.
John Ford, Detroit, 4 musk melons.
Loren Moore, York, 1 dish quinces.
Mark Norris, Ypsilanti, 1 dish quinces.
" " 1 dish grapes.
H. P. Crouse, Hartland, 1 bottle pickled cucumbers.
J. Holmes Agnew, Ann Arbor, 1 dish peaches.
Bela Hubbard, Detroit, 9 varieties pears.
" " 2 varieties quinces.
O. Derby, Niles, 1 doz. quinces.
S. G. Pattison, Marengo, 10 specimens seedling peaches.
J. G. Welch, Plymouth, 31 varieties apples.
John Falconer, Manchester, 2 varieties apples.
J. S. Bagg, Detroit, 2 varieties muskmelon.
Loren Flint, Novi, 11 varieties apples.
Peter Doolittle, South Nankin, 2 varieties apples.
Wm. Baby, Chatham, C. W., dish of peaches.
J. Brown, Battle Creek, 12 quinces.
Mr. Frost, Rochester, N. Y., lot of peaches.

FLOWERS.

J. H. Morrison, Detroit, 2 Cacti.
Wm. Neldrett, Ann Arbor, collection cut flowers.
do do 1 bouquet.
Hubbard & Davis, Detroit, collection of cut flowers.
do do do dahlias.
B. G. Stimson, do do green house plants.
do do do roses and phloxes.
do do do dahlias.
do do 3 bouquets.
do do 1 floral design.
do do basket cut flowers.
Rev. Geo. Duffield, do collection of dahlias.
Mrs. Geo. Duffield, do 1 round bouquet.
do do Pyramid of dahlias.
John Ford, do collection of dahlias.

John Ford, Detroit, collection of green house plants.
do do round bouquet.
do do flat bouquet.
Messrs. Mixer & Co., do 1 floral design.
do do 1 bouquet of dahlias.
do do 1 hand bouquet.
do do 1 bouquet miscellaneous flowers.
de do collection green house plants.
do do do dahlias.
do do do verbenas.
do do do cut flowers.
Mrs. J. C. Holmes, do 3 bouquets.
Mr. Frost, Rochester, N. Y., collection of dahlias.
do do do cut flowers.
Wm. Adair, Detroit.
Jerch. Brown, Battle Creek, dahlias.
Mason Branch, Stockbridge, cockscomb.
Mrs. R. B. Norris, Ypsilanti, bouquets.

.VEGETABLES.

Joseph Voorhies, Van Buren, table beets.
Wm. Maiden Plank Road, double parsley, blood beets, orange carrots, onions, seed corn.
Linus Cone, Troy, parsnips, turnips and onions.
D. S. Deane, Canton, 3 varieties of squash.
David Weeks, Hamtramck, stack of onions.
D. D. Gillett, Sharon, 3 pumpkins.
J. C. Holmes, Detroit, 3 varieties squash.
Jereh. Brown, Battle Creek, sqashes, cabbage, beets.
" " tomatoes, parsley.
Wm. Ten Brook, Adrian, ½ peck Lima beans.
John S. Bagg, Detroit, table beets, pumpkins, onions, 3 varieties of tomatoes, ½ peck Lima beans, 1 bunch double parsely, 12 ears of 12 rowed corn, 12 of 10 rowed, and 12 of 8 rowed, ½ peck of long pink eyed potatoes, ½ peck of round pink eyed potatoes, and ½ peck round white potatoes.

H. Walker, Detroit, beets, Valparaiso squash.

George Chamberlain, Redford, onions, beets, &c.

J. Meredith, Detroit, cabbages, 3 varieties tomatoes.

George Crabb, " cabbage, onions, carrots, parsnips, beets, salsify.

A. Bradner, Dearborn, pumpkins.

P. Roberts, Lasalle, California peppers.

" " 1 sack of sage.

" " 1 sack of prepared sweet corn.

John Ford, Detroit, 2 vegetable eggs, 1 bunch parsley, 3 heads of cabbage, collection of table vegetables.

H. Stevens, Redford, 1 bunch of beets.

Bela Hubbard, Detroit, 2 varieties tomatoes.

do do 3 varieties beets.

do do parsnips, carrots.

do do Lima beans, pumpkins.

do do sweet and yellow dent corn.

do do cauliflowers, potatoes.

do do peppers, black radish.

O. B. Blackmar, Moscow, Lima beans.

A. Wood, Detroit, 1 bunch of onions.

B. G. Stimson, Detroit, 1 doz. sweet potatoes.

" " ½ doz. red cabbage.

Clark Besley, Troy, lot of ruta baga.

P. J. Loranger, Detroit, winter squash.

Peter Doolittle, South Nankin, lot of onions.

G. V. N. Lothrop, Detroit, 4 vegetable eggs.

Hubbard & Davis, Detroit, Aberdeen turnips.

" " tomatoes, pop corn.

" " Stowell's evergreen sweet corn.

Simeon Havens, Trenton, 7 varieties potatoes, viz: Irish cup, western red, Melocoton, Carter's, Merino, mountain June, Bloomers.

J. S. Miller, Detroit, 5 heads red cabbage.

A. Stockwell, Redford, onions and turnips.

John Kirk, Dearborn, specimens of 1,000 head cabbage.

Jacob Perkins, York, ½ peck potatoes.

" " 1 sweet pumpkin.

FLOUR, SEEDS AND FIELD CROPS.

H. K. Farrand, Colon, 1 bushel white berried blue stem wheat.

F. S. Finley, Ann Arbor, 1 bushel blue stem wheat.

Myron Gates, Plymouth, 1 do do

Ambrose Burr, do 1 do do

O. D. Richardson, Pontiac, specimen of millet.

Wm. Congdon, Plymouth, 1 bushel blue stem wheat.

J. Brown, Battle Creek, 1 dozen ears Flint corn.

de do 1 do Dent "

J. B. Springer, Livonia, oats and seed corn.

Wm. Beesly, Waterford, 1 bushel white Sicilian spring wheat.

N. D. Bingham, Clarkston, Crystal Flint wheat.

do do specimen of 8 rowed yellow corn.

A. Knapp, Novi, 1 bushel blue stem wheat.

H. E. De Garmo, Ann Arbor, 1 bushel seed oats.

do do 12 ears corn.

C. T. Wilmot, do 1 bushel oats.

Mark Norris, Ypsilanti, 2 bbls. flour, "Eagle mills."

O. Derby, Niles, 1 bushel wheat.

Matthew Rowen, White Pigeon, 1 bushel wheat.

G. Ten Brook, Adrian, broom corn.

Jesse Truesdail, Whitmore Lake, 1 bushel Flint wheat.

L. Flint, Novi, sample blue joint wheat.

Jacob Perkins, York, seed corn.

Edward Boyden, Webster, 1 bushel Boyden wheat, produced from Italian spring wheat.

John Cupit, Rochester, 1 bbl. flour "Rochester city mills."

George Paddock, Commerce, 1 bbl. flour, "Strait's mills."

Matthews & Beach, Pontiac, 1 do

J. Cadet, Shelby, 1 do "Shelby mills."

Mead & Co., Detroit, 2 do "Detroit city mills."

H. I. Going, Pontiac, 1 do

Linus Cone, Troy, Egyptain wheat, chicken cone.

A. Knapp, Novi, 1 bushel blue stem wheat.

MISCELLANEOUS ARTICLES.

Wm. Phelps, Detroit, 3 cases confectionery.

Wm. Fletcher, " boot and shoe lasts.

John Daines, Birmingham, lot of drain tile.

Henry Blake, Detroit, a Treatise of Faith, printed 1629.

" " ornamental pen.

C. F. Kneeland, " 2 hot air furnaces, registers, and stove for the same.

Isaac McNeil, Newark, N. Y., self-acting regulator to adjust a uniform height of the water in steam boilers.

J. T. Wilson, Jackson, 1 bekstead fastening.

John Beard, Clyde, 5 white pine plank.

J. G. Leuchers, Detroit, 1 box refined tallow candles.

E. Price, Jr., " 6 sides upper leather, Hemlock tanned.

" " 6 calf-skins, "

" " 6 " Heberts patent.

Cyrus A. Chipman, Rochester, 50 samples merino wool.

Martin Rich, Juno, Wisconsin, tail block and dogs for saw mill.

Ladue & Eldred, Detroit, 11 pcs. belting.

" " 12 sides harness leather.

" " 6 sides bridle "

" " 12 " tap "

" " 2 doz. calf-skins.

" " 12 pcs. cordovan.

" " 12 sides sole leather.

H. Sangster & Co., Buffalo, N. Y., 6 patent spring carriage house lanterns.

" " " 6 pat. sp'g barn or hand lanterns.

" " " 6 " R. Road & stemb't "

" " " 6 " hand lanterns.

" " " 2 " Jenny Lind lanterns.

" " " 1 " railroad signal "

" " " 1 steamboat " lamp.

" " " 1 vessel signal lamp.

" " " 1 propeller " showing all the different lights required by law.

Henry Miller, Detroit, chewing tobacco and cigars.

Edward Shepard, Detroit, 1 large gass lamp.

J. R. Smith, Adrian, melodeon.

Messrs. Sawyer & Giddings, Milford, 1 water drawer.

Walter Wright, Adrian, 1 fleece French merino wool.

Wm. Ten Brook, " 1 model house for drying fruit.

Hugo Fidelity, Detroit, cyphering machine.

Mrs. J. Palmer, " 3 bottles currant wine.

T. H. Armstrong, Detroit, 2 cases containing regalia for free masons odd fellows and sons of temperance.

C. & P. Mellus, Detroit, circular, cross cut and wood saws.

A. C. Hubbard, Detroit, 3 bottles currant wine.

" " 1 bottle vinegar.

" " 1 bottle grape wine.

L. P. Hurd, Marshall, 2 set grave stones.

David French, Detroit, 1 bbl. plaster for agriculture.

" " 1 bbl. plaster for stucco work.

W. H. Buckland, Howell, 1 bottle wine.

E. VanZandt, Detroit, 5 colors stucco washed brick.

W. J. T. & Levi Wilson, Brooklyn, Ohio, specimens of fire proof paint.

George Clark, Detroit, 12 specimens of fish from Detroit river, viz: Herring, large and small, black bass, rock bass, sun fish, pike, yellow birch, pickerel, mullet, dog fish, sheep head, bull head.

Mrs. Titus Dort, Dearborn, 1 bottle currant wine.

REPORTS

Of Viewing Committees—1851.

CATTLE.

REPORT OF COMMITTEE ON CLASS I—SHORT HORNS.

The committee having discharged the duty assigned them, report that the whole number of cattle entered in class 1st, was thirty-seven. The number of premiums recommended, twenty-four, viz:

Best bull, 5 years old or over, No. 87, B. W. Phillips, Lagrange, Cass county, medal and $8 00

This bull is a red Durham, of beautiful color and proportions. The committee deem him a fine animal and worthy the attention of breeders of cattle.

The number of short horns exhibited that were three years old and under five, was four, three of which were exceedingly fine animals, forming so close a competition, and embracing so many excellent points, that your committee found great difficulty in fixing a preference, but finally awarded the first premium to No. 249, a nearly white bull, owned by Ira Phillips, of Romeo, Macomb county, medal and $8 00

2d best, No. 89, a grey bull of fine qualities, owned by David Vickery, of Gidley's Station, ... 7 00

3d best, No. —, a pure white bull, ... 5 00

Of two year old bulls, there was but one entered, No. 252, a spotted grey Durham. Though but one, your committee, in view of the fine qualities of the animal, deem him worthy of the first premium, owned by S. G. Pattison, Marengo, Calhoun county, ... 8 00

In yearling bulls there was a fair competition and some very fine specimens.

The first premium is awarded to No. 6, a dark red and grizzly Durham, owned by Silas Sly, of Plymouth, ... $5 00

2d premium No. 257, a red and white bull, "Lord Byron," of large size and good qualities. His sire was the imported bull "Splendor," dam "Victoria," a fine blood cow, imported by Thos. Weddle, Livingston county, N. York, H. E. De Garmo, Ann Arbor, ... 3 00

The 3d premium to No. 184, a pure white bull of fine qualities, well calculated to improve that kind of stock in our State, owned by David M. Uhl, Ypsilanti, ... 2 00

Of Durham bull calves, three only were exhibited, and it is due to the exhibitors to say, that they were all good animals.

The 1st premium is awarded to No. 12, Silas Sly, Plymouth, ... $3 00

2d premium, to No. 51, J. B. Ward, Farmington, ... 2 00

3d do No. 378, David Williams, Royal Oak, Transactions.

First premium on short horn cows, over 3 years old and under 5, is awarded to No. 205, owned by George E. Pomeroy, Clinton, Lenawee Co., ... $8 00

2d best, No. 259, J. Mulholland, Monroe, ... 7 00

3d " 378, David Williams, Royal Oak, ... 5 00

First premium on short horn heifers 2 years old, No. 9, Silas Sly, Plymouth, ... 5 00

2d prem., No. 30, Edward Belknap, Henrietta, Jackson Co., .. 3 00

3d " 31, do do .. 2 00

Of yearling heifers, the first premium is awarded to No. 258, O. B. Blackmar, Moscow, Hillsdale county, ... 3 00

2d prem., No. 10, Silas Sly, Plymouth, ... 2 00

The whole number of heifer calves was five, most of them very fine animals.

Best, No. 255, an exceedingly beautiful calf, of the grizzly color, raised and exhibited by S. G. Pattison, Marengo, . $3 00

2d best, No. 205, Geo. E. Pomeroy, Clinton, ... 2 00

3d " 11, Silas Sly, Plymouth, Transactions.

In class 3—Herefords, no animals were exhibited; which circumstance your committee very much regret, as they consider the Herefords a very beautiful and symmetrical class of animals.

In conclusion, your committee while they regret that the competitors were so few, and believing that our farmers and breeders of good stock might, much to their own interest and your gratification, greatly augment the number; yet they are fully convinced, and are happy to express the feeling that this department of stock has been greatly improved, both in number of animals exhibited and in quality, since your first fair, and that your association may well say "*avancer,*" to those engaged in this department, for the signs of the times fully show that a few more days of care, and years of attention will fully realize their fullest hopes and highest temporal interests.

All which is respectfully submitted.

BENJ. PIERSON,
N. A. BALCH,
A. L. HAYS,
Committee.

REPORT OF COMMITTEE ON CLASS II. AND IV.—DEVONS AND AYRSHIRES.

The committee on Devon stock, respectfully report that they have discharged the duties assigned them, and that in the performance of that office they were governed by the pedigree furnished, the general appearance, the symmetry of form, and developement of those points indicating hardiness of constitution, purity of blood, easy keeping, and early maturity of the animals, together with their adaptation to the purposes for which they are designed. Much of the stock exhibited was of a superior quality, and your committee are satisfied that it is such as it is represented to be, viz: of the pure Devon blood, and we consider it an honor to the discrimination, judgment and enterprise of the gentlemen owning it. Your committee have adjudged the following premiums:

Best bull 5 years old and over, No. 62, F. V. Smith, Coldwater, Branch Co., medal and.................... $8 00

2d best bull 5 years old and over, No. 207, Geo. E. Pomeroy, Clinton, Lenawee Co.,........................ 7 00

3d best bull 5 years old and over, No. 279, Elijah Leland, Quincy, Branch Co., ... $5 00

Best bull 3 and under 5 years old, No. 188, S. M. Bartlett, Lasalle, Monroe Co., medal and 8 00

2d best bull 3 and under 5 years old, No. 237, O. B. Blackmar, Moscow, ... 7 00

3d best bull 3 and under 5 years old, No. 280, John Allen, Coldwater, ... 5 00

Best bull 2 years old, No. 63, F. V. Smith, Coldwater, ... 8 00

2d do 246, G. P. Bennett, Jackson, ... 5 00

Best bull 1 year old, No. 72, E. Butterworth, Coldwater, ... 5 00

Best bull calf, No. 70, F. V. Smith, ... 3 00

2d do No. 242, O. B. Blackmar, ... 2 00

Best Devon cow, 5 years old or over, No. 64, F. V. Smith, medal and 8 00

2d Devon cow, 5 years old or over, No. 65, F. V. Smith, ... 8 00

Best cow, 3 years old and under 5, No. 66, F. V. Smith, ... 8 00

2d do do No. 71, E. Butterworth, Coldwater, ... 7 00

3d do do No. 218, O. B. Blackmar, 5 00

Best heifer 2 years old, No. 67, F. V. Smith, ... 5 00

Best heifer calf, No. 68, F. V. Smith, ... 5 00

2d do No. 69, do 2 00

3d do No. 199, E. M. Crippen, Coldwater, Transactions.

But one Ayrshire bull was exhibited, and your committee thought he was not a superior animal of that stock; having, however, some good points, and as an inducement to the propagation of the stock, the committee recommend the award of the third premium. This bull, No. 197, is owned by H. King, of Detroit. 5 00

All which is respectfully submitted.

W. H. MONTGOMERY,
WM. D. CANFIELD,
O. M. ROOD,

Committee.

REPORT OF COMMITTEE ON CLASS V.—GRADE CATTLE.

The committee on grade cattle, after examining carefully the animals allotted to our class, with but few attested statements from owners as to the amount of cross or improved blood, or the manner of keeping, for at least, the last four months, have acted in accordance with their best judgment as to possessing such points as are generally deemed evidences of good milkers, size, color and general form, symmetry and developements of points indicative of hardy constitutions, report the following list of premiums:

Best bull 5 years old or over, No. 48, F. Gardner, South Lyons, Oakland Co., medal and $8 00

Best bull 3 ys. old and under 5, No. 151, A. Spaulding, Grass Lake, Jackson county, ... 8 00

Best bull 2 years old, No. 220, Andrew Y. Moore, Schoolcraft, Kalamazoo Co., ... 8 00

2d best 2 years old, No. 171, J. Freeman, Manchester, Washtenaw county, ... 5 00

3d best bull 2 years old, No. 299, Luther Shaw, Chesterfield, Macomb Co., ... 3 00

Best bull 1 year old, No. 195, J. D. Yerkes, Northville, Wayne Co., ... 5 00

2d best bull 1 year old, No. 342, Wm. Camfield, Mt. Clemens, Macomb Co., ... 3 00

Best bull calf, No. 226, A. Y. Moore, Schoolcraft, ... 3 00

2d do 173, J. Freeman, Manchester, ... 2 00

3d do 202, Sam'l Blackwell, Novi, Oakland Co., Transactions.

Best cow 5 years old or over, No. 183, D. M. Uhl, Ypsilanti, Washtenaw county, medal and 8 00

2d best cow 5 years old or over, No. 211, A. R. Chapman, Novi, ... 7 00

Best cow 3 and under 5 years old, No. 146, David C. Vickery, Parma, Jackson Co., ... 8 00

2d best cow 3 and under 5 years old, No. 223, Andrew Y. Moore, ... 7 00

3d best cow 3 and under 5 years old, No. 35, J. Milham, Kalamazoo, ... 5 00

Best heifer 2 years old, No. 111, Chas. E. Stewart, Kalamazoo, ... $5 00
2d best heifer 2 years old, No. 173, J. Freeman, Manchester, 3 00
3d do do 224, Andrew Y. Moore, ... 2 00
Best heifer calf 1 year old, No. 226, Andrew Y. Moore, ... 3 00
2d do do 260, J. Mullholland, Monroe, 2 00
3d do do 226. We found 3 calves in this lot belonging to Andrew Y. Moore. Transactions.

The committee were much pleased to see so full and fine a collection of animals in the class of grades, and in several instances deliberated for some time which to select, and they feel confident that the breeding of even such stock is not only meritorious, but most prove profitable.

M. FREEMAN,
AMOS HAMLIN,
M. CLAWSON,
Committee.

CLASS VI. & VII.—WORKING OXEN AND STEERS.

The committee on working oxen and steers, make the following report:

Best yoke working oxen, No. 153, 7 years old, John Starkweather, Ypsilanti, Washtenaw Co., medal and........$10 00
2d best yoke working oxen, No. 154, 4 years old, John Starkweather, ... 10 00
Best yoke of steers, 3 years old, No. 150, S. A. Randall, Norvell, Jackson county, ... 10 00
2d best yoke of steers, 3 yrs. old, No. 215, John Renwick, Novi, Buel's Farmers Companion and................ 5 00
Best yoke steers, 2 years old, No. 155, John Starkweather, .. 6 00

Said cattle are grade, mostly Durham.

ISAAC J. VOORHIES,
H. S. HOLCOMB,
LEWIS T. MILLER,
Committee.

CLASS VIII., IX. & X.—FAT CATTLE AND MILCH COWS.

The committee on fat cattle and milch cows, beg leave to make the following report:

Best fat cow, fed on hay and grass alone, after one year old, No. 104, owned by Ambrose Burr, of Plymouth, Wayne county, ... $5 00
2d best fat cow, fed as above, No. 182, D. M. Uhl, Ypsilanti, 3 00
1st best milch cow, No. 1, F. S. Finley, Ann Arbor, Allen on Domestic Animals and 8 00
2d best milch cow, No. 343, Thos. Bigley, Detroit, ... 5 00

STEPHEN ELDRED,
L. MAYNARD,
GEORGE CLARK,
Committee.

CLASS II—FOREIGN CATTLE.

The committee on foreign cattle respectfully report that they have discharged the duty assigned them, and awarded premiums as follows:

Best short horn bull, 4 years old, No. 308, owned by Isaac Askew, of Amherstburgh, C. W., diploma, and $8 00
Best short horn cow, 4 years old, No. 352, owned by John Hamilton, Amherstburgh, C. W., diploma and 8 00
Your committee would respectfully call attention to Durham bull No. 24, 3 years old, owned by David Brooks, of Avon, Livingston county, N. Y., ... 5 00
Also, Durham bull No. 261, 6 years old, owned by John Amer, of Gosfield, C. W., they recommend a premium of 3 00
Bull calf No. 27, 3 months and 11 days old, owned by David Brooks, ... 1 00
Heifer calf No. 26, 3 months and 23 days old, David Brooks, .. 3 00
do 25, 6 months old, do 2 00

All these animals we think worthy of discretionary premiums, and therefore refer them to the consideration of the executive committee.

Your committee regret being compelled to say that the exhibition of foreign stock is inferior in numbers to that of preceding years. What may be the cause your committee would not presume to say; but would venture to suggest to the executive committee the adoption of some measures to induce the introduction of the best stock from abroad both for exhibition and sale. This may be done by a wider range of premiums or such other course as will best promote the object. Thus disseminating more rapidly throughout the State one of the surest means of agricultural wealth and comfort.

GEORGE E. POMEROY,
Chairman of Committee.

HORSES.

CLASS 1.—FOR ALL WORK.

The undersigned, your committee on horses, class 1, for all work, beg leave to report, that in the execution of that duty, they make the following report:

Best stallion 4 years old or over, No. 117, "Victor," 12 years old, owned by Cheney Hill, of Richfield, Genesee Co., medal and $8 00

2d best stallion 4 years old or over, No. 294, "Green Mountain Morgan," owned by John Parker, Kalamazoo, ... 8 00

3d best stallion 4 years old or over, No. 250, "Sir Henry Duroc," 9 years old, owned by J. Hammond, Spring Arbor, Jackson Co., Youatt on the horse, and 3 00

Best brood mare, with foal at foot, No. 122, Willard White, Southfield, Oakland Co., medal and 8 00

2d best brood mare, with foal at foot, No. 45, George Clark, Lapeer, ... 8 00

3d best brood mare, with foal at foot, No. 185, D. M. Uhl, Ypsilanti, Youatt on the horse and 3 00

Best stallion 3 years old, No. 193, Gelba Knapp, Albion, Calhoun Co., ... 7 00

2d best stallion 3 years old, No. 77, D. Macomber, Baltimore, Barry Co., ... 5 00

3d best stallion 3 years old, No. 265, John Eckles, Plymouth, Wayne Oo., ... 3 00

Best mare 4 years old, No. 357, C. W. Green, Farmington, .. $5 00

Best stallion 2 years old, No. 145, E. N. Fairchild, Marion, Livingston Co., ... 3 00

2d best stallion 2 years old, No. 267, Wm. Anderson, Utica, Macomb Co., Youatt on the horse.

3d best stallion 2 years old, No. 203, Samuel Harring, Kalamo, Eaton Co., American Veterinarian.

Best mare 2 years old, No. 186, D. M. Uhl, Ypsilanti, ... 3 00

2d do do 75, Joseph Tireman, Greenfield, Wayne Co., Youatt on the horse.

Best stallion 1 year old, No. 123, Willard White, Southfield, 3 00

2d do do 180, James Clisby, Quincy, Branch Co., Transactions and American Veterinarian.

3d best stallion 1 year old, No. 76, Joseph Tireman, Greenfield, Transactions.

Best mare 1 year old, No. 46, George Clark, Lapeer, ... 3 00

Your committee have had only one case of any difficulty in deciding on the merits of the several horses presented to them, and that was in their choice between the "Green Mountain Morgan," and the stallion "Victor." Their decision in favor of Victor was made on account of size; he being the best adapted as a horse for all work.

GEORGE C. MUNROE,
DANIEL B. HIBBARD,
DANIEL C. WATERMAN,
Committee.

CLASS II & IV—DRAUGHT AND MATCHED AND SINGLE HORSES.

The undersigned, a committee on draught and matched and single horses, report as follows:

MATCHED HORSES.

Best matched horses, No. 43, H. O. Bronson, Jackson, ...$10 00

2d do No. 323, Austin Wales, Erin, Macomb county, ... 8 00

3d do No. 108, R. Lee, Novi, Oakland county, . 5 00

4th do No. 306, H. Bogart, Novi, ... 3 00

SINGLE HORSES.

Best single horse, No. 356, M. Rugg, Cassopolis, Cass Co., bay horse 7 years old, ... $500

2d do No. 121, black horse, 5 years old, F. W. Backus, Detroit, Youatt on the Horse, and . 3 00

3d do No. 91, sorrel horse, H. R. Johnson, Detroit, 3 00

4th do No. 78, bay mare, 3 y'rs old, Walter Brewster, Battle Creek, Calhoun Co., ... 2 00

DRAUGHT STALLIONS.

Best draught stallion, No. 79, "Michigan Cock Fighter," Isaac Schram, Grand Blanc, Genesee Co., medal and .. 8 00

2d do No. 277, Findley McHardy, Almont, Lapeer Co., stallion, 10 years old, "Black Sampson," 8 00

3d do No. 181, James Clisby, Quincy, Branch Co., stallion, 7 years old, ... 3 00

For draught stallion, No. 329, S. M. Lathrop, Plymouth, "Plow Boy," 8 years old, ... 8 00

Best draught stallion, 3 years old, No. 213, Horace Smith, Tekonsha, Calhoun county, ... 7 00

The committee would recommend a span of matched colts, two years old, No. 80, belonging to Isaac Schram, of Grand Blanc, as worthy of notice, and award a premium of Youatt on the Horse.

The committee in judging upon matched and single horses, in arriving at conclusions, have been governed more by the size, strength, action and symmetrical proportions, than by the external beauty, and in single horses they have paid particular attention to the action exhibited in the horse.

In regard to draught horses, the committee have been governed by the power, weight and size, and have awarded premiums to that class of horses that were not overgrown in height.

W. H. COLEMAN,
GEORGE CLARK,
VAN HOVENBURGH,
Committee.

CLASS III—BLOOD HORSES.

The committee on blood horses beg leave to report as follows:

Best stallion 4 years old or over, No. 374, P. L. Carter, Jackson, sorrel horse, "Glencoe," medal and $8 00

2d do No. 321, "Telegraph," John Hamilton, Flint, ... 8 00

3d do No. 218, "Bucephalus," A. Y. Moore, Schoolcraft," Youatt on the Horse, and 3 00

BROOD MARES FOUR YEARS OLD OR OVER.

Best brood mare 4 years old or over, No. 141, 8 years old, F. E. Eldred, Detroit, medal and 8 00

2d do No. 360, B. F. Spalding, Dowagiac, ... 8 00

3d do No. 370, 5 years old, Chas. Sly, Farmington, Oakland county, Youatt on the horse, and 3 00

THREE YEAR OLD STALLIONS.

Best stallion 3 years old, No. 212, J. L. Graham, Three Rivers, St. Joseph county, ... 7 00

2d do No. 305, J. Swan, Birmingham, ... 5 00

THREE YEAR OLD MARES.

Best mare 3 years old, No. 245, Geo. Chamberlain, Redford, 5 00

TWO YEAR OLD STALLIONS.

Best stallion 2 years old, No. 39, "Black Hawk," M. L. Cole, Climax, Kalamazoo county, ... 3 00

2d do No. 103, Myron Gates, Plymouth, Youatt on the Horse.

EBER ADAMS,
G. W. HOWE,
Committee.

SHEEP.

CLASS I.—LONG WOOLED. CLASS II.—MIDDLE WOOLED.

The committee on Sheep, class 1 and 2, in compliance with their duty, have made all the examinations necessary to satisfy their minds in reference to the superior merit of said animals and award the following premiums to wit:

CLASS I—LONG WOOLED.

Best buck over 2 years old, No. 137, Leicester, Washington Heath, of Plymouth, Wayne Co., medal and $5 00

2d best buck over 2 years old, No. 164, Cyrus Stone, York, Washtenaw Co., ... 5 00
3d best buck over 2 years old, No. 96, Philip Mims, Detroit, 3 00
Best pen of 5 ewes, No. 136, Leicester, W. Heath, medal and 5 00
do 5 ewe lambs, No. 162, Cyrus Stone, York, ... 5 00

CLASS II.—MIDDLE WOOLED.

Best middle wooled buck over 2 years old, No. 232, Southdown, W. Whitfield, Waterford, Oakland Co., medal and 5 00
2d best middle wooled buck over 2 years old, No. 232, Southdown, Wm. Whitfield, ... 5 00
Best buck 1 year old, No. 233, Wm. Whitfield, ... 5 00
Best pen of 5 buck lambs, No. 234, Wm. Whitfield, ... 5 00
Best pen of 5 ewes, No. 235, Wm. Whitfield, medal and... 5 00
2d do No. 81, Wm. Maiden, Plank Road, Wayne Co., ... 5 00
Best pen of 5 ewe lambs, No. 236, Wm. Whitfield, ... 5 00
2d do do 82, Wm. Maiden, ... 3 00

The committee would state that the competition was exceedingly limited, there not being enough of the different kinds to receive the full list of premiums offered.

There was but one fat sheep exhibited, and it not being highly meritorious, no premium was awarded.

In class 1, all the sheep were of the Leicester breed, and remarkable fine specimens.

A lot consisting of 1 buck, 3 ewes, 10 ewe lambs and 2 buck lambs, had just arrived from the fair of the New York State Agricultural Society, held at Rochester, N. Y., and owned by C. Stone, Esq,, of York, Washtenaw county, Michigan. They show purity of blood, fine size, and symmetry of form.

Class 2, middle wooled, were all of the Southdown breed, showing great beauty and purity of blood.

Respectfully submitted by

ANDREW Y. MOORE,
A. J. KEENEY,
C. A. CHIPMAN,
Committee.

CLASS III, MERINOES.—CLASS IV, SAXONS.

The committee on Merino and Saxon sheep make the following awards:

Best merino buck over 2 years old, No. 287, Harris Newton, Novi, Oakland county, medal and..................$5 00
2d do No. 40, E. T. Lovell, Climax, Kalamazoo county,... 5 00
3d do No. 18, E. Arnold, Dexter,..................... 3 00
Best merino buck 2 years old or under, No. 23, G. W. Gale, Ypsilanti,.................................... 5 00
2d do No. 159, J. A. Austin, Plymouth, Wayne Co.,..... 3 00
3d do No. 159, do do do 2 00
Best pen of 5 buck lambs, No. 271, N. Dickinson, Romeo,.. 5 00
2d do do No. 190, Hiram Taylor, Romeo,.. 2 00
3d do do No. 38, J. Milham, Kalamazoo, Transactions.
Best pen of 5 ewes, No. 18, E. Arnold, Dexter, Washtenaw county, medal and.......................... 5 00
2d do No. 268, Ira H. Butterfield, Utica,.............. 5 00
3d do No. 300, Alonzo Henry, Borodino, Wayne county, American Shepherd and...................... 2 00
Best pen of 5 ewe lambs, No. 269, N. Dickinson, Romeo,.... 5 00
2d do do No. 270, do 3 00
3d do do No. 288, Harris Newton, Avon, Transactions.

SAXONS.

Best Saxon buck over 2 y'rs old, No. 124, D. D. Gillet, Sharon, Washtenaw county, medal and................$5 00
2d do No. 115, G. M. Simonson, Royal Oak,............ 5 00
3d do No. 129, J. P. Gillet, Sharon,.................... 3 00
Best Saxon buck 2 years old or under, No. 130, J. P. Gillet,. 5 00
2d do do No. 128, D. D. Gillet,. 3 00
Best pen of 5 ewes, No. 135, J. P. Gillet, medal and........ 5 00
2d do do No. 126, D. D. Gillet,................ 5 00
3d do do 131, J. P. Gillet, Amer. Shepherd and.. 2 00
Best pen of 5 buck lambs, No. 128, D. D. Gillet,............ 5 00

Best pen of 5 ewe lambs, No. 127, D. D. Gillett, ... $5 00

All which is respectfully submitted.

EDWARD SAWYER,
Chairman of Committee.

CLASS V.—NATIVE AND GRADE SHEEP.

Best buck over 2 years old, No. 41, a buck 4 years old, ¾ merino and ¼ Saxon, owned by E. T. Lovell, Climax, Kalamazoo Co., medal and $5 00

2d best buck over 2 years old, No. 228, Wm. Ten Eyck, Dearborn, ... 5 00

3d best buck over 2 years old, No. 57, 3 years old, C. A. Green, Troy, ... 3 00

Best pen of 5 ewes, No. 227, Wm. Ten Eyck, medal and ... 5 00

2d do 274, John Kirk, Dearborn, ... 5 00

Best pen of 5 buck lambs, No. 290, Wm. Whitfield, Pontiac, 5 00

SHEPHERD'S DOGS.

Best shepherd's dog, No. 134, owned by D. D. Gillet, of Sharon, Washtenaw county, ... $5 00

2d best shepherd's dog, No. 133, by J. P. Gillett of Sharon, American Shepherd.

In class No. 5, there were but few entries; the quality as grade sheep was good. Your committee would recommend the practice of crossing a merino buck with our native ewes, thereby producing an animal combining a hardy constitution, with a fleece sufficiently fine for most domestic purposes, also affording to the butcher's stall a desirable quality of mutton.

There were but two shepherd's dogs on the ground. They were well trained to perform the part assigned them, evidence being given of sagacity and a readiness to obey the instructions of their masters.

This animal, doubtless the most valuable of his race to the farmer, should grace his domicil, to share his prosperity or adversity, always being his constant friend.

All of which is respectfully submitted by

JOHN STARKWEATHER,
GEORGE MACOMBER,
H. K. FARRAND,
Committee.

CLASS VI—FOREIGN SHEEP.

The undersigned committee, having carefully examined all the foreign sheep offered for premium, report as follows:

We found no specimens of long wooled, middle wooled or Saxony sheep, consequently our attention was confined to French and Spanish merinos. In this department the exhibition was quite large and the quality very superior. The committee found it very difficult to decide which were the most superior animals. They recommend the following premiums:

Best merino buck, No. 14, French merino, A. L. Bingham, of West Cornwall, Vt., ... $5 00

Best pen of 5 ewes, No. 284, J. Bingham, Shoreham, Vt.,... 5 00

No person presented the number of 5 buck lambs, but your committee recommend a premium of $3 00 for NO. 16, owned by F. S. Gale, of Bridgport, Vermont. This we consider an extraordinary buck lamb, superior in size and fleece and richly entitled to a premium. Your committee would speak in high praise of bucks in pens No's. 19, 20 and 21, as a superior lot of sheep, and recommend a discretionary premium to the exhibitor, R. R. Wright, Waybridge, Vermont.

We would also recommend a premium to No. 168, being 19 yearling bucks, presented by Chas. Rich, of Lapeer, Michigan. We think them very fine animals, and worthy a premium.

J. SHEARER,
J. B. BLOSS,
J. S. PEABODY,
Committee.

SWINE.

We, the undersigned, committee on swine, award the following premiums:

Best boar over 2 years old, No. 44, owned by R. B. Merrill, of Battle Creek, Calhoun county,........................ $5 00

2d do No. 47, owned by Jereh. Brown, Battle Creek,.... 3 00

Best breeding sow over 2 years old, 194, a Leicester sow, owned by G. Knapp, Albion, Calhoun county,........... 5 00

Best lot of pigs 5 months old, No. 247, G. P. Bennet, Jackson, $5 00
2d do do No. 295, Doct. J. C. White, Detroit,... 3 00

JACOB SUMMERS,
WM. TEN EYCK,
Committee.

POULTRY.

The committee, upon examining the colleotion of poultry, find various breeds exhibited under one entry, designated as No. 208, consisting of eight coops; also other lots under their appropriate numbers. They award the premiums as follows:

Best lot of Dorkings, No. 208, Doct. M. Freeman, Kalamazoo, $3 00
do Polands, No. 208, do do 3 00
do large fowls, " do do 3 00
do poultry owned and exhibited by one person, with statement furnished and verified. consisting of 12 varieties, No. 208, Doct. M. Freeman, Bement's American Poulterer's Companion and................... $4 00
Best lot of turkies, No. 83, Wm. Maiden, Plank Road, Wayne county,.................................... 3 00
Best lot of geese, No. 110, John Griffith, Farmington,...... 3 00
To a lot of beautiful white Bloomer Bantams, No. 187, T. Wiley, of Detroit, we recommend a premium of....... 1 00
In lot No. 208, was a lot of Seabright Bantams, a new variety, exceedingiy small, perfect in symmetry, and clean legged; to these we recommend a discretionary premium of.................................... 1 00
No. 353, a foreign lot of Dorkings, exhibited by James Dougall, of Windsor, C. W., very fine and worthy a premium, 2 00
No. 281, a beautiful lot of wild turkeys, exhibited by Amos Mead, Plymouth,.................................... 1 00
No. 353, a lot of fancy pigeons, James Dougall, C. W.,.... 1 00
" 359, a lot of very fine geese, C. Perkins, York,....... 1 00

All of which is respectfully submitted, by

A. B. MOORE,
Chairman of Committee.

FARM IMPLEMENTS.

CLASS I. AND II.

Your committee on farm implements, class 1 and 2, would respectfully report that they have carefully examined all the articles presented for their consideration, and make the following awards:

Best farm wagon, No. 18, E. H. Davis, Kalamazoo,........ $5 00
Best corn cultivator, No. 61, spring steel teeth cultivator, B. B. Morris, Pontiac,.............................. 3 00
Best corn stalk cutter, No. 72, F. F. Parker & Bro., Detroit, 5 00
Best straw cutter, No. 28, D. O. & W. S. Penfield, do 3 00
We would recommend a discretionary premium to straw cutter, No. 86, Moses Rogers, Ann Arbor, Transactions.
Best corn and cob crusher, by horse power, No. 22, J. T. Willson, of Jackson,.......................... 5 00
No. 20. "Hydraulic Churn," Augustus Day, Detroit, dip.
No. 17. "Premium Churn," J. B. Tillinghast, Graham's Station, Meigs Co., Ohio, committee recommend a premium of.................................. 2 00
Best grain cradle, No. 12, a light, beautiful article, R. Simmons, Farmington,.......................... 2 00
Best six hand rakes, No. 41, D. O. & W. S. Penfield, Detroit, 2 00
Best hay and manure forks, No. 95, Ellithorp & Co., Detroit, 4 00
Best clover huller, No. 90, A. A. Holmes, Tecumseh,...... 5 00

EDWARD SMITH,
Chairman of Committee.

CLASS III. AND IV.

The committeee on farm implements and machinery, class 3 and 4, report as follows:

No. 1, Moore's improved seed planter and seed drill, a very good and compact machine, exhibited by Messrs. Nixon & Voorhies, of Adrian, Lenawee county, diploma and. $3 00
Also to Messrs. King & Walker, of Lancaster, Pa., for a specimen of Moore's improved seed planter, exhibited by them, ... 3 00

No. 2, an eighteen horse power high pressure engine, a very simple and beautiful article, manufactured by Messrs. Johnston, Wayne & Co., Detroit, ...Diploma.

No. 5, tile machine, for manufacturing tile for under-draining and other purposes, a very useful article, exhibited by John Daines, of Birmingham, Oakland county, ...Diploma.

No. 6, a sewing machine, presented by Messrs. Hallock & Raymond, Detroit. This is one of the latest improved machines, beautifully finished, and the work performed by it is excellent, ...Diploma.

No. 9, a stump extractor, a very simple and powerful machine, Paul Coffin, Byron, Shiawassee county, ...Transactions.

No. 13, cider press and grinder, a very good article, C. H. Bennet, Plymouth, Wayne county, ...Transactions.

No. 14, grain drill—this is a very beautiful article, great taste has been displayed in its construction, Beckford & Huffman, Macedon, N. Y., ...Diploma.

No. 19, Gate's patent dies for cutting screws on metal—this is an invention of great merit and usefulness, Gates & McKnight, Chicago, Ill., ...Diploma.

No. 21, improved railroad horse power—this power possesses some valuable improvements, which do not appear in many of the horse powers on this principle, Messrs. D. O. & W. S. Penfield, Detroit, diploma and $5 00

No. 23, an overshot thresher and separator, a good article, Messrs. Penfield, ... 5 00

No. 25, a sawing machine, an article valuable for its simplicity, Messrs. Penfield, Detroit, ...Transactions.

No. 26, a feed mill. A good and simple article for grinding corn in the ear.

No. 30, vegetable cutter. An article worthy the notice of farmers. Messrs. Penfield, ...Transactions.

No. 32, two water rams. These are valuable improvements in the method of raising water, ...Premium.

No. 37, two cornshellers, which do the work with ease and rapidity. Messrs. Penfield, ...Diploma.

No. 53, cooking stove. This is an arttcle patented by Stuart, and has a wide spread reputation for economy and durability. Messrs. Penfield,..................... Diploma.

No. —, hot blast cooking stove. For its many and valuable qualities has attained a reputation seldom if ever equaled by any other stove. M. Howard Webster, Detroit,...Diploma.

No. 54, Smith's ventilating smut machine. This is a good article and worthy the consideration of millers. Messrs. Penfield, Detroit,............................Diploma.

No. 64, three platform scales, 1 letter press, 1 dentist's lathe, and 3 cast steel stamps, all good and valuable articles, manufactured and exhibited by D. E. Rice, Detroit,...Diploma.

No. 65, Hinkston's cast iron wheel cultivator with improved steel teeth, a new invention, D. Hinkston, Clarkston, N. York, ...Diploma.

No. 70, a new cylender grain drill. This is an article possessing superior excellence in workmanship and beauty of finish. Aaron Palmer, Brockport, N. Y.,...........Diploma.

No. 75, Clark's patent portable merchant mill. This is a valuable improvement. T. M. Clark, Lancaster, Pennsylvania, ...Diploma.

No. 77, a model board fence. This fence possesses great advantages over the ordinary board or rail fence. John Brewer, Ypsilanti,............................Diploma.

No. 80, Hydraulic engine and pumps. This engine was manufactured for the city of Detroit, by Messrs. DeGraff & Kendrick, and while it reflects great credit upon the manufacturers, for their skill and workmanship, for its beauty of finish, the ease and certainty with which it performs its work, it is also an honor to the city; Messrs. DeGraff & Kendrick,............................ Medal.

Nos. 81, 82, 83, 84 and 85. Five stationary high pressure steam engines, on cast iron frames, ranging from 15 to 60 horse power, all of which are of superior workmanship and beauty. The arrangement of their several parts display good taste and judgment in uniting beauty, compactness and durability. DeGraff & Kendrick.

No. 88. 1 dozen ox-bow fastenings; a valuable improvement, Messrs. Penfield, Detroit,........................Diploma.

No. 91. A miller's or inspector's brand. This is a newly invented article, and possesses superior advantage over the brands now in use; J. M. Holbrook, Detroit,........Diploma.

No. 72. A two horse endless chain power, and a one horse endless chain power, both manufactured by Wheeler, Mellick & Co., of Albany, N. Y., and exhibited by F. F. Parker & Bro., Detroit.

1 threshing machine, 1 corn and seed planter, 1 saw mill, 1 Burrell's iron corn sheller, 1 southern corn sheller, all good and valuable articles, and all exhibited by Messrs. Parker & Bro., Detroit, diploma and................ $5 00

No. 104, a mulley harvest rake, a very light, handy and convenient article, Sam'l Stanton, Plymouth,............Diploma.

No. 117, Virginia Reaper. This is the celebrated reaping machine which has been received with great favor in this country and in Europe, and is deserving of consideration by the American agriculturist. The improvements added to this machine within the last year, render it in our opinion the most useful farm implement coming under our notice; C. H. McCormick, Chicago, Ill.; we therefore award to it the premium of a medal and..... 10 00

No. 87. Emery's seed planter, by Messrs. Penfield,........ 3 00

No. 120. 1 seperator and horse power, on the sweep principle, Peabody & Munroe, Albion, Calhoun Co.,......Diploma.

No. 20. Bartle's self-acting regulator for adjusting and maintaining at a uniform heighth, the water in steam boilers. Isaac McNeil, Newark, Wayne Co., N. Y.,..........Diploma.

A. A. WILDER,
P. R. ADAMS,
Committee.

PLOWS AND PLOWING.

The committee on plows and plowing, beg leave to report, that want of time prevents them from making a full and extended report, and they will for the present, merely state to whom they have awarded premiums:

To A. Wattles, of Troy, Oakland county, 1st premium for plowing, ...$10 00

No. 298, James Moore, Almont, Lapeer county, Gardner's Farmer's Dictionary and 7 00

No. 349, A. Knapp, Novi, Oakland county, ... 5 00

No. 292, J. J. Joice, of Plymouth, the committee recommend a discretionary premium.

The quality of plows not tested by the committee, but there was a large number exhibited, of superior workmanship.

LINUS CONE,
Chairman of Committee.

The executive committee make the following award upon plows:

No. 7, L. H. Hubbard, Mt. Clemens, Macomb county, for cast iron beam plow, ...Diploma.

No. 8, sod plow, O. C. Mosher, Jackson, ... "

" plow for old land, " " .... "

No. 15, sub-soil plow, James H. Fellows, Manchester, ... "

No. 63, J. S. Smith, Plymouth, sod plow, ... "

No. 97, iron beam plow, Albert Wilber, Rives, Jack. Co., .. "

No. 98, double mold board, one-horse plow, Albert Wilber, Rives, Jackson county, ... "

BUTTER, CHEESE, SUGAR, HONEY AND BEE HIVES.

The committee on butter, cheese, sugar, &c., have examined the articles presented in their departments, and noted their awards opposite the numbers. There were but few cheeses of one year old and over, but several new cheeses equal, in the opinion of your committee, to any made at the best dairies in the State of New York.

BUTTER.

Best lot of butter made from 5 cows, in thirty consecutive days, No. 53, James Smith, Detroit, ...$7 00

No. 47, best 15 lbs. made at any time, E. Cross, Redford, ...3 00

No. 48, 2d do do Clark Beardsley, Troy, Transactions and 1 00

No. 13, 3d do do J. B. Springer, Livonia, ... 2 00

No. 36, 4th best 15 lbs. butter made at any time, W. Dennison, Troy, 7th Vol. Michigan Farmer.

No. 7, best 10 lbs. made in June, Mrs. Isaac I. Voorhies, of Waterford, Oakland county,........................ 3 00

No. 31, 2d best 10 lbs. made in June, Mrs. Titus Dort, Dearborn, Transactions and........................... 1 00

CHEESE.

Best cheese 1 year old and over, No. 20, Luther Lapham, Farmington,....................................$5 00

2d do No. 15, W. Gillet, Rome, Lenawee county,........ 3 00

An excellent cheese, No. 25, F. H. Murry, Farmington,...... 3 00

3d best cheese 1 year old, No. 42, L. Moore, York, Washtenaw county,.................................... 2 00

No. 55, two very excellent cheeses, made without pressing, Mrs. Titus Dort, Dearborn,....................Premium.

HONEY.

Best 10 lbs. honey, No. 22, Amos Mead, Plymouth,.........$3 00

2d do do No. 28, A. Stockwell, Redford,.......... 2 00

3d do do No. 37, S. Grodfrey, Paw Paw, 8th vol. Michigan Farmer.

Best bee hive, No. 37, S. Godfroy,...................... 3 00

MAPLE SUGAR.

Best 10 lbs. maple sugar, No. 27, D. White, Northville,......$5 00

3d do do No. 43, Loren Moore, York, 8th vol. Michigan Farmer.

AUSTIN WALES,
Chairman of Committee.

—

DOMESTIC MANUFACTURES.

CLASS I.

The undersigned, committee on domestic manufactures, class 1, respectfully report that they have examined with as much care as circumstances would allow, the various articles presented for premiums.

The number of qualities to be taken into consideration in goods of this kind, has often rendered a judgment difficult, but they have en-endeavored to ascertain which article combined the greatest amount of skill with excellency of materials, and have decided accordingly. They would likewise state that they met with another difficulty, the remedying of which they would respectfully suggest before another fair. In such goods as fulled cloths, coverlets, &c., the manufacturer by profession, and the farmer's wife come into competition, while it is impossible that the latter can make as handsome an article as the former, though the actual quality may be as good. But it is believed that this Society peculiarly intends to foster "home" productions of this sort; and therefore your committee think that it would be best to offer two setts of premiums, one for manufacturers, the other for agriculturists. Among the goods presented are some for which no premium was offered, and yet which were considered worthy of distinction. These will be found mentioned at the end of the list; and we would respectfully suggest that diplomas be presented to the exhibitors. Indeed, some extra encouragement of this kind would appear to be more necessary from the fact that so small a number of manufactured articles are exhibited, while those present are generally most creditable to our State, and speak well for the ability and industry of our citizens, yet we are convinced that a much greater number might have been presented, and every increase of manufacturers among our agricultural community, which helps to consume our produce and absorb our labor at home, is to be hailed as a general benefit.

Best pr. woolen blankets, No. 102, George Chamberlain, Redford, diploma and $4 00

2d best, John Gray, Dearborn, transactions and 2 00

Best 10 yds. flannel, No. 91, A. Taylor, Kalamazoo, diploma and 4 00

2d best, No. 116, Edward Chase, Rose, Oakland county, ... 3 00

Best 10 yds woolen cloth, No. 91, A. Taylor, Kalamazoo, diploma and 4 00

2d best, No. 146, C. Cornwall, Ann Arbor, transactions and 2 00

Best hearth rug, No. 203, Miss J. Tooker, Superior, Washtenaw county, ... 3 00

2d best, No. 163, J. Farrier, Ypsilanti, transactions.

3d do 212, James Smith, Greenfield, 8th volume Michigan Farmer.

Best 10 yards rag carpet, No. 124, Miss Julia Ann White, Northville, diploma and $3 00
2d best, No. 164, Mrs. A. Bradner, Dearborn, ... 2 00
3d " 135, David Thomas, transactions.
Best pr woolen knit stockings, No. 19, Mrs. Mark Norris, Ypsilanti, transactions and 1 00
2d best, No. 139, Mrs. A. C. Hubbard, Detroit, ... 1 00
Best pr woolen knit socks, No. 118, Mrs. John Palmer, Detroit 2 00
2d best, No. 141, Mrs. J. H. C. Garvin, Washington, ... 1 00
Best pr woolen knit mittens, No. 83, Edward Sawyer, Grand Blanc, ... 1 00
Best woolen coverlet, No. 86, John Thomas, Oxford, diploma and 1 00
2d best, No. 167, James Bailey, Troy, transactions.
Best pc satinet, No. 93, Hibbard & Davis, Milford, diploma.
Best sample woolen yarn, No. 198, Mrs. O. P. Stout, Troy, .. 1 00
" worsted yarn, No. 85, Mrs. Jabez Warner, Plymouth, ... 1 00
Best pair of worsted stockings, No. 173, W. Dennison, Troy, 2 00

In addition to the above, we would recommend the following discretionary premiums:

The general good quality of the whole lot of cloths and cassimeres offered by the Kalamazoo company, is deserving of a diploma and $5 00
No. 167, 10 yds "home-made" fulled cloth, James Bailey, Troy, ... 2 00
No. 33, a pair of black silk stockings, and specimens of the silk before dyeing, both beautiful articles, G. W. Collins, Farmington, ... 3 00
No. 130, lot of woolen yarn, Mrs. Francis Leslie. Dearborn, 1 00
One fleece clean washed French merino wool, weighing 14 and 9-16th lbs., Walter Wright, Adrian, Lenawee county, Silver Medal.
No. 130, a kint woolen jacket and comforter, Francis Leslie, Dearborn, transactions.

No. 172, ten articles children's lambs wool dresses, &c., very beautiful, John Clay, Germantown, Pa., diploma.

No. 116, a very beautiful woolen shawl, made in this State, in imitation of the "Bay State" shawls—it deserves honorable mention, Edward Chase, Rose, Oakland Co., Downing's Cottage Residences.

Messrs. Raymond & Nall, of Detroit, exhibited some very fine specimens of shawls of American manufacture, . . Diploma.

This firm also exhibited some very beautiful samples of carpets.

L. Beecher, of Detroit, exhibited some splendid specimens of carpets.

No. 7,	bed quilt,	Mrs. Hiram Arnold, Ann Arbor, ...	Premium.
" 26,	"	Mrs. Linus Cone, Troy, ...	"
" 103,	"	Mrs. Mary Christie, Detroit, ...	"
" 123,	"	Mrs. M. Beebe, Southfield, ...	"
" 147,	"	Mrs. Wm. Lowe, Farmington, ...	"
" 195,	"	Mrs. J. Boutwell, Ann Arbor, ...	"
" 208,	"	Mrs. Moreland, Manchester, ...	"

CHAS. FOX,
Chairman of Committee.

CLASS II.

The committee on domestic manufactures, class 2, respectfully report, that of the specified products committed to their care no specimens were presented in the once prominent department of domestic industry, comprising linen and tow cloth, linen diaper and linen thread. This is doubtless owing to the improved quality and cheapness of the cotton substitutes, through the application of machinery and steam. It affords an additional illustration of the revolutionizing power of such agencies over the tastes, habits and occupations of society.

The same defect is also to be noted in regard to the specified article of woven hosiery, a branch of industry that for some reason has never found much favor with our native citizens, although the consumption of the products of the foreign knitting frame in the United

States is very large. May we not hope that the characteristic ingenuity of our citizens will e'er long reduce this manufacture and its mechanism to a form better adapted to their taste and to a profitable investment of labor and capital.

Of knitted cotton and linen stockings a small but very excellent and beautiful assortment was presented, some distinguished by the substantial usefulness of their fabric and some by the taste and beauty of their texture. Among them your committee have distinguished the following as worthy of special notice and reward, viz:

No. 82, best cotton stockings (clocks) Edw. Sawyer, Grand Blanc, ... $2 00

No. 149, best Lisle thread stockings, J. B. Bloss, Shiawassee, .. 1 00

No. 141, cotton stockings, Mrs. J. H. C. Gargin, Washington, .. 1 00

No. 83, best linen stockings, Edward Sawyer, ... 2 00

All which is respectfully submitted.

JOSEPH PENNY,
Ch'n Committee.

CLASS III.

The committee on domestic manufactures, class 3, would make the following report:

On examination we find that overcoat No. 192 is entitled to the first premium, Geo. Common, Detroit, diploma and .. $4 00

No. 128, dress coat, Eagle & Elliott, do do ... 2 00

do best pantaloons, do do do .. 2 00

do 2d best overcoat, do do 3 00

No. 166, a very beautiful embroidered vest, Hyde & Satchell, Detroit, Medal.

No. 211, best hats, F. & C. H. Buhl, Detroit, ... 2 00

No. 199, best ladies' shoes, Smith & Tyler, Detroit, ... 2 00

No. 162, best calf boots, A. Boor, Detroit, ... 3 00

No. 11, best boot and shoe lasts, Wm. Fletcher, Detroit, Diploma.

No. 131, a dress coat made by Mrs. Alice W. Butterfield, of Utica, Macomb Co., Dowing's Fruit and Fruit Trees and . 1 00

B. F. EGGLESTON,
Ch'n Committee.

CLASS IV.

The uudersigned, committee on domestic manufactures, class No. 4, beg leave to make the following very brief report:

No. 12 embraces a lot of drain tile, which from its appearance, cheapness and durability, we think it entitled to a premium; we therefore award the maker, John Daines, of Birmingham, a premium of a diploma and Transactions.

No. 19. Hot air furnaces, registers and stoves for do., C. F. Kneeland, Detroit, ... Diploma.

Nos. 69 to 77 inclusive are all articles manufactured and presented by Messrs. H. Sangster & Co., Buffalo, N. Y., consisting of various steamboat, rail road, and other lamps, of beautiful and ingenious workmanship, we therefore award a .... Diploma.

No. 106. An open, one seat buggy, of excellent construction and finish, for which we award the first premium, Moses Stanfield, Jackson, diploma and $5 00

No. 209. Open buggy, Vreeland & Sperry, Ann Arbor, 2d premium, ... 3 00

No. 179. A superb two horse carriage, manufactured at Rochester, New York, by Mr. Cunningham and presented by George Heron, of Detroit, ... Diploma.

No. 187. 2 horse shoes, J. B. Brumfield, Plymouth, ... Diploma.

No. 188. 2 horse shoes, Thomas Hall, Detroit, ... 1 00

Nos. 216 to 230. All articles of elegant cabinet furniture, of foreign manufacture, presented by J. W. Tillman, of Detroit, ... Transactions.

No. 232, consists of three top buggies, manufactured at Toledo, Ohio; also one 2 seated carriage, all very beautiful in appearance, and novel and ingenious in construction, L. M. Butts, ... Diploma.

Nos. 3 and 4. A two horse carriage and 1 2 seated rockaway. Both are splendid and ingenious in their formation, manifesting superior skill, which our limits will not enable us adequately to describe, Messrs. Paton & Co., Detroit, diploma and 10 00

No. 234, called a mammoth chair, it might well be denominated nonsuch, presuming that nothing of the kind has heretofore been presented to the vision of Wolverines, and for the exhibition of such masterly skill, perseverence and patience, in devising and constructing that wonderful, matchless and beautiful article, we award the maker, Professor Barnes, of Port Huron, a prem. Transactions.

No. 22, bedstead fastenings. A valuable article, as bedsteads thus fastened could be instantly taken down and removed from the room in case of fire, J. T. Willson, Jackson, Diploma.

No. 49 to 55 inclusive, consisting of several varieties of leather, Messrs. Ladue & Eldred, Detroit,..... ....Diploma.

Your committee are aware that in many cases, justice has not been done to exhibitors, as we are not sufficiently informed in relation to many articles belonging to our department.

JOHN BURCH,
WM. W. CALKINS,
Committee.

PAINTINGS, DRAWINGS AND DAGUERREOTYPES.

The committee on paintings, drawings and dagurreotypes, report that the best specimen of animal painting in oil, is

No, 12, cattle scene, by L. T. Ives, Detroit,............... $3 00

Best specimen of oil painting, by Michigan Artist, No. 65, "The Mizers," F. E. Cohen, Detroit,............... 3 00

They recommend the following, as possessing great merit, and worthy of praise:

No. 8, an India ink painting, Miss Sarah E. Smith, Detroit,.. 1 00

" 36, two landscape drawings in monochromatic, Mrs. Mark Norris, Ypsilanti,.................................. 2 00

No. 76, a fruit piece, John Atkinson, Detroit,............Diploma.

" 57, monochromatic painting in rustic frame, the whole work executed by the exhibitor, Miss Standish, Romeo, Macomb couuty,..........Downing's Cottage Residences.

No. 45, a picture frame, gilt, and carved with great taste and skill, John Atkinson, Detroit,....................Diploma.

No. 52, a design for a hotel, Lum and Ford, Detroit, ... Diploma.
" 53, do church, do do .... "
Best specimen of daguerreotype, No. 44, M. Sutton, Detroit, 2 00
2d best do do 59, O D. Moore, " 1 00

F. E. COHEN,
HENRY LEDYARD,
BENJ. FOLLETT,
Committee.

NEEDLE, SHELL AND WAX WORK.

The committee appointed to examine the various articles in the department of needle work, &c., would present the following report:

That from the great number of articles to be examined they found it very difficult to make the full examinations so as to compare perfectly. Yet they are satisfied of having done the work assigned them as carefully and impartially as could possibly have been done under the circumstances, and present the result of their decisions as follows:

No. 2. one straw bonnet. In this article there was no ocompetition. This bonnet is finely wrought by Mrs. Jacob Hendrickson, Pontiac, ... 3 00

No. 1, embroidered mantilla or cape. This article presents evidence of much careful and ingenious needle work. We recommend to the maker, Miss Clara Davis, of Ypsilanti, a premium of 3 00

No. 92, a very neatly embroidered pair of suspenders. These show fine taste and very neat execution, ... 1 00

No. 6, a crochet quilt. This is a very neat article, shows much and very careful labor, 1st premium, Mrs. A. Fowler, Detroit, ... 2 00

No. 11, one needle work wall basket, beautifully wrought by Mrs. Griffin, Detroit, ... 1 00

No. 15, a beautifully wrought scripture scene, showing fine taste and much work, Mrs. C. W. Williams, Detroit, ... 2 00

No's 16 and 17, a very fine group of neatly wrought articles showing much taste and labor, Miss Gillman, Detroit, .. 2 00

No. 18. one satin embroidered pin cushion, Miss D. Brown, Ypsilanti, ...$1 00

No. 35, one satin smoking cap, embroidered; a most exquisite piece of needle work, and while the committee feel obliged to recommend it for the first premium in needle work, they cannot but regret that so much taste, beauty and skill should be bestowed upon an article to be used for so unworthy a purpose; Miss C. H. Spencer, Ypsilanti,. 3 00

No. 21 and 22, knit bed spreads or counterpanes, of most excellent quality, both showing great skill and labor. To the latter number we recommend the premium, E. M. & A. H Moyles, Detroit,........................ 2 00

No. 23, 2 silk bonnets, very neat, beautiful workmanship, S. Freedman & Bro., Detroit,........................ 2 00

No. 24, a neat piece of worsted work, deserving much credit, Miss Mary S. Palmer, Detroit,.................. 2 00

No. 25, 2 pieces of fancy bead work, finely wrought, showing much taste, Mrs. P. E. DeMill, Detroit,........... 1 00

No. 73, 1 silk patch work table cover, very carefully wrought, showing patient labor and much skill, Mrs. Riddet, Detroit, ... 3 00

No. 30, a tapestry work table cover, wrought with great skill, a very beautiful article, Mrs. P. Desnoyers, Detroit,... 2 00

No. 31, 1 crochet lamp mat, the first work of a little girl, and is worthy of praise, as showing both industry and genius, Miss A. Clark, Detroit,.................. 1 00

No. 34, one shell work chair, an article wrought with great care. The committee found it somewhat difficult to decide between several very neat articles of this class, yet they felt obliged on account of the neatness and great accuracy, as well as the very substantial character of the work, to recommend for a premium, Mrs. Ellwood, Detroit, ... 2 00

No. 20, 1 smoking cap, neatly wrought by Miss C. Davis, Ypsilanti, ... 1 00

No. 37, 1 beautifully embroidered handkerchief, also an embroidered collar, both very neat, to the latter the committee recommed a premium, Mrs. A. C. Hubbard, Detroit, ... $2 00

No. 39, 1 very neat purse and one crochet shawl, very creditable to the maker, Miss M. W. Ballard, Ypsilanti, ... 1 00

No. 40, one very beautifully wrought fire-screen. This article the committee learn, was wrought by a lady aged 61 years. and without spectacles. The committee confess themselves greatly at a loss to decide between this and another piece of needle work, yet they felt obliged to yield the palm to the other; still the committee deem this article as worthy of great praise, and recommend a discretionary premium, Mrs. J. N. Elbert, Detroit, ... 3 00

No. 41, a very handsome group of fancy needle work, exhibiting fine taste and workmanship, Mrs. Caton, Detroit, .. 1 00

No. 42, one embroidered chair and four ottomans, very beautiful, and tastefully wrought, Misses Baldwin, Detroit, .. 3 00

No. 48, three cases gentlemen's shirts—they show very neat workmanship in a very useful branch of home industry, Mrs. E. Thomas, Detroit, ... 3 00

No. 48, one sofa cushion, one pair of ottomans, and one sample of fancy work, Mrs. R. W. Van Fossen, Detroit, ... 1 00

No. 49, one piece of gilt frame worsted work; this is a very beautiful piece of work of the kind, being a representation of Bolten Abbey, and does very great credit to the genius and industry of the lady who wrought it, Mrs. Charles Piquette, Detroit, ... Downing's Cottage Residences.

No. 50, a shell work gothic church, constructed by a daughter of Michigan, now in California, of very beautiful workmanship—the committee regret that it has been somewhat damaged on its way hither, by which its beauty is somewhat impaired—we recommend a premium, Mrs. H. B. Withington, Lasalle, Monroe county, ... 1 00

No. 56, a shell work cottage—this article is beautifully and tastefully arranged, and deserves great credit, Mrs. R. W. Withington, Lasalle, Monroe county, ... 2 00

No. 60, this consists of one doll bedstead, very beautiful, also two boxes of shells very tastefully and beautifully arranged, Mrs. B. B. Kercheval, Detroit. Encyclopedia of Domestic Economy.

No. 65, 1 pink crochet hat, a very neatly wrought article, showing genius and industry, Mrs. F. Finley, Detroit,...$1 00

No. 68, a case of ornamental hair work, set in jewelry. These articles are exquisitely wrought and show much taste and skill, Mrs. Lemcke, Detroit,................ 2 00

No. 69, two large ottomans of very fine workmanship, worthy of much praise, Mrs. F. Buhl, Detroit,................ 2 00

No. 70., one hour glass stand, a very neat article, Miss Cornelia M. Ayers, Detroit,............................ 1 00

No. 27, one silk patch work table cover, a beautiful article, exhibiting taste and industry, Mrs. H. E. De Garmo, Ann Arbor,.................................... 2 00

No. 74, a very prettily wrought worsted camp chair, Miss Duffield, Detroit,.................................. 2 00

No. 75, one hour glass stand. The competition between this and No. 70 is so close that the committee do not pretend to judge, Miss Ayres, Detroit,.................... 1 00

No. 78, eleven pieces of ornamental needle work, showing much labor and taste, Mrs. J. Starkweather, Ypsilanti,.. 2 00

No. 180, a worsted work table spread and one work bag, very neat, Mrs. Wm. Hale, Detroit,.................... 1 00

No. 45, a cotton quilt, Mrs. Isaac J. Voorhies, Waterford,... 1 00

No, 46, two pine burr baskets and one port folio, neatly wrought, Miss Mary N. Smith, Detroit,.............. 1 00

No. 82, white cotton quilt, very neatly wrought, Edw. Sawyer, Grand Blanc,............................... 2 00

No. 86, two fly brushes, neatly wrought of peacock feathers, very beautiful, Mrs. B. G. Stimson, Detroit,............ 2 00

No. 30, one pair lamp mats, worsted, neatly wrought, G. W. Collins, Farmington,............................ 1 00

No. 49, one worsted and silk sofa pillow, crochet work, Mrs. Chas. Piquette, Detroit,........................... 1 00

No. 51, worsted embroidery, Mrs. A. B. Matthews, Pontiac, 1 00

No. 59, a show case of fancy articles, consisting of 2 lamp mats, 1 slumber cushion, 1 net toiled cushion, 1 wall basket, 5 pair child's socks, 2 small scarfs, 1 comforter, 1 crochet work bag, 1 crochet bag for soiled muslins, &c. This is a very beautiful group of articles, and the committee think well deserves the premium for the best variety of worsted work, Mrs. Babe, Detroit,.......... Diploma.

No. 58, 1 camp chair, although not quite in season for competion, yet the committee noticed it and cannot but regret that it was not entered in season, a beautiful article, by Miss Sarah Bingham, Detroit,.................... $1 00

No. 138, 1 silk patch work chair, containing 3,206 pieces, beautifully wrought, by Mrs. A. A. Fish, Detroit, Downing's Cottage Residences.

No. 108, a very neat shell work quilt, Mrs. A. W. Comstock, Port Huron,.................................. 1 00

No. 109, a cotton patch work quilt, an article not only beauiifully designed but very finely executed, Mrs. E. Force, Manchester, ... 2 00

There are many other articles which the committee would be very happy to mention with commendation, as showing taste, skill and industry, but cannot, for want of time.

All of which is respectfully submitted,

J. A. BAUGHMAN,
Chairman of Committee.

FRUIT.

The committee on fruit respectfully report that they have examined the fruits presented for exhibition and award the following premiums:

No. 34, best and greatest variety of good table apples, Jas. H. Fellows, Manchester,.......Hovey's Fruits and Fruit Trees.

No. 23, 2d best and greatest variety of good table apples, Cyrus A. Chipman, Rochester,.......................$5 00

No. 48, 3d best and greatest variety of good table apples, A. Streeter, Bruce, Thomas' Fruit Book and............. 2 00

No. 38, best 10 varieties table apples, Albert Terry, Avon,.. 5 00

No. 78, 2d do do Wait Peck, Manchester,.. 3 00

No. 125, best 6 winter varieties table apples, J. G. Welch, Plymouth,..............Hovey's Magazine of Horticulture.

No. 25, 6 good winter varieties table apples, Isaac W. Ruggles, Pontiac,.............Hovey's Magazine of Holticulture.

No. 44, 2d best 6 good winter varieties table apples, Wm. Ten Brook, Adrian,........Downing's Fruits and Fruit Trees.

No. 29, best and greatest variety of pears, J. C. Holmes, Detroit,............Fruits and Fruit Trees, with colored Plates.

No. 81, 2d best and greatest variety of pears, B. G. Stimson, Detroit,..........................1 vol. Horticulturist.

No. 10, best 10 specimens peaches, Rev. Chas. Fox, Grosse Isle,..$3 00

No. 33, best 6 specimens seedling peaches, D. D. Gillet, Sharon,.. 2 00

No. 28, best 12 quinces, J. Simmons, Farmington,.......... 5 00

No. 18, 2d 12 do Linus Cone, Troy.......Hovey's Magazine.

No. 75, 3d 12 do Thomas Keal, Ann Arbor,...Mich. Farmer.

No. 30, best and most extensive collection of grapes grown in the open air, J. C. Holmes,......................$5 00

No. 86, 2d best and most extensive collection of grapes grown in the open air, C. C. Trowbridge, Detroit, Allen on the Grape and.. 2 00

No. 86, best dish of native grapes grown in the open air, Catawba, C. C. Trowbridge,......................... 2 00

No. 30, 2d best dish of native grapes grown in the open air, J. C. Holmes,......................Allen on the Grape.

No. 104, best 4 specimens of muskmelon, John Ford, Det.,..$2 00

No. 108, 2d best do John S. Bagg, Detroit,..Mich. Farmer.

The committee recommend the following discretionary premiums:

No. 27, a very fine collection of foreign grapes grown in the open air, F. E. Eldred, Detroit,............Allen on the grape.

No. 81, a very fine collection of foreign grapes under glass, B. G. Stimson,........................Allen on the grape.

No. 22, a very fine dish of foreign grapes, Black Prince, grown in the open air, Francis Raymond, Detroit, Transactions.

No. 47, a jar containing nine very large and fine yellow Alberge peaches, T. Butterfield, Brooklyn, Thomas' Fruit Book.

Best sample of currant wine, Mrs. J. Palmer, Detroit, Downing's Cottage Residences.

No. 28, 2d best sample of currant wine, A. C. Hubbard, Detroit, ... Country Residences.

Best sample cider vinegar, A. C. Hubbard, Detroit, ... Transactions.

Best sample wine vinegar, H. Walker, Detroit, ... Transactions.

Your committee would particularly notice a very fine collection of fruits from Messrs. Frost & Co., of Rochester, N. Y., for which they recommend a .... Diploma.

No. 138, two dishes of very fine peaches from Wm. Baby, of Chatham, Canada West, ... Transactions.

The show of apples was good, several lots being so nearly equal in specimens of good quality, that some to which no premium was awarded, would have been recommended for discretionary premiums but for the number of inferior kinds with which they were mixed, viz: Chesbro Russet, Pennock's Red Winter, Black Gilliflower, &c. These varieties should not be offered at our fairs.

JAMES DOUGALL,
J. W. SCOTT,
A. P. COOK,
Committee.

FLOWERS.

The committee on flowers respectfully report, that after making as careful an examination of the extensive collection submitted for inspection, as their limited time would admit, they award the premiums as follows:

No. 81, the best and greatest variety and quantity of cut flowers, B. G. Stimson, Detroit, ... $3 00

No. 114, best and greatest variety of dahlias, Mixer & Co., Detroit, ... Beck's Botany of the U. S.

No. 80, 2d best, Wm. Adair, Detroit, ... 1 00

" best 12 dissimilar blooms, Wm. Adair, Hovey's Magazine of Horticulture.

No. 101, 2d best 12 dissimilar blooms, John Ford, Detroit,.. $1 00
" 102, best single dahlia, John Ford,.................. 1 00
" 80, best and greatest variety of roses, Wm. Adair, Detroit.. 2 00
No. 81, best six varieties of phlox, B. G. Stimson,......... 2 00
" 115, best and greatest variety of verbenas, Mixer & Co., 2 00
" 81, best collection of green house plants owned by one person, B. G. Stimson, Detroit,.................... 3 00
No. 81, Best floral design, B. G. Stimson,..........Horticulturist.
No. 111½, 2d best floral design, Mixer & Co., Detroit,...... 2 00
" 103, best hand boquet, flat, John Ford, Detroit,....... 2 00
" 121, 2d best " " Mrs. J. C. Holmes, Detroit, 1 00
" 111½, best round boquet, Mixer & Co.,..Beck's Botany of U. S.
" 81, 2d " " B. G. Stimson, Detroit,....... 1 00
" 119, best and most beautifully arranged basket of flowers, Miss Lucinda Walker, Detroit,................ 2 00

The collection of dahlias was large, very fine, and of superior bloom. Your committee found the task of deciding for premiums very difficult, where all were so beautiful and so little inferior to each other.

No. 101 was a fine collection.
" 137, a collection from A. Frost & Co., Rochester, N. Y., were superb—we recommend a premium of.......... $1 00
Also a pyramid of Dahlias was elegant. Among the miscellaneous flowers, No. 112, from Mixer & Co., was a beautiful boquet,............................... 1 00
No. 102, was a fine collection of green house plants, containtaining many choice and rare varieties,............. 1 00
From Mrs. Doct. Smith, of Toledo, Ohio, a flat bouquet, very beautiful and very tastefully arranged,.............. 1 00
No. 19, 2 mammoth Cacti, from J. H. Morrison, Detroit,... 1 00

The committee remark that they were compelled to pass by flowers in some instances, in consequence of the neglect of those presenting them, not having entered them according to the rules prescribed by the Society. Your committee suggest that it be made a rule of the Society, that each person presenting flowers for a premium shall also present a written list with names and numbers of the articles presented. Until this is done, the decisions of judges must be more or less erroneous.

It is desirable in all miscellaneous collections that the number and names should accompany the flowers; and it is surprising to the committee that florists and gardeners do not pay more attention to this part of the subject.

All of which is respectfully submitted.

D. C. WALKER,
MRS. R. R. NORRIS,
Committee.

VEGETABLES.

The undersigned, committee on vegetables, award the following premiums:

No. 18, best white table turnips, Linus Cone, Troy,.........$1 00
No. 20, three best crook neck squash, D. S. Dean, Canton, Johnston's Agricultural Chemistry.
No. 20, best round squash, weighing 83 lbs. each, D. S. Dean,.. 1 00
No. 32, second best pumpkins, D. D. Gillet, Sharon,........ 1 00
No. 50, twelve best ears seed corn, J. S. Bagg............. 2 00
No. 51, best and largest pumpkins, J. S. Bagg, Gardner's Farmer's Dictionary.
No. 52, twelve best table beets, J. S. Bagg,................ 1 00
No. 55, best 12 tomatoes, do 1 00
No's 61, 62 and 63, second premium for seed potatoes, J. S. Bagg.............Gaylord & Tucker's American Husbandry.
No. 124, best seed potatoes, B. G. Stimson,............... 1 00
No. 105, six best stalks celery, John Ford,...........Transactions.
No. 105, best and greatest distinct variety of vegetables raised by the exhibitor, John Ford,...................... 5 00
No. 107, 2d best 3 heads of cabbage, John Ford,...Mich. Farmer.
No. 118, twelve best carrots, Bela Hubbard, Detroit,........ 1 00
No. 118, twelve best parsnips, Bela Hubbard,............... 1 00
No. 120, best ½ peck lima beans, O. B. Blackmar, Moscow,.. 1 00
No. 123, best bunch of onions, A. Wood, Detroit,........... 1 00

No. 19, best ½ peck table potatoes, N. D. Bingham, Clarkston,$2 00
No. 133, best vegetable eggs, G. V. N. Lothrop,............ 1 00
No. 134, six very fine Aberdeen turnips, Hubbard & Davis, Detroit,..................................Transactions.
No. 135, three ears Stowell's ever green sweet corn, Hubbard & Davis,............................Transactions.
No. 87, best three heads cabbage, George Crabb, Hamtramck,..2 00
Simeon Haven, of Trenton, exhibited seven varieties of very fine potatoes, which have severally resisted the rot. These potatoes were not entered in time for a premium. We would, however recommend that a premium be awarded him,............Johnston's Agricultural Chemistry.

All of which is respectfully submitted by

S. H. PRESTON,
JOHN CHAMBERLAIN,
Committee.

MISCELLANEOUS ARTICLES.

The committee on Miscellaneous Articles respectfully report that they have examined with as much care as time has permitted, the articles assigned to them. Their duties are the more onerous and difficult, from the fact that in most instances their examinations have been confined to single specimens without the opportunity of comparing with different articles of the same class. They have also to award discretionary premiums only, as the executive committee cannot affix beforehand premiums to articles not enumerated, because not foreseen. Doubtless many articles worthy of a premium will be found which are not herein recommended, for your committee not being advised of the amount which the executive committee can thus appropriate, have feared they might trespass upon funds otherwise disposed of.

A few articles upon the schedule are not mentioned, for the reason that they were not found, while others needed explanations from the exhibitors to enable your committee to do full justice to their merits. In general it may be remarked that the articles presented afford gratifying testimony that in ingenuity and mechanical skill, we shall soon rival older States and more densely populated communities; competing successfully with their accumulated capital and long experience.

The committee recommend the following premiums:

No. 8, 3 cases of confectionary, affording evidence that Michigan need not go abroad for this article of luxury, Wm. Phelps, Detroit, ... $3 00

No. 15, ornamental pen, tastefully wrought, Henry Blake, Detroit, ... Transactions.

No. 17, riding saddle, worthy of favorable notice, Joab Polhemas, Marshall, ... 2 00

No. 22, corn and cob mill, appears well adapted for the purpose designed and worthy the attention of every farmer, J. T. Wilson, Jackson, ... 5 00

No. 23, passenger railroad car. The style and finish of this article, it is believed, is not excelled by any State in the Union. The mechanics whose united skill produced it are worthy the highest measure of praise that this society can bestow. It is recommended that a suitable testimonial be directed by the executive committee to the Michigan Central Railroad, ... Diploma.

No. —, specimen of book binding, which for strength and durability we recommend a premium, A. Richmond, Detroit, ... Diploma.

No. 193, another specimen of book binding, for beauty and elegance of finish we recommend a premium, Friend Palmer, Detroit, ... Transactions.

No. 263, a Spanish Jack, considered a good specimen, and to the attention of the executive committee is respectfully suggested the encouragement that is needed, Isaac Cozzens, Detroit, ... 5 00

No. 87, a great variety of chewing tobacco and cigars, very creditable to exhibitor, Henry Miller, Detroit, Diploma & 3 00

No. 90, large gas lamp, merits favorable notice, Edw. Shepard, Detroit, ... Diploma.

No. 92, melodeon, a good instrument and worthy of encouragement, J. R. Smith, Adrian, ... 3 00

No. 94, water drawer, should be further tested to show its superiority as an article of pure necessity, Seaver & Giddings, Milford, ... $2 00

No. 12, twine wound corn brooms, an excellent article, John Hutchins, Southfield, ... 2 00

No. 95, wire and twine wound corn brooms, very good, Alexander Wattles, Troy, ... 1 00

No. 110, cyphering machine, no explanation given to the committee, they therefore refer it to the curious, Hugo Feality, Detroit, ... Transactions.

No. 351 and 363, specimens of deer, ruined by the bondage of civilization, shut out of their forest home, they lose much of their beauty.

No. 132, spring lounge, N. Flattery, Detroit, diploma and.. 2 00

No. 133, thread reel and ball, an ingenious and useful article, Peter Cheld, Detroit, ... Transactions.

No. 134, case of jewelry manufactured by Messrs. Valentine & Cruwell, Detroit, highly creditable to the manufacturers, and worthy of the Society's highest premium, Valentine & Cruwell, Detroit, ... Medal.

No. 137, upright piano, sweet tone, the manufacture is purely domestic and highly creditable to the manufacturer, H. Schoonmaker, Detroit, diploma and 1 00

No. 237, a 7 octave piano, of superior finish and tone, of great power and compass, worthy of the State, J. Bloynk, Detroit, diploma and 1 00

No. 100, wire cloth, fine specimens of a most useful article, Wm. Snow, Detroit, ... Diploma.

No. 142, grave stones, L. P. Hurd, Marshall, ... 2 00

No. 144, light fancy double harness; when such are made in Michigan we need not go elsewhere. J. P. Booth, Fentonville, diploma and 3 00

No. 161, Banning's body brace, too well known to need commendation, Doct. Rose, Detroit, ... Diploma.

No. 165, 2 bbls. plaster, an excellent article, David French, Detroit, Transactions and 2 00

No. 168, sole and upper leather, Geo. Kirby, Detroit, Transactions.

No. 169, Spanish saddle, a well manufactured article, and is deserving a favorable notice, Orrin Derby Niles, ...Diploma.

No. 170, cow bells, a good article, O. Starr, Royal Oak, Transactions.

No. 171, a case of gold pens, a great variety and worthy of favorable notice, Chas. Piquette, Detroit, ...Diploma.

No. 174, a jar of cream candy, Benj. Lee, Detroit, Transactions and 1 00

No. 178, artificial teeth—a splendid and necessary article, Doct. L. C. Whiting, Detroit, ...Diploma.

No. 181, telegraphic instruments and electrical apparatus, J. A. Bailey, Detroit, ...Medal.

No. 182, electro magnetic instruments, C. Crosman, Detroit, .M'dal.

" 183, ingot copper—fine specimens of the finest copper known to the commercial world, J. R. Grout, Detroit, Transact's.

No. 184, wigs and hair work, unsurpassed, Wm. Tate, " ..Diploma.

" 186, carriage whips. L. S. Noble, Detroit, ... 1 00

" 189, specimens of whittling, displays ingenuity that shows its maker capable of achieving more valuable ends, A. Smith & Sons, Birmingham, ...Transactions.

No. 191, fancy-colored morocco, good specimens, and deserving favorable notice, John Ladue, Detroit, ...Diploma.

No. 196, pressed brick, remarkable for smoothness and beauty, Chas. H. Wood. Detroit, ... $3 00

No. 197, electro plated ware, appears to be well manufactured, and ought to be encouraged, Judson Rockwell, Detroit, ...Diploma.

No. 200, Yankee knife sharpener, a good domestic article, Elisha Tyler, Detroit, ...Transactions.

No. 201, miniature copper tea kettle, made out of a single cent—the apprentice of ten months, only needs to cultivate his ingenuity by persevering efforts, to become a boss mechanic, Henry J. Hopkins, Detroit, ... 2 00

No. 202, lot of cabinet furniture, of fine workmanship and finish, highly creditable, Howard, Watkins & Co., Detroit, ...Diploma.

No. 206, an extensive variety of cabinet furniture, tastefully and thoroughly manufactured, Stevens & Zug, Detroit, Diploma.

No. 207, a case of silver ware, splendid, Geo. Doty, " "

No. 210, farmers' baskets, a good article, J. Perkins, York.

No. 177, Lyons Katharia, Jamaica ginger, and an assortment of perfumery, equal to any foreign article and worthy American patronage, L. S. King, New York, ... Diploma.

No's 214 and 215, rifle and shot guns. These specimens of Michigan manufacture exhibit mechanical skill and workmanship unsurpassed, John Sutherland, Ann Arbor 3 00

No. 232, stucco washed brick. Should this article prove as valuable as it promises, it will be of great importance, and a great saving in the more expensive paints, E. Van Zandt, Detroit, ... 2 00

No. 236, fire proof paint, W. P. & Levi Wilson, Brooklyn, Ohio, ... Diploma.

No. —, a box of mineral paint discovered sixty days since in Sharon, Washtenaw county, by John Townsend. It bids fair to rival the Ohio article. It should be thoroughly tested, and the attention of the society is especially invited to it. If it proves capable of preserving our buildings from fire as well as water, it cannot be too highly prized, ... Diploma.

No. 47, fifty samples of merino wool. These being very superior specimens, fine and uniform, it is recommended that a premium be awarded the owner on presentation of evidence that the samples were each from different animals; the committee presume they are, though the fact is not stated, Cyrus A. Chipman, Rochester, Oakland county, ... Medal.

No. 48, tail blocks and dogs for saw mills. These are recommended as a valuable improvement, Martin Rich, Juno, Dodge county, Wisconsin, Transactions and 2 00

No. 18, a side saddle; the quilted and worsted work covering is worthy of special notice, and we recommend a premium, Joab Polhemus, Marshall, ... 2 00

No. —, a beautiful specimen of book binding, C. A. Hedges, Lansing ....Diploma.

No. —, case of confectionary, N. Aldrich & Son, Detroit,.Diploma.

No. —, regalia for Odd Fellows, Sons of Temperance and Free Masons, T. H. Armstrong, Detroit,.............Diploma.

No. 238, twelve varieties of fish from Detroit River, by Geo. Clark, Detroit,.................................Diploma.

CHAS. G. HAMMOND,
MARK NORRIS,
E. G. MORTON,
Committee.

GRAIN, FLOUR AND SEEDS.

The committee appointed to examine grain, flour and seeds, would respectfully report:

No. 19, winter wheat, crystal flint, N. D. Bingham, Clarkston, Oakland Co.,.................................. $5 00

No. 16, best spring wheat, Wm. Beesley, Waterford, Agricultural Chemistry.

No. 11, best sample of oats, J. B. Springer, Livonia, Colman's Tour.

No. 186, best bbl. of flour from the least quantity of wheat, viz: 4 18-60 bushels being used, C. W. Chapel, Utica, 5 00

No. 121, best bbl. of flour without regard to the quantity of wheat used, Mead & Co., Detroit City Mills,.......... 5 00

No. 44, second best bbl. of flour without regard to quantity of wheat used, John Cupit, Rochester, Oakland Co.,.. 3 00

No. 80, 3d best bbl. of flour without regard to quantity used, George Paddock, Commerce, Oakland Co.,......Mich. Farmer.

Other samples of wheat and flour highly creditable to the producers were entered for competition, and the committee regret, that under the rules of the Society, they could not award more premiums, for they deemed them not only creditable to the producers, but also to the State.

All of which is respectfully submitted.

JOHNSON NILES,
F. S. FINLEY,
HENRY C. KIBBEE,
Committee.

ESSAYS.

The committee, to whom were referred the agricultural essays offered for premiums, report as follows:

In awarding a premium for an agricultural essay, this Society virtually endorses it as being, in its opinion, sound in theory, judicious in its practical recommendations, and faultless, at least, in its phraseology and style. Endorsed with this character, the essay is entered upon the permanent records of the transactions of the society, and is published to the world in the numerous agricultural journals, as worthy of the practical attention and adoption of the agricultural community. An essay, therefore, to be entitled to a pre-eminence, should, in the opinion of the committee, rank far higher than an ordinary, hastily prepared article for ephemeral newspaper publication. It should be more like a judicious, well studied, and deliberately prepared Agricultural Address; its subject matter practical and useful; its parts well and systematically arranged, and its style, plain, grammatical, free from technical terms and ambiguous phrases, and clearly expressive of the precise idea intended to be conveyed. With these general views, the committee have examined the Essays submitted to them as follows:

On agricultural fences and enclosures.

On the stave business.

On the management of sheep.

On the cultivation of potatoes.

On the cultivation of corn.

On the cultivation of wheat.

On the cultivation of fruit.

On the rotation of crops.

Most of these Essays have considerable merit, and might be regarded as respectable articles for agricultural newspapers, where they would go forth as the individual ideas and suggestions of the writer alone. There is but one of the number, however, which the committee feel justified in recommending for a premium, and that is upon "agricultural fences and enclosures." This Essay is not as full and complete as the committee would desire, but it is on a subject of much practical importance, and respecting which little has been written and published for general information, they would recommend that it be awarded a premium of $15.

All of which is respectfully submitted.

J. S. BAGG,
A. C. HUBBARD,
CHA'S BETTS,
Committee.

REPORT OF THE EXECUTIVE COMMITTEE.

ON FARMS AND CROPS.

At the session of the committee, held at Jackson, Dec. 9th, 10th and 11th, 1851, the following premiums were awarded:

To Peter J. Van Vleet, of Macon, Lenawee county, the first premium on wheat crops, he having raised 276 24-60 bushels on 5 acres and 3 rods, medal and............ $7 00

To D. M. Uhl, of Ypsilanti, Washtenaw county, the first premium on corn crops, he having raised 1,132 bushels of ears on seven and one-fifth acres,.................. 8 00

To Garret Ten Brook, of Adrian, Lenawee county, the 1st premium on crop of broom corn, being 736 pounds to the acre,.................................... 3 00

To Charles T. Wilmot, of Ann Arbor, the 1st premium on oat crop, he having raised 150 bush. on two acres, Colman's Tour.

To George Clark, of Lapeer, for best farm report from Lapeer county,......Diploma, and 1st and 2d vols. Transactions.

To F. W. Curtenius, of Grand Prairie, Kalamazoo county, for best farm report from Kalamazoo county, Diploma, and 1st and 2d vols. Transactions.

To W. H. Miller, of Moscow, Hillsdale Co., for best farm report from that county, Diploma, and 1st and 2d vols. Transact's.

To Samuel Rapplege, of Adrian, Lenawee county, for best farm report from that county, Diploma, and 1st and 2d vols. Transactions.

To George Clark, of Lapeer, for farm report, Diploma and 2d vol. Transactions.

COMMUNICATIONS.

NEAT CATTLE.

BY D. M. UHL.

J. C. Holmes, *Secr'y:*

Dear Sir—I am in receipt of yours of the 7th inst., in which you request my opinion respecting the grade cattle, which are a cross of the Durham and Native, exhibited by me at the last fair of the State Agricultural Society.

In reply, I can only say that for the dairy I think them second to no stock whatever. Their working qualities, taking their weight and size into consideration, I think equal to any variety of cattle with which I am acquainted, whether regard is had to their capacity for draught or endurance.

For beef, I consider them decidedly superior to any other variety in respect both to their fattening qualities and the appearance of the flesh in the market.

Their reputation in eastern markets is well established and they command good prices and ready sale.

As many farmers express a preference for red Durham cattle, I beg leave to say that I am decidedly partiai to Durhams of any color; but farmers should, in searching for red Durhams, avoid being imposed upon by a cross of the Durham and Holderness cattle, as many of his breed pass for red Durham.

Ypsilanti, 14th January, 1852.

DEVON CATTLE.

BY GEORGE E. POMEROY.

J. C. Holmes, Esq.:

Dear Sir—I brought into Michigan two years since, two pure Devon Bulls, one four years, the other coming two years old. The young bull I wintered as I did my other young stock. The Spring showed a vast difference—the Devon was far ahead of my Native stock.

I also brought an imported Durham cow, and from careful observation, I have come to the conclusion that the Devons keep on full one-fifth less feed than the Natives; and *I know* full one-half less than the Durhams. My opinion is, that the Devons, of all cattle, are *the* cattle for a new country. They feed well at the straw stack and thrive on marsh hay; and while they are small eaters, they are *good* feeders, and they stand the cold weather better than any cattle I ever saw, and are less dainty, and better feeders than any cattle I ever saw, and the calves thrive on *crusts* of bread and *hay* tea. I am this winter raising one on skimmed milk and bread, in fact, anything but good feed; and one on a fine heifer, taking all the milk she wants—and the one by hand keeps an even pace with the one that sucks. Again, they are uniform in color, being a deep red, and are finely made, close and compactly built, quiet, and always gentle. I was offered $10 each, for my calves, at three days old.

As we get older, and acquire wealth, we shall all seek the Devon, to get oxen, that we can make weigh 5,000 lbs. Let some of your friends three years hence, call at my farm and see my heifers and steers, all deep mahogony red, with clean buff noses, and every pair of steers matched. I flatter myself I produce some of the best milkers in any country.

I have put down over a mile of brush under-drain, of which I hope to give you some account another year.

Yours truly.

Clinton, January 12th, 1852.

DEVON CATTLE.

BY F. V. SMITH.

J. C. Holmes, *Sec. Mich. State Ag. Society:*

Sir—Yours of 7th January, requesting from me an article upon Devon Cattle, their value for labor, dairy and beef, as compared with other breeds, my mode of keeping, together with a description of my stock, I received in due time, and should have answered sooner had not my time been entirely taken up with a business requiring an unusual degree of attention.

The Devons, although among the oldest of the improved breeds of cattle, are not very numerous in this country, yet it is supposed we have some of the best specimens that exist. They have generally been great favorites, but have at times been supplanted in the estimation of breeders by others claiming some particular excellence over them. At these periods the pure blood has been lost by mixing with other breeds, except in the hands of a few who have tenaciously adhered to it in its purity, believing that when time should test the matter, they would stand second to no breed in point of profit for food consumed.

They have for the last few years been fast growing in public favor. Many importations have been made from some of the best herds in England, which in addition to those bred in this country, without alloy, furnish in different sections an occasional herd that may be regarded as fair specimens of the breed. The excellence of this breed of cattle does not consist in their great size, as that is not desirable with the usual qualities attached to it. They are of medium size, the most favorable to symmetry and perfection of form, are very vigorous and hardy, retain their shape and good qualities, when put upon ordinary feed or even neglected. Being of medium size, they are not as likely to degenerate, lose their good points and grow mis-shapen, as animals that have attained great size by high breeding, when they receive only the ordinary care of the farm yard, or bred by persons who have not well studied the principles involved. I believe this principle will hold good in all the domestic animals that medium size is most favorable to a full and harmonious developement of all the parts, and when we would increase the size more than ordinary care and diligence must be exercised, or we sacrifice some valuable point to obtain it.

The Devons are universally underrated by persons not acquainted with them in judging of their size and weight, their delicacy and fineness, together with their perfect symmetry, give them a light and active appearance. The bulls of some breeds are the largest and most showy part of the herd, but the reverse is the case with the Devon. The bulls are much inferior in size to the oxen, and comparatively smaller than the cows. For improving our native stock they are admirably adapted, imparting their color and characteristics in an eminent degree. It is frequently difficult to distinguish a half or three-fourth bred animal from a full blood, so strongly do they partake of their characteristics when only moderate pretentions are made to blood. Their color is uniformly a beautiful red, varying from a rich cherry to a dark mahogany, except the bush on the end of the tail, which is always white in the older animal, but dark while calves, (the change usually commences from eight to twelve months old,) with frequently a white udder and strip under the belly, and occasionally a shade of white around the outer edge of the ears. Their white bushy tails are very remarkable and a sure test of the blood which never fails in a pure Devon, and usually runs with the blood to a very great extent. The nose and rings around the eye are of a bright clear orange, which lights up the countenance and gives it that look of intelligence which they actually possess. The head is small and well set up, the eye prominent and bright, the horns are slim, clear, smooth and handsomely curved, and yellow at the root while young, neck tapering from the shoulders to its attachment to the head, round in the band, full in the cheek, with a projecting brisket, straight on the back, tail small and tapering with a white bushy tuft; small in the bone, firm and clean in the limb, skin yellow and soft, hair fine and frequently curled, which is most common with the bulls, whose neck head and shoulders are generally covered with soft curly hair, which has been described as resembling ripples upon the surface of the water. Their uniform color and appearance renders them easy to match for labor, for which they excel all other breeds. The ox attains as much size as is consistent with activity, they are excellent travellers, intelligent, docile and tractable, and notwithstanding their excessive fineness, they possess great strength and are remarkable for enduring heat and fatigue. Their slim horns handsomely curved upward, their beautiful rich color, straight clean limbs, straight backs, round barrels, bright and keen looking eyes, gives them a truly noble appearance, and make a team not only of utility, but one which pleases the fancy and taste.

Compared with other breeds, great difference of opinion exists as to their value for the dairy. One person has been acquainted with an extraordinary Native cow, and he comes to the conclusion that our Native cows excel all others. Another has known a Devon cow that was a remarkable milker, and he believes that the Devon stand unrivalled. Another has known a short horn that gave an unusual amount of milk, and he decides in their favor for the dairy. Others have been acquainted with cows of each of these breeds that were inferior milkers, and they judge the whole breed are worthless. From such slight testimony as this from isolated individuals, whole breeds are judged and prejudices formed for or against. I believe very much depends upon the purposes for which cattle have been bred. Some families of short horns are without doubt excellent milkers, while others which have been bred expressly for beef without reference to their milking qualities, are worthless for the dairy and hardly supply milk sufficient to nurse their young. The same is true of the Devon though not to so great an extent;—more uniformity exists with them, but still a great difference in different families. I think they will just about milk with the best of our native cows as to quantity, but it is always very rich, yielding a great proportion of butter and cheese. I believe as a breed they will not suffer in comparison with any except it may be with some of those who claim no other recommendation than their superior milking qualities which they certainly ought to possess, for every other excellence seems to have been sacrified to obtain it. The estimation in which they are held by those who have tested their milking qualities may be judged from a challege given by Mr. Bloomfield, manager of the late Lord Leister's estate at Holkhom, to milk forty pure Devons to be selected out of his own herd, against the same number of any breed in the kingdom, without finding a competitor.

Their qualities for making beef, are a great inclination to fatten, a rapid increase for the food consumed; small in the bone, with little offal. The fine quality of their beef and its unusual proportion in the most desirable parts, and containing greater weight in the same superfices than any other breed. They are eagerly sought after to drive to distant markets both for store and beef. Their capacity for traveling enables them to stand the drift with less loss than other breeds.

In localities near market, where beef only is the object, with a mild climate, luxuriant pasturage, and where early maturity and large size are wanted, they may be excelled by other breeds. But in a climate changeable as ours, with our scanty pasturage and severe winters, I believe no other breed so well adapted or will prove as profitable. If beef is the object, the female as well as the male will acquire as much size as is profitable in a grazing animal, if the propagating qualities are taken from them. And the various uses for which cattle are bred in this State, I am satisfied that none combine so many excellencies, or will yield as much labor, beef and dairy for the food consumed.

My stock has descended from various importations; the first was bred by Messrs. Garbutt & Beck, of Wyoming county, N. Y., whose original stock was obtained from the Hon. Rufus King, of Long Island, imported from the herd of the Earl of Leister. They afterwards made cross upon this stock from the importation of Messrs. Pattison & Caton, Baltimore, and subsequently by a bull imported from the herd of Mr. Davy, (known as the Dibble bull,) by a Mr. Vernon, of Genesee county, N. Y. I purchased last season a bull and heifer, each two years old, last spring, bred by Mr. Davy, North Moulton, Devonshire, England, imported in October, 1850, by Mr. Gapper, of Richmond Hill, Toronto, C. W. They are animals of great promise, and I expect from them a decided improvement on my old stock, which have suffered some for the want of fresh blood, proper crossing and breeding together, where too nearly related.

My mode of keeping does not differ materially from the ordinary way. I design always to keep my stock in good condition, and if I can avoid it, never allow them to get fat or poor. I consider it very injurious to breeding animals or their progeny, to carry too much flesh. I stable my stock in winter, and they run to pasture in summer. My feed in winter is hay, straw, corn-fodder, and bran to cows giving milk. Hay is my main dependence. I have some 25 feet of stable floor, made of three inch scantling, upon the plan mentioned by Editor of the Michigan Farmer, in his correspondence, as having seen in England. It works admirably—is a great saving of labor in cleaning stables, besides being much more cleanly.

Coldwater, Feb. 6th, 1852.

DURHAM CATTLE.

BY ISAAC ASKEW.

J. C. Holmes, *Sec. Mich. State Ag. Society:*

Sir—In compliance with your request, I herewith submit a short article relative to Durham cattle.

The first principle in which the improver of cattle wlll find the means of attaining his object, is the well kown rule that "like will beget like." A rule which applies with equal force of truth to good and bad animals. That breeder's attention and skill will be best rewarded, who so manages the breeding of his stock that the good shall predominate in a greater degree. The fact is evident that great attention should be paid in the selection of the bull. The breeder will do well to seek in his animal a proper form, viz: wide and deep in the girth around the heart, large, broad and level loins, extending their width well forward; good, hooped ribs, setting close, leaving but little space between them; the hips deep; flank full and low, allowing as little light as possible between the thighs; legs short, the bone not being so fine as to indicate delicacy of constitution. There are other points of perfection, which being more dependent upon taste and fancy, and such as the eye, when in search of a good form, will readily notice, and presuming sufficient consideration has been gven to form, it will be sufficient to state, that a thin hide, covered with much hair, full and soft to the touch, generally indicates an extremely mellow flesh, and a quick feeder; but such cattle are rather apt to be delicate in constitution, and are not best adapted to resist the flies in this section of country.

In my short experience with the short horn cattle, I find that the red and white Durhams are best adapted to this western country. They are more hardy, and endure the extremes of both heat and cold, far better than the white Durhams; but I am free to acknowledge that as far as my experience goes, the white Durhams are the best milkers, but the red and white make the best beef, and I am sure they winter on less feed than the white. They also make the best cross with the Native cow, for general purposes. Admit they are not all superior milkers, they will always bring a good price from the butcher; but in most instances they are good milkers. As regards their feeding qualities, if they are allowed open sheds in winter, they will keep in good condition on corn-fodder and wheat straw, with plenty of salt and water.

I have a few three year old grade Durham heifers, crossed with Ayrshire; they are rather inclined to be small, but fine in the neck and head; they are excellent milkers, both in respect to quantity and quality, and are well adapted for dairy purposes. They are good graziers, and winter well, but I think the full blood short horn Durham is the most profitable breed of cattle for the State of Michigan. Your clover lands, with your improved state of cultivation, will feed them rapidly and profitably.

Amherstburgh, April, 1852.

WORKING CATTLE.

BY JOHN STARKWEATHER.

J. C. Holmes, Esq.:

Herewith I send you a statement respecting the stock which I exhibited at the last annual fair. The working oxen were seven years old last spring; they are quarter Durham and three quarters Native breed, and girth in ordinary working condition 7 5-12 feet and weigh 3,660 lbs. They are perfectly orderly and kind in the yoke and in the pasture.

My red and white matched steers, four years old last spring, are three-quarters Durham and quarter Native breed. They girth each 7 2-12 feet and weigh 3,420 lbs. When in ordinary condition they are not easily surpassed in qualities that constitute good working oxen.

My two year old steers are of the same breed as those last mentioned, of white and light roan colors. They girth each 6 7-12 feet and weigh 2,370 lbs.

The foregoing cattle have been reared in that scanty process of feeding the Michigan farmer generally follows. Nothing but the excellence of the breed entitles them to consideration. They have grown up in spite of indifference to care and thrift.

The farmers in this place and vicinity are much indebted to Messrs Jno. W. Van Cleve, who, as early as 1838, introduced some fine animals of the short horn Durham breed, from which the improvements in neat cattle have mostly originated. We just begin to appreciate the efforts he so generously bestowed in time and money to improve the breed of stock, and hope that much more good will yet result from the noble enterprise.

Yours, most respectfully.

Ypsilanti, Oct. 2, 1852.

FRENCH MERINO BUCK, OWNED BY WALTER WRIGHT, OF ADRIAN.

["YOUNG NAPOLEON."]

SHEEP—FRENCH MERINOS—BY WALTER WRIGHT.

J. C. HOLMES, Esq., *Sec'y Mich. State Ag. Society:*

DEAR SIR—Yours of the 7th January is received, wishing me to "furnish for the pages of the third volume of the Transactions of the State Agricultural Society, an article upon Sheep, describing bucks that shear the big fleeces, my opinion respecting the variety I deem the most profitable for the farmer of Michigan to keep; also my mode of feeding, keeping, and such matters as I deem of importance."

Notwithstanding I have paid more attention to the purchasing of wool since I have been in Michigan, than to wool growing, yet I have done more or less of the latter for the last twenty years or more. I am, therefore, exceedingly anxious that the wool-growing interest in this State should be more generally appreciated by our farmers than heretofore, and gladly would I contribute my mite for that, as well as all other branches connected with the farmers interest or prosperity. And in complying with your request, I can only give you my own limited experience, which may not be of any value, yet in the hope that I may say something that will excite wool growers in this section more particularly, (as well as other parts,) to improve their flock, and also to manifest my willingness to do what I can in sustaining and encouraging you in your arduous duties which you have performed with so much success and honor to yourself and the Society which you represent. There has been a great want of interest in wool growing in Michigan, particularly in the improvement of breeds, until within a very short time—the greater part of our farmers keeping the native or long legged races, with wool as coarse as dog's hair; others, again, keeping a mongrel Saxon, which I consider still worse, not having constitution, size, or wool enough to pay for keeping; others, still although but few, keep the Pauler, or Spanish merino, which for profit is without doubt much preferable to the others in all respects. I have no doubt, indeed, I know that a most valuable cross of the improved French merino can be obtained from this variety. One reason, amongst others, that this state of things has remained so long among us, is that wool dealers have been in the habit of putting men and boys in the markets who know no more about the quality of wool than an Arab or a Hottentot, and when I have urged upon farmers the importance of improving their flocks, the reply has been, "why, what is the use? these blockheads that buy our wool make no discrimination, we get just as much (or nearly so) for ours, whether washed, soaked or unwashed, as those do who have a fine article, and well washed." And so long as eastern wool dealers continue to put such men in market to buy wool, this state of things will continue. One would think that the enormous high prices paid the last few years for the coarse wools, would put wool dealers on the lookout to find the cause. In no instance, that I know of, has anything been lost on fine, well washed wool. But a change in this respect, I confidently believe will be speedily brought about; and when farmers have inquired whether they had not better turn their attention to growing coarse wool, I have endeavored to dissuade them from doing so, and encouraged them to continue to improve their flocks by selecting (for a cross) the best bucks that can be found, having reference to size and form of sheep, quality and quantity of wool, and that the time is not far hence in which they will be amply rewarded for all their pains and expense. And what more natural than that I should recommend them to the variety of sheep which I have chosen to improve from. And this brings me to your first request, viz: "the bucks that shear the big fleeces."

The fleece of wool for which I received the first prize of a medal valued by the executive committee at fifteen dollars, (but which I value at more than twice that amount,) at the State fair at Detroit, Sept., 1851, is from my French buck two years old, bought of J. D. Patterson, of Chautauque county, New York, and is a descendant of a French buck imported from France by J. A. Taintore, Esq., of Hartford, Connecticut. His wool was thoroughly washed on his back; one week after which he was sheared, his fleece weighing 14 pounds and 9 ounces.

Now I do not pretend to say that this fleece weighs more than any fleece taken from a merino sheep, but I do say and believe that there never was a fleece of merino wool exhibited at the State, or the World's fair, more free from oil, gum or dirt than mine was. I have no doubt (unless something should befall him) that another year the same buck's fleece will weigh considerable more than it did this year.

And now for my reasons for believing these are the most profitable sheep for the farmers of Michigan to keep. They are of a much larger size than any of the fine wooled sheep, frequently weighing from one hundred and seventy-five to two hundred and fifty pounds. Having a much hardier constitution, their wool is about as fine, more compact, and much longer, covering their entire frames, body, legs and head, and what is very remarkable their wool is of good staple even to the hoofs.

I have no doubt from the limited experience I have had in this variety of sheep, that I can raise a flock by a cross with my Spanish ewes that will average from seven to eight pounds per head.

My flock of sheep at present number not far from three hundred, a part of them pure blooded Spanish merinos; a few full blood Saxons; the rest are a grade sheep. Having so many varieties, I have experimented a little in crossing. From the Saxons, I did not obtain any improvement of consequence, but from my Grade and Spanish I obtained a cross that is perfectly satisfactory; and although I paid a large sum for my buck, I feel repaid amply and with interest in my first year's crop of lambs. Their wool is much improved in fineness and length and very even over the whole body, and more compact than the ewes were.

My manner of keeping sheep in the summer as well as winter is, to keep as few in a flock as possible, changing often, salting at least once a week, always keeping tar in their troughs, and occasionally mixing a little sulphur with the salt. During winter sheep should have a good shelter from the storms, plenty of good (I prefer clover) hay twice a day, and from five to ten bundles of oats in the sheaf at noon for every hunered sheep. The oats should be cut when quite green and lay in the swath a few hours to dry; then bound up and set in bunches to cure. This course will prevent the shelling of the oats, and the straw is much more valuable for food.

A manger formed of rough boards, the length of the board, and about three feet wide, without bottom or top, and with an opening the length of each side, sufficiently wide to admit the head without the body of the sheep, is indispensable with economy in feeding sheep, and will save at least one quarter of the food, and many pounds of wool. By observing the above rules, I have seldom lost any sheep during either winter or summer; but this winter is an exception. In many flocks they are dying off in scores—some farmers have lost one-third of their flocks—I have lost several. There seems to be no apparent cause. The symptoms when first discovered, are drooping, a great weakness, and gradual failing, sometimes lingering for a month or more, eating as usual every day, until so weak they cannot stand, then generally die in two or three days, eating however, until the last. No treatmant as yet has been of any avail. I have a small flock that were kept on a very dry peice of land the past summer, none of which are as yet affected with the disease. I attribute it to a species of the rot, or something very nearly resembling it, caused no doubt by the extreme wet season that we have just experienced. I consider a wet summer more to be dreaded than a dry one for stock of all kinds, and doubly so for sheep. Wet pastures are unhealthy for sheep.

Enclosed (at your request) I send you a likeness of my buck.

Adrian, Feb. 17, 1852.

MERINOS.

BY G. W. GALE.

J. C. Holmes, Esq.

Dear Sir—I will try and answer your questions according to the best of my ability. In the first place, "what do I consider the most profitable breed of sheep for Michigan farmers? All will agree that the breed which will bring in the most dollars and cents from a given amount of food, but all might not agree as to the breed of sheep to do it with. For one I would take the right class of merinos, in preference to all others. Although we have different classes of sheep in other breeds, as well as in that, yet you may take an inferior or common breed, and there is not so much danger from mismanagement in breeding. It is here and in keeping that the great secret lies, if a difference occurs in the same breed. Let us look a little; we want them for wool, tallow and mutton—not like the horse—we do not require great muscular action, but only a good constitution, free from disease, and a disposition to feed and be quiet. By some it is said that animals require food according to their size; but this is not true. In men we frequently see small men most voracious eaters, and large ones of quite the contrary. Just so in animals. How is all this? Simply in their physical organization. Physiologists say where the secreting organs predominate, that the man or animal is lymphatic. Have we not all seen men both large and fleshy, and yet very small consumers of food, and as the saying is, too lazy to work off their fat. Now just such a temperament I would choose in an animal that was not for purposes of labor. Is it mutton or wool we want? I think no candid man would say a mutton sheep for Michigan, yet a few flocks of mutton sheep might be profitably kept near Detroit, and some of our large villages, but for all to go to raising mutton would be unprofitable in the extreme. If we had a London market it would alter the case materially; then we would say Leicesters, Southdowns, &c. But we have no such market, and we must depend on a foreign market for some time to come. It is therefore wool and not mutton on which we must chiefly depend for profit.

Some say "I can raise more pounds of coarse wool, and it will bring as much into five or ten cents per pound." But we do not endorse that doctrine at all. I have yet to see the coarse wooled sheep that will shear as many pounds of wool as the right sort of merinos. I do not say every fine wooled sheep is right—far from it—and until farmers understand breeding much better than at present, I verily fear the number of poor sheep will increase.

I will give you a history of a flock close by me. Six years ago, the owner purchased two bucks, and kept them with the same ewes for five years. His first and second crop of lambs were an improvement. When the young stock began to raise lambs, many of them died. Those that lived were very sharp on the back, narrow breast, a sinking down of the neck from the shoulders, and every appearance of a light constitution. True, a little was gained in fineness, but at a bitter expense. The next cross from the same bucks on the same flock (on the second crop of lambs understand) produced the most sorry race of animals I have ever seen; indeed, not one-fourth of them lived, and those that did live never came to one-half the size of the original stock. Their faces and ears were entirely pealed off, as if they had been sun-burnt. The owner sold them for 50 cents per head; thus you see by in and in breeding you produce a high nervous temperament, which is known by a thin skin and a light fleece. I must say that breeding from near blood relation has always produced this same effect. If I wish to bring out some particular point more fully, I choose from both sides those that have the point required most fully developed.

I have no doubt but in and in breeding will produce fineness, and at the same time destroy constitution. If you wish a good merino sheep, choose one wooled all over, as sheep men say, and not one with bare legs and face. This is one of the first failings, and the difference is considerable, where you get a pound or two of wool from those parts generally bare in common sheep. Look well to the form and skin; a fine velvety hide, with some thickness, is far preferable to one of those paper skins, as thin skins and heavy fleeces do not go together. Finally, choose those sheep that have the most fine wool on the poorest part of the fleece, and I will warrant the better part to increase in the same ratio.

Wool on the sheep should open white and oily, but entirely clear from anything like yellow chunks of yolk or gum, as some term it. The action of the sun on the oil will produce a crust on the out ends of the wool which should not extend into any great depth. One of the greatest faults of farmers in breeding fine sheep is, they breed them too young and too old. They never should rear lambs short of two years, (and three would be better,) they will then produce strong stock that will endure the vigor of our climate, and when they fail from age, which is seldom before twelve or fifteen years, they should be thrown one side, as not being fit for breeders.

Another common error is the excessive use of stock bucks. A stock buck should be well fed and cared for, and never turned loose with the flock, as is the common practice. A much better way is to limit the number per day and at stated intervals, so as not to have the sire in any way reduced. Then we may expect a crop of lambs that will be something alike.

You ask me about my management of keeping sheep. I feed almost every thing, hay, oats, straw, corn stalks, wheat, shorts, oats, &c. I sometime ago fed corn as the grain part, but do not like it. It is too concentrated and binding for sheep. If obliged to feed it, I would get it ground, cob and all, to give it more bulk. Corn produces too much inside fat and too much contraction. I have been the loser of a good many lambs by it, and have lost none, or nearly so, since I left it off. Neither do I believe it possessed of the qualities that produce wool. I have fed roots with the best of success, and intend to try carrots. From what I saw in Vermont, last spring, I think they are ahead of any root in existence for sheep. Carrots and wheat shorts, mixed together, are as hearty food as sheep should ever have where wool is the object.

From the first of November to the first of May, I consider it very important to keep sheep out of the drenching rains. Many diseases arise from bad colds, and a loss to the fleece is sure to follow. I never suffer my sheep to get wet in the winter. As a condiment of health, use one part sulphur, one of ashes, two of salt, mixed and put where they can have free access to it. Water in winter is indispensable if sheep are kept on dry food. It is a disputed point whether they should have water in summer or not, but I do know that they will drink freely in dry weather when there is little or no dew.

My flock has increased within four years from three to over five lbs. Some of my best sheared but a little short of fifteen lbs.; but these were a select few. When I first commenced improvement some of my brother farmers told me I could never get a flock that would average five lbs. each. Having attained this, I shall mark a much higher figure and have no doubt of attaining it in time. In so doing I shall exercise care not to get the fleece any thing but fine wool. I know it is said by many that a heavy fleece must not be fine, but on the contrary I have seen very heavy ones fine, strong, elastic wool, and I never saw a strong fibre on one of those very light dry fleeces. The wool gets weather beaten and is not naturally strong, partaking in some degree from the source whence it sprung. A feeble constitution always did and always will produce a light fleece, but it does not always follow that a good constitution will produce a heavy fleece. The organization may not be right to produce it. It may run all to flesh like some of the mutton breeds, and it is plain to be seen that mere contentedness of disposition is not all, for many of the mutton breeds have it to a high degree and do not produce a fine heavy fleece. All sheep of that class have coarse skins and wool of the same character; thus it becomes necessary to have a high degree of refinement, but not that kind that destroys constitution by materially lessening arterial actions. Extremes of all kinds should be avoided, especially in feeding; over feeding is one of the worst of evils, if it be on grain, as good constitutions may be destroyed in this as well as in many other ways.

Ypsilanti, Jan. 15, 1852.

SAXON BUCK, OWNED BY D. D. GILLET, OF SHARON, WASHTENAW COUNTY.

["MATCHLESS."]

SAXON EWE, OWNED BY D. D. GILLET.

["REFINA."]

SAXON SHEEP—BY D. D. GILLET.

J. C. Holmes, *Sec. of the Mich. State Ag. Society:*

Dear Sir—In compliance with your request, I place at your disposal an account of my flock of pure blooded Saxon sheep, now numbering two hundred and fifty, all of which were either purchased at different times, of that most indefatigable wool grower, the late D. A. Mills, of Manchester, in this county, or were raised from those purchases on my own farm, I am in the receipt of a letter from G. J. Barker, of Grand Rapids, Kent county, a co-partner with Mills in his purchase, under date of February 9th, 1852, in which the history of the origin of Mills' flock is thus detailed:

"The largest share of the ewes at the time of Mills' decease, were from the flock of S. C. Scoville, of Salisbury, Litchfield county, Connecticut, or rather their descendants crossed by bucks from the flock of Mr. Church, of Oneida county, N. Y. I think Mr. Scoville informed me that he purchased his from Humphries (query, Henshaw's, D. D. G.) importation, at Boston, which must now be from twenty-five to thirty years since; but for the last ten years before Mills and my purchase of him, (I think eight years ago,) he had crossed from bucks purchased from H. D. Grove, of Hoosic, Rensselaer county, N. York. Said Grove was bred in Saxony, a Saxon shepherd, by a rich uncle, and imported his flock himself, from the then three best flocks in Saxony. Said Mills had also a small lot of ewes of said Church, Oneida county. If you have the list of premiums that have been awarded at the New York State fairs for the last six or eight years on Saxony sheep, you will see that Mr. Church generally has obtained a large share of the first. Scoville took the first premium at the last New York fair, on a foreign Saxon buck.

"The last purchase of bucks by Mills for his own use, was of Douglass Clark, Esq., of Dutchess county, N. York. He told me about eight years since, that he was ahead of Scoville on weight of fleece, having crossed with bucks from the east part of Connecticut, of whose importation I have forgotten, but they were pure Saxon."

Herewith I send you sketches of a buck and ewe, taken as you will see by W. R. Wheeler, a young artist of Manchestsr. They are lower than the Saxons originally imported are represented in those cuts which I have seen, have a round and, as compared with the Spanish merino, rather long body, are well covered over the entire surface of the body with wool of a superfine quality, are as hardy as any breed of sheep with which I am acquainted, enduring our severe winter storms with as little shelter, as little care, and as little food in proportion to their live weight (the average live weight of my grown sheep is about sixty-five pounds) as any of the ovine race. I am not unaware that this last position is strongly controverted by Vermont sheep peddlers and their western echoes; and that a great many changes are being wrung upon the "over delicacy of the Saxons," "the hot house plants," etc. But all my reading and experience, as well as the experience of this neighborhood in which they have been kept for years on the same farms, and in many instances in the same flocks with Merinos, Natives and all the different grades of the three varieties, tend toward the conclusion above expressed. My own flock for the last seven years has numbered from one hundred to two hundred and fifty, kept in one of the bleekest and most exposed situations in the country, almost without shelter, on the most ordinary fare, and my annual loss has not averaged two per cent. But sufficient on this point to indicate the reasons that have suggested my conclusions.

In the preparation of my wool for market, I have hitherto taken considerable pains to have it *pure* and properly done up, having for several years past sent to an eastern market where *condition* as well as other excellencies is somewhat appreciated. As the result of my care in this particular, I give you an extract from the letter of Messrs. Blanchard & Co., (the gentlemanly proprietors of the Kinderhook wool depot, of whom I asked for instructions in regard to the matter,) acknowledging the receipt of my last years' clip, and endorsing a statement of the grades of the same, which is as follows:

KINDERHOOK, Aug. 2, 1851.

D. D, GILLET, Esq., *Sharon Mich.:*

DEAR SIR—Your three bales of wool grade thus:

Super Super.		Super.		Extra.		Prime.		No. 1.		
216	—	152	—	53	—	25	—	5	=	441 lbs.

We can give you no information in regard to washing your sheep or putting up and sacking the fleeces. As we never received a clip of wool shorn in this State equal in condition to yours, we would rather send our wool growers out to Michigan to learn from you, than to endeavor to impart instruction to one whose clip was all it should be.

Very respectfully,

H. BLANCHARD & CO.

Compare the grading of my clip as given above, with that of Joseph Barnard, of Hopkinton, New Hampshire, as given in the American Shepherd, at page 399, and which is cited by the author of that work as one of those American Saxon flocks which rivalled the German, and you will see at a glance that the quality of my wool is superior. In order to a correct idea of the average weight of fleece, I ought perhaps to say, although I apprehend that no one is more fully convinced of the effect which good or gross keep exerts to increase that weight; yet my flock has generally been in ordinary, or low condition, and never in high; has always had a great proportion of ewes and lambs, and clips an annual average of from two and a half to two and three quarter pounds of pure wool, which has sold at the Kinderhook depot, for the last three years, for from forty-two to sixty-six cents per pound, the last clip averaging sixty eents per pound. Now when the relative weight of the Saxon to his only wool producing rival, the Merino, (sixty-six to eighty-eight pounds,) and the relative difference of the value of the fleece per pound are considered, the conclusion seems unavoidable that the former may be at least as profitable as any rival they have, so long as *wool* determines the matter of profit.

Sharon, Washtenaw Co., Feb., 1852.

LEICESTER BUCK, OWNED BY J. SHEARER, OF PLYMOUTH, WAYNE COUNTY.

LEICESTER SHEEP—BY J. SHEARER.

HON. J. C. HOLMES, *Sec'y of the Mich. State Ag. Society.*

DEAR SIR—In answer to your inquiry of my mode of keeping sheep, their adaptation to the climate of Michigan, and a correct drawing of my long wooled buck, will be most cheerfully complied with. As to the mode of keeping my Leicester Sheep, they require nothing more of *housing* than common sheds in winter, and shade of booths or trees in summer, which prove congenial. Their food principally consisting of grass and hay, of timothy and clover; they are fond of turnips; chop feed is good for them in small quantities, say half a pint to a sheep, daily, as near as may be, with plenty of water. As to their adaptation to our climate, it suits them admirably—as much so as any climate in the world—their constitution is remarkably good and healthy. They fatten early and kindly, on the least food of any variety of sheep, maturing at two years old. They possess a quiet disposition, and afford friendly society, good nurses, and particularly fond of their young, and with the common grade sheep, the cross proves of the best order.

In this vicinity, there are some two hundred lambs, several of which have been sold at five dollars per head, weighing some one hundred pounds at six months old—live weight. By crossing the wool is not as long, but more compact, and very much is added to the constitution of the grade variety. The wool of the full blood is long, fine and soft, maintaining its length and fineness all over the body, and is in great demand for the manufacturing of muslin delaines. Their form, as will be seen by the above drawing, is full in the breast, broad straight back, light in the chest, with small, fine bones, comprising all the profitable qualities essential to the market, manufacturer and producer—in the good quality of its mutton, amount of tallow and wool, with early maturity.

All of which is respectfully submitted for your consideration.

Plymouth, January 26, 1852.

POULTRY.

BY DOCT. M. FREEMAN.

Mr. J. C. Holmes, *Sec'y Mich. State Ag. Society.*

Dear Sir—I am induced to comply with your request from various considerations. The pleasing interest which I have taken for thirty years in propagating the domestic fowl, has become still more gratifying, as a much greater opportunity is now offered to test new varieties; also by seeing a similar spirit of improvement and fancy manifested for the past two years, by a large number of citizens throughout the different States, in that too long neglected branch of domestic economy, and willingly contribute my mite of experience in many varieties towards inducing farmers and villagers to look more to their interest in this important branch of profit and economy. I also feel confident in saying the time is fast coming when an account of the poultry yard will stand as an important item in the farmer's account of yearly sales.

The disposition of hens is so natural to lay eggs and rear chickens, that notwithstanding the neglect they generally receive, such as the want of comfortable accommodations, sufficiency of food, &c. &c., still, if statistical information was furnished from the market of any large town, it would be shown, even under such circumstances, that no branch of the farming interest, pay so large a profit for the amount of labor and capital invested, and now the fact has been clearly ascertained from experiments, (at least in my mind) that there are certain varieties of domestic hens that produce during the year a much larger number of eggs than are furnished by other varieties. And again, that certain varieties will furnish for the table, with the same outlay and trouble, twice the quantity of food that can be obtained from others, and to those qualities may be so combined by judicious crossings, as to possess large size, fine form, hardiness, and free laying properties, quiet disposition, with beauty of plumage, all which are peculiarities of the domestic fowl, and to me a most pleasing occupation, well worthy the attention, and should receive the consideration of every farmer and prudent house-wife, otherwise an important branch of profit, convenience and luxury is neglected.

My own experience for many years, fully warrant the above remarks, though I have not tested many of the new popular breeds. The varieties which I have kept, furnish very different results as to their productiveness, hardiness, size and laying properties. The tests were as fairly conducted as I know how, taking into consideration all circumstances which were made to bear as equally as possible upon each, of which I have kept a strict account, which shows that there are as great differences among different varieties of fowls, as in any of the various species of domestic animals.

In early life, fancy induced me to devote considerable attention to the habits, constitution and peculiarities of the domestic fowl, not then knowing but one kind, viz: the common dunghill fowl, save a particular breed kept for fighting.

One of the sources of enjoyment consisted in crossing them, so as to produce fancy colors, but circumstances forced upon me conclusive evidences that a great difference then existed between different yards, through my section of country, which was no doubt owing to a sprinkling of game blood, as I have since found that even a mixture of the game variety makes free layers and very hardy; and now, within a few years, importations have been made which show an improvement in many poultry yards, quite equal to that of any of the imported animals.

The varieties which I have since experimented with have been the common fowl of the country—English games, Polands, Malays, Dominico, Game and Dominico, mixed. The Emigrant's white hen, with red crest and muffle, made up by careful selections for several years, and cross on a large size game cock, making the variety called Kent Co. Dorkings, Shanghae, and Royal Cochin China, most of which I have crossed by way of experiment, some of which did not succeed well, others exceeded my expectations, statements of which have been given at considerable length with comments and remarks in the Michigan Farmer, at different times, and could now only give a condensed repetition of those communications, had I not this last season greatly improved my varieties in some respects, by procuring a pair of imported Shanghaes, and a pair of Cochin Chinas, which, as far as my experience of one season warrants me, I can truly say the Shanghae importation I have are second to none, as respects size and symmetry, hardiness and laying properties. Their eggs are not large but exceedingly rich, possessing great vitality. The chickens are remarkably hardy, and grow rapidly. I did not receive them until the first of June, but hatched upwards of fifty chickens, and have never seen one of them droop in the least. The last brood now, Nov. 22, are still unfledged, and run out in the snow and mud, appearantly with less inconvenience than the early chickens of other breeds. The imported pair were unexpectedly sent to Wisconsin. I did not weigh them, which I now regret, confidently believing they would exceed twenty pounds in weight. However, I since weighed a cock chicken, when four months and one day old, and he fully balanced six pounds and three ounces. The chickens when first hatched are uniform in their color, (an evidence of pure blood,) covered with long, coarse, dark brown down. The pullets feather full and early, their legs are short and yellow, set wide apart, remarkably heavy in the thigh, broad on the back with deep full breast, with but little variation in the color of the plumage, viz: a grayish brown. The cocks generally lose the down before the feathers start on the body, which however soon burst out, and in a few days, entirely covers them, and are uniformly a dark red, and more upon the legs, which are lightly covered with soft downy feathers. The quill and tail feathers in both are very short. I have never seen them attempt to fly as high as an ordinary fence. Their disposition is very mild and quiet. The hen showed great perseverence in setting, and when removed to the chicken coop, would brood at night and stormy days, all the chickens that she could get under and around her. My Cochins are from a stock imported last year, and were exhibited in New York and invited comparison. Their color when in full plumage, is a greenish black, of a metalic lusture, equal to the finest specimens of peach-orchard coal. They as yet have not done well with me. They had the rape when I received them in July, which they communicated to my different yards. The imported cock was selected, but he died a few days after, still he weighed fourteen pounds. I then obtained one of his chickens, which, when full grown, I think will equal its sire. The few chickens which I have had, have proved hardy, feather early, with white specks on their breasts, which at the first, moderately disappear. I have crossed them on the Malay, which bid fair to become very large and hardy. I permitted a game hen to pass in and out of the Cochin's yard, which her size admitted, and in due time brought forth a brood. The hen and most of the chickens were carried off by a fox; the balance were left to cater as they could by way of experiment. They grew well, evincing the hardiness and appearance of the games. The Dorkings should not be lost sight of in our fondness for new or fancy varieties; they certainly may be classed among the best laying variety of large eggs, compact, hardy form and good size. A cross with the Shanghaes bid fair to make in appearance at least, as we would suppose a large fine form, hardy variety, and am of the opinion will be great layers, both breeds being hardy, and first class as layers.

When eggs are the principal object, I would prefer the Dorking or a cross on the game, and again repeat that I have ever found the games, either in their purity or crossed, valuable as layers. The objection to full bloods is removed, (viz: their fighting propensity,) by mixing them with other breeds. My favorite Kent Co. are half game. There has never been a fight among them when chickens. They were above medium size, (until reduced by breeding in and in,) very hardy, great layers, the best of nurses and not surpassed by any in their rich gaudy plumage. Crossed on the Dorking forms a variety larger than either their parentages, finely formed, remarkable free layers of large eggs. Their plumage is a white ground, shaded with a lemon or cinnamon hue. I think I have tested them fairly, and deem them well worthy a conspicuous yard in my choice collections. The Dominico I have found a good variety and when mixed with the game are a very valuable breed for eggs.

The Polands are not favorites of mine, though they lay freely. The chickens are difficult to raise, and not a choice kind for the table. There are several new varieties which I have not tried, some by importation and others from crossings which are respectively recommended for size, as good layers, and beauty of plumage by different breeders, and each be sought after by their admirers, but as utility is the great object with most persons, and generally but one variety is kept, and that too for both eggs and the table, I would recommend to such persons to select such varieties as all breeders agree in possessing the important requisites, viz: the Dorking and Shanghae, and if they should possess a little spice of fancy in plumage, I can honestly recommend a mixture of the Kent Co. and Dorking next, without the fear of lessening any of the desirable qualities.

The Malays are not as good layers as the above named varieties, still their eggs are large and of rich quality. Year before this I had but fourteen hens of improved varieties. I purchased eighty common fowls, and kept a strict account of all the eggs laid in February, March, April and May. This year I yarded forty-six hens, consisting of popular breeds, such as I have alluded to, and kept this year also a strict memorandum of all the eggs collected during the same months as the year previous, and on comparing them I find I collected rather more than twice the number of eggs this year than I did the year before, with twice the number of hens, and last year, too, I kept my fowls under more favorable circumstanees than I did this year.

In addition to the above statement, circumstances have recently come to my knowledge which enhances the importance of keeping fowls, and too, the plan of yarding them. A gentleman lately informed me that he hereafter would keep them if they did not lay at all. Having occasion to build a poultry yard, he enclosed a part of his plumb orchard, and those trees which were within the yard have since borne yearly the finest fruit of their kind and those without were utterly useless. The fruit was invariably destroyed by the Curculio, which statement corrobarates a circumstance which occurred with my plumb trees this year. Those trees which my chickens had access to were loaded with perfect fruit, while the main orchard protected from the chickens did not yield a sound plumb. If such results prove generally the case it is a discovery of great importance, which I shall fully test by making large yards and filling them with plumb trees.

McCORMICK'S REAPING AND MOWING MACHINE.

J. C. Holmes, *Sec. Mich. State Agr. Society:*

Herewith I hand you a cut and description of my Reaping and Mowing Machine. A reaping machine being now almost universally regarded as a necessary implement to the farmer (and not much less so than the plow, because of the high price of labor at the season of the harvest, and the great importance of the work to be done in a short time,) they should investigate carefully the principles of different machines.

My machine is adapted and warranted to cut all kinds of grain and grass in the very best manner, and as is generally well known, operates with a *sickle edge* and *reel,* combined with a seat or stand for the raker, so that the grain is delivered at the side of the machine, and the binding made entirely independent of the cutting. It is sim-

McCORMICK'S REAPING AND MOWING MACHINE.

ple, of light draught, not liable to get out of order, and the sickle will cut 100 to 300 acres of grain, without a second grinding. That the *sickle* is the *true cutter,* I think does not admit of a doubt, while the *reel* is indispensable to the successful and satisfactory operations of the machine through the harvest. Though the farmer may put his reaper together by the above cut, I give the following

DIRECTIONS.

I. Lap the in-hound and cross piece which has two holes for the axle, for high or lwo stubble, Lower hole placed on the axle for higher stubble.

II. Bolt the angular board (S, in the cut) marked thus ——, to its place with the 8 inch bolt on the back end, which rests close to the platform.

III. Put the axle in Place.

IV. Place the wheel frame on the other end of the axle *lapping* the finger beam above or below—suiting the *higher* or *lower* cut at Q.

V. Put on the main brace D, marked thus ═, one side of driver's seat on same bolt.

VI. Separator W., (marked O,) to its corresponding mark, there is a little wooden pin, to be removed to give place to a bolt.

VII. Dividing iron, marked thus ⋈. If the reel be required *very* low, this iron is taken off, for small wheat.

VIII. Side board, (with brace nailed to it,) marked U., to its place, removing a small block to give place to it.

IX. Two small posts erected and cloth bar on top marked T.

X. Reel bearer V.

XI. Reel shaft Z to its place, arms in with blocks, (on each end) *forward* as the reel turns, and after braces are in and tightened by a cross *pin* in the one with long tenon (there is one brace for each end with long tenon) the boards are then nailed on the blocks. See *d* in cut. The boards have to be strained into a shape that will fit, as the arms are not parallel. Each of these boards having a block on the end (and the end without the block is next the wheels) put on, so as to pass $\frac{3}{4}$ of an inch from the angular board. Right and left nuts on reel gudgeons. See that reel turns free.

XII. If fingers do not fit right, they can be knocked out with a punch, and with the wedges trained right with very little trouble.

XIII. The small ground wheel has a third and lower sett by bolting the block on the *upper* instead of the *lower* side, and the side next the horses may be raised or lowered some two inches by changing the position of the tongue, which may be done if required, by boring another hole in the tongue. Open side of small ground wheel outward.

XIV. The square washer placed on left end of the axle to gear deeper, and other washers of leather can be added.

XV. Put in the driver (9 or 12 for lower or higher stubble) and bolt on the guard (small cast-iron piece) to keep it in place, see that the guard does not tighten on the driver. The block *in* the driver to be tightened on the crank, when it wears, and made very tight at first.

Boats having beviled heads belong to corresponding places, heads of tongue bolts to the left. The 8 inch bolts (in the box) goes through the back end of angle board (S. S. in the cut,) cross piece and in hound C. The 5¼ bolt through cross piece and in hound near the same place. The two 7 inch bolts through in hound and cross piece near the axle.

OPERATION.

1. Run the reaper on hard ground, and see that it works free, as it may be tight at first. Work the reel as high as the smallest grain will admit of, as it runs lighter to be higher. The band will stretch at first and must be taken up, and the end kept tied, or it will fasten and be injured in the cogs.

2. Grind sickle with a *short* bevil, on the smooth side when required, (but not often.) To be a little broken will not injure it, but may be evidence of good temper.

3. The foregoing cut shows *exactly* how the raking should be done. The raker can have a little intermission between the sheaves, and then with a *very quick sweep*, (catching by the heads) pull the sheaves round, leaving the butts next the grain.

Keep the nutts (washers only put on where bolts are changed, because not so liable to lose nutts) and keys tight, and oil well.

When iron and wood work together, very little oil is best, and I will add, that long experience has convinced me that the main axles of a reaper are better to run in wood, (having slow motion,) than metal—

nothing to get loose, and if *ever* they wear to require it, it will only then be necessary to put boxes in, and the axles will be perfect.

By the shipper the cogs may be geared deeper or shallower, or put entirely out of gear when not cutting.

The strip on the angle board may be raised or lowered to suit the reel. As the reaper is not wholly put together at the factory in some cases very slight trimming at the laps may be found necessary.

If the band flies, it will be owing to its stretching, or the shaft not level, or the pulleys not being fair.

Chicago, Ill., 12th Jan., 1852.

CHEESE.

BY MRS. D. DORT.

J. C. Holmes, Esq., *Sec'y Mich. State Ag. Society:*

Dear Sir---Pursuant to the requirements of the premium list of said society, I send herewith a statement of the manner of making *cheese without pressing* for which a premium was awarded to me at the society's fair for 1851.

The milk is curded and the rennet prepared in the same manner as is prescribed in the essay of A. L. Fish, Esq., of Herkimer county, New York, which will be found in the Patent Office Report for 1848, pages 620 and 621. When the curd is properly prepared as therein directed, it is put into a clean dry linen or cotton bag, a little warmer than it should be when put to the press, and pressed down slowly as hard as can be done with the hand, so as to make it solid—it can be in just such shape as the maker may desire; I think the cone or sugar loaf the best shape—and then hung up to drip and dry, in about the same temperature as is fit for drying and curing pressed cheese. No more than fifteen or twenty pounds should be put into a cheese of this kind. They cure much sooner than when pressed, and are very rich, as none of the oil passes off as it does under the power of the press. They need no care in drying, as the cloth is a perfect shield against flies, and prevents them from cracking. When cured, they are very convenient for family use.

I have made but a few in this manner, and they have cured and kept as well as those that I have pressed of the same weight, and are much richer; besides it saves a great deal of care and labor in turning and greasing to prevent injury by flies.

Dearborn, January, 1852.

MAPLE SUGAR.

BY DEXTER WHITE.

J. C. Holmes, *Sec'y Mich. State Ag. Society:*

Dear Sir—It must be evident to every one that liquid filtrated through solid maple timber cannot be otherwise than pure. It also must be evident that sap caught in old half-decayed wooden troughs with a liberal infusion of leaves and dirt, impart a great impurity to the sap. Rainwater, decayed vegetable matter, &c., add chemical ingredients to the sap; is troublesome to extract, and injurious to the quality if not removed.

Consequently, cleanliness is the principal secret in making nice maple sugar—cleanliness both in vessels, kettles and every thing else pertaining to the business. My buckets are mostly tin and are a cheap and neat article. Sugar made from sap caught in such vessels cannot be otherwise than clean, and if no dirt of any kind gets into it, the consequence is I have nice sugar. If I wish to have extra nice, I do not boil it so long but that it will drain; consequently if there is the least impurity or dark color in it, it drains out. Sometimes a wet cloth wraped around a cake (except the bottom) helps to make it white.

The syrup I let stand and settle, strain through a flannel strainer, and cleanse with milk and eggs. Four eggs, well beat with about four quarts of milk, is sufficient to cleanse syrup for a hundred weight. Sweet does not rust anything. Tin as well as wood should be painted on the outside, and when done using them for the season serve them as a dairy woman does her milk pans, and they are sweet and clean for use again.

Novi, Nov. 14, 1851.

BEE HIVES.

BY GODFREY & MOON.

J. C. HOLMES, Esq., *Sec'y Mich. State Ag. Society:*

DEAR SIR—We sent you for exhibition "Chrystal Bee Palace," and now an abreviated description of the peculiarities of our invention:

Imprimis. The manner of securing the boxes or apartments together by grooves.

2d. Simplicity of arrangement, being the most convenient form.

3d. The *arrangement of the slides* for the purpose of intercommunication and removing the bees from one apartment to another, and for the purpose of removing honey and replacing *other* apartments; and,

4th. The *self-acting door,* which affords better protection to the bees from their natural enemies.

Not having the specifications and drawings here, we shall be obliged to delay further description for the present, but will forward at as early date as possible.

Very truly yours, &c.

Paw Paw, Sept. 27, 1851.

HOUSEHOLD FURNITURE,

BY SAMUEL ZUG,—[OF THE FIRM OF STEVENS & ZUG.]

J. C. HOLMES, Esq., *Sec'y Mich. State Ag. Society.*

DEAR SIR—In compliance with your request that we would furnish you an article having reference to the woods used in our business, and such other matters in connection with it as would be suitable for publication with the records of your Society. We have thought a short history of Furniture, in connection with other matter would not be uninteresting to the intelligent readers of the records.

The history of Furniture, necessarily short, will be found abundantly full to afford an idea of the progress of refinement and luxury at home and abroad.

As nations advanced in civilization, the quality of household furniture improved. We learn from the sculpture upon the tombs of the Egyptians, that the realm of the Pharoahs had advanced to some degree of elegance in such matters, and there are figures taken from Egyptian tombs, showing a lady seated in a chair, shaped in modern style. The fact that no braces were used to the legs of the chair gives evidence of the skill of the Egyptian cabinet maker, in making them of sufficient strength without. The Greeks borrowed their style from the Egyptians, and improved upon it, and in time gave it to the Romans. Our knowledge concerning the furniture of the Greeks and Romans, is mostly had from their sculpture, and the works of their classic writers. The excavations at Pompeii, and Herculaneum, have enriched the museum at Portici with many splendid articles of ancient furniture, and show that the ancients applied the fine arts to that branch of industry, and combined the ornamental with the useful in a high degree. Nor were they unaware of the use and value of the more expensive kinds of wood. When luxury and refinement were at their heighth in Rome, tables made of the rarer woods were sold for enormous prices. Cicero, we are told, gave a sum equal to $35,000, for a Cyprus table, and we read of others fully as valuable. It is probable that these tables are inlaid with ivory and gold, and profusely ornamented with carving. Pliny warmly commends the beauty of the maple. The art of veneering was well known at that period.

The style of furniture in every age and nation partakes of the character of the people. When Europe sank into barbarism, furniture lost its beauty, and only rose to its dignity as civilization progressed. The architecture of a people has been held to mark their progress—and the style of furniture, as a general thing, is compatible with the architecture.

In England, the country of our immediate progenitors, furniture made slow advance to perfection of shape or construction. The continual wars of the houses of York and Lancaster kept the country so unsettled that the castles and palaces of the wealthy displayed little of domestic comfort, and much less of luxury. The close of these by the coronation of Henry VII., paved the way to a new era. To the bloody Henry VIII., the country was indebted for a movement in this branch of the mechanic arts. He employed Holbein, the celebrated Swiss artist, during his leisure time, to make designs for all kinds of furniture, which surpassed anything before in use. But this spirit was not enlarged during immediately succeeding reigns. The furniture, specimens of which are still extant, was manufactured in Holland or Flanders, or after the style of those countries. The legs of chairs and tables, and the posts of bedsteads, were straight and decorated with tinsel ornaments, and in the later Flemish modes, scroll work and carving were introduced. But this furniture which is now mostly known as Elizabethan, was by no means of the perfect kind. The chairs were straight backed, and therefore devoid of comfort; and the bedsteads, from their profusion of projecting points, and the draperies with which they were loaded, became resting places for dust.

In the modern imitations, while the general outline is preserved, these objections are sought to be obviated, and the carvings are far more appropriate and graceful.

As before stated, furniture and architecture are generally accordant; and hence, when Grecian architecture was revived in Europe, displaying the Gothic, the furniture was also changed, and partook of both styles—mingling the twisted column, the grotesque tracery and the ornaments of the Gothic, with the rectangular fashion of the Greek. The architecture of the time was a corruption of the Roman, rendered still worse by ill-judging artizans, who had neither originality nor purity of taste.

During the reign of Louis XIV., when luxury reached its height at the court of France, was introduced that rich and ornate style, possessed of much originality and beauty, which is known as "Louis Quartoze," and which is now one of those supreme in public taste. It is marked by elaborate carvings, mostly in bold relief, and filled with graceful curved lines; and these are shown to still greater advantage by heavy and gorgiously colored damask, which harmonizes with the style of the wood work. During the succeeding reign, the furniture underwent some change—the carving becoming more delicate and intricate. With Louis XVI., another change came, and a rage for the antique caused an alternation in furniture to correspond with the altered taste.

In England, however, the introduction of mahogany caused a still greater change. From the vivid color of the wood, and the high polish it readily attained, it seemed to have acquired no enrichment.

by carving. Tables, cabinets and bedsteads were made; and the art of wood carving seemed to be almost forgotten. But when furniture had sunk to its greatest point of plainness, insipidity and bad taste, a wealthy English gentleman, Mr. Hope, endeavored to revive the art of cabinet making by filling his mansion with furniture, modelled by himself after the antique. It was with great difficulty that he could find workmen capable of executing his designs. By perseverance, and at great expense, he succeeded, and afterwards had drawings made of the articles thus executed. These were engraved and published in large folio volume in 1807.

The present style of furniture is that of Louis XIV, Louis XV, and an improved Flemish—while many of our first families decorate their houses with a style embracing the best points of all these, with enrichments from the Gothic, the Elizabethan and even the Oriental. But there is a growing taste for entirely new styles of furniture, and this keeps designers busily at work to gratify it.

THE MATERIALS EMPLOYED.

The principal material of which furniture is made, is one or other of those costly and valuable woods so familiar to every housekeeper, and it is very important, especially in this climate, that the wood employed should be well dried, or as it is technically called, *seasoned*, otherwise it will shrink or warp. When it is required of curved forms, it has mostly to be cut out of the solid wood, making a deal of waste and adding to the expense; though it can sometimes be bent—as is mostly the case in Russian furniture—into various forms, and there retained by drying. This process has, however, a serious objection in the variations of our climate, which renders work of this kind little to be depended on, and forces our cabinet makers to the more expensive but certain mode of manipulation.

Veneering, which is the application of the more expensive parts of wood, in their plates, to a surface of inferior woods, is applicable to all flat and highly curved surfaces; and if well done, and the baser wood beneath properly seasoned, can be relied on to stand very well, even sudden changes of temparature. If not judiciously done, and by the most skillful workmen, it is apt to warp, crack, split or blister, and is then repaired with great difficulty; sometimes, indeed, in the cheaper kind of ware, the common woods are stained in imitation of the more expensive, and thus a handsome article, though not so durable, is produced at a small price.

Furniture is occasionally inlaid. When this is done with wood it is known as *Marquetry;* when done with shell or metal it is called Buhl.

Almost as much depends upon the nature of the wood of which furniture is made, as upon the workmanship, and hence it is important that housekeepers should have a general knowledge of this—a knowledge, the chief points of which we intend to communicate in as brief a space as possible. To go into a detailed history of all woods, would occupy more space than we can spare, but we can speak fully of the principal, and summarily of the other kinds.

Woods differ in color—some being of a reddish hue, as cedar, cherry, birch and mahogany; some of a brown or chocolate hue, as walnut; some of a yellow color, as oak, maple, pear-tree, box, English walnut, and satin-wood; some black as ebony, and others of two distinct colors, as rosewood. No less different are they in hardness, durability, and their capacity to receive a lustre. Much of the different qualities of wood depends upon its structure. The woods used with us for furniture is invariably of the kind denominated by botanists *oxogenous*, that is, the sort which increases its thickness by an annual addition of a circular layer. These layers being composed of soft wood on one side, and hard on the other, we are readily enabled to count them by the lines exhibited when a section is made. Sometimes thin plates of hard wood also radiate from the centre, and cross these concentric circles, producing the fine effect called by most of our trade, the silver grain. In close grained wood, such as the pear-tree, these concentric circles cannot be distinguished. Other peculiarities in woods are their curls, knots and waves, or, as in one variety of maple, a succession of spots, from the color passing off in waves, producing a fine effect.

Of all the foreign woods, mahogany has for a long time been the most employed. It was accidentally brought to the notice of the cabinet maker about the beginning of the 18th century, although it is mentioned so far back as the 16th, having been used in 1597, at Trinidad, in the repairing of one of Sir Walter Raleigh's ships. The circumstances attending its introduction for household use, are quite interesting, and are detailed by Philips, in his work on Pomiculture. It appears that at the time spoken of, a famous physician of the day, Dr. Gibbons, was building a house in London.

His brother, who commanded a vessel in the West India trade, having to make the return voyage in ballast, brought over some logs of mahogany, which he thought might serve as lumber for the Doctor's house. The carpenters tried it, but finding the wood to be too hard for their tools, condemned it to the woodpile. Some time after that, Mrs. Gibbons desiring to have a candle-box, her husband sent for his cabinet-maker, Mr. Wollaston, and desired him to have one made from the mahogany which lay in the garden. The cabinet-maker found the same fault as the carpenters, but made the box. When the latter was finished, it looked so well that the Doctor ordered a bureau of the same material. When this was brought home, its lustre and color attracted his admiration so forcibly, that he invited his friends, among whom was the Duchess of Buckingham, to see the novelty. The Duchess was so delighted that she begged some of the wood, and ordered Wollaston to make her a bureau. The wood at once became fashionable, and so did the cabinet-maker, who first used it, and both progressed steadily in popularity, from that time out.

There are two kinds of mahogany in use—one of these is inferior, as a general thing, and is known as bay-wood. It comes from the British Colony of Honduras; the other is called Spanish, and comes from Cuba and Hayti. The latter is hard, with a close, silky grain, and fine pores, and is beautifully curled, mottled and waved. It takes a light polish, and is admirably adapted for all good furniture. The Honduras mahogany is lighter and softer, with a more open grain, and evident pores. It does not take so high a polish, and is more apt to retain dirt than the closer grained sort.

Rosewood, or *pallisandre*, which is the next most commonly used of the foreign woods, but far more expensive than mahogany, is brought from Brazil. From its hardness it is difficult to manufacture, which adds somewhat to the cost of furniture formed of it. Its color is of deep brown, upon brownish red, and is arranged in elongated double cones. It will fade, however, if exposed long to sunlight; and no matter how well seasoned, or how carefully varnished, the lustre will, in time grow dim—owing, it is thought, to an oil in the wood, which no seasoning can thoroughly evaporate. The appearance of the wood is exceedingly rich, and a *fauteuil* or sofa of this material, carved *en arabesque,* and covered with heavy silk or satin damask, forms a most attractive and substantial piece of furniture. The sofa and tete-a-tete frames we exhibited at the fair last fall, and which were so much admired, were manufactured from this wood, the *pallissandre* or rosewood.

The woods of our own country are equal to almost any in the world, and are rapidly becoming generally introduced into use. Among them, walnut (of which we have an abundance in our own State,) holds the first place. It is of a dark brown, with a firm, and in good specimens, a very close grain, and takes a high polish. Logs are found as fully variegated with curls and knots as the mahogany. The tree from which it is obtained, the *Inglans forcina,* is one of the most picturesque of any in our national forests, being different in its character of foliage from the oak, with which it is mostly surrounded, rising frequently to the height of eighty feet, with its high head, broad gnarled arms and rough black bark, it has a most weird-like appearance. It is subject to a disease, a sort of tumor, accumulating upon it, which sometimes extend many feet in length. This puts us occasionally in possession of the most elegant wood ever used in furniture, and called the burled walnut, or sometimes the tortoise shell wood. The best specimens of the walnut we get in this State, are mostly shipped to Boston, where it is in great vogue, and when manufactured sell for as much, and sometimes more, than mahogany.

Our native forests are rich in various oaks, some of which possess great beauty. Unfortunately, many of these are so liable to warp, that it is almost impossible to use them, no matter how thoroughly they have been seasoned. The white oak, a most noble looking wood, possesses this fault to such an eminent degree, that it is unfitted for furniture, except where the article can be composed of small pieces. Our black oak stands better, and is sometimes very handsomely curled.

The maples, of which we also have an abundance in our own State, both the curled and birds-eyed varieties, the latter resulting from a vagary of nature in the sugar maple, are woods which receive a high polish, and are hard and durable, but their yellow color does not seem agreeable, and they are little used in the better kinds of moveable furniture. It is, however, sometimes used in making fine toilet or bedroom sets, but mostly for ordinary post bedsteads.

The native cedar, the wood of the *Juniparus Virginiana*, which is of a pleasant red color, and might well be used for the interior of drawers and wardrobes and for chests for packing away blankets, because of its agreeable odor, which keeps away moths and other insects. It is soft and light, but next to the catalpa, the most durable of woods, not even excepting the locust and the live oak, of Florida.

Our native oak, from its toughness, is useful in making frames that are to be entirely covered or upholstered.

The roots of the birch, beech and cherry are good examples of woods; but those we have hitherto enumerated, are the main reliance of the cabinet maker.

But we must not forget the cherry, also an excellent wood, and mostly used for dining room sets. The best specimens of cherry in our own State are to be found in the valley of the St. Joseph River. That which grows in the eastern counties is generally small and filled with gum specs.

But, sir, we fear you think this will occupy too much space in your journal, and will therefore close by adding directions for keeping good furniture in order.

If there should be any paint or grease on the surface, remove it by gently rubbing it with a woolen rag, moistened with spirits of turpentine; and then, with a soft rag rub the latter entirely off. Take a soft sponge wet with clean cold water, wash well all over, then wipe with a soft chamois or very fine buckskin, which has been wet with cold water and wrung dry as can be done.

After this, if any places be scratched or bruised, apply a very small quantity of boiled linseed oil to the injured places, and then rub it off with a piece of silk. Be careful to leave none on the furniture as it will unite with the varnish and make a bad and rough surface. Should the furniture be discolored by water, the linseed oil thus applied will restore the color.

Shun all receipts for polishing old furniture. There are none sold, that we know of, calculated to do it the least good. If your furniture should need varnishing, take finely ground pumice stone, with a piece of cloth well wet with water, and rub the surface, being careful not to rub through to the wood. Then, with a flat varnish brush, put on a coat of very fine copal varnish. If it does not look well after the first application, renew in three or four days the same process.

Furniture should be cleansed with water as above directed every fall and spring, by which means it will always look well.

Detroit, January 12, 1852.

WHEAT.

P. J. VAN VLEET'S STATEMENT.

J. C. Holmes, *Sec'y Mich. State Ag. Society:*

Sir—I give below a statement of a field of wheat, upon which I have applied at the State Fair for a premium.

The wheat all threshed, except the rakings—which by good judges is laid at twenty-five bushels on eleven acres and eleven rods—is 519 58.60 bushels.

I plowed on the last of May and first of June, 1850, and turned to common. Harrowed once, the middle of July. On the first of August shut in 125 sheep for one and a half days. On the first of September cultivated the whole once, and a part of it twice. Began sowing at noon of the 9th September, and finished on the 10th, harrowing both ways. When the wheat was nicely up I rolled it with a heavy roller. About the first of April, 1851, sowed 457 lbs. of plaster. The field lay in clover sod, and I applied about 14 loads of long manure per acre.

Product upon 5 acres and 3 rods,	276 24.60	bush.
" " 6 acres and 8 rods,	243 34.60	bush.
Total,	519 58.60	bush.

The variety of wheat raised was the Virginia May or Soules Wheat.

Macon, Sept. 19, 1851.

State of Michigan, County of Lenawee:

On this 4th day of November, 1851, personally appeared before me, the above named Peter J. Van Vleet, and made oath that the foregoing statement is true according to the best of his knowledge and belief.

HENRY DARLING,
Justice of the Peace.

BOYDEN WHEAT.

BY LUTHER BOYDEN.

J. C. Holmes, *Sec'y Mich. State Ag. Society:*

Sir—I herewith send you one bushel of a new variety of wheat of my own producing, having had it in progress for the last eight years. Of this wheat, I sowed last year, on a clover sod of once plowing, seven acres. The yield was thirty-five bushels to the acre, weighing sixty-four pounds to the bushel. The test in grinding was a barrel of flour from 4 9-60 bushels, and the bakers here say it is the best flour they ever used. I state these for facts, and I feel confident that this is the very best sample of wheat ever produced in this State.

I remain very respectfully yours.

Webster, Washtenaw Co., Sept. 22, 1851.

CORN.

STATEMENT OF DAVID M. UHL.

J. C. Holmes, *Sec. Mich. State Ag. Society:*

The ground upon which I raised corn the past summer was green sward, sandy soil, opening plain land. I covered this land with coarse barnyard manure, not thick, and immediately plowed the same under, rolled the ground with a heavy roller, and dragged it three times. I planted the large grained eight rowed white and yellow corn, in hills three feet and ten inches apart, about the first of May. The plowing being done the last of April preceding.

As soon as the corn came up, it was plastered on the hill, and hoed once—after that kept the ground clean with the cultivator. I measured and harvested seven and one-fifth acres of the corn, in November last, and had on the same, eleven hundred and thirty-two bushels of ears of sound corn, being at the rate of 157 bushels of ears of corn to the acre.

This corn yields more than one bushel of shelled corn to two bushels of ears.

Ypsilanti, Dec. 9th, 1851.

—

I saw the above described corn while growing, I also saw the corn after it was harvested. It was a remarkable good yield of corn, and noticed as such by the neighbors.

GROVE SPENCER.

OATS.

CHARLES T. WILMOT'S STATEMENT.

J. C. Holmes, *Sec'y Mich. State Ag. Society:*

Dear Sir—My farm lies on what is called the South Ypsilanti road, one mile south east of the city of Ann Arbor. The land upon which I raised the oats I offer for premium, has been cultivated some twenty years. The soil is of a gravelly loam; it was manured with about 25 loads of long manure to the acre last year, and planted with corn and potatoes. It was green sward and turned over the fall before. I plowed the land the latter part of last April, and went over it with a cultivator. I then sowed about two and a half bushels of common white oats to the acre and covered them with the cultivator. Soon after they were up, I sowed a peck of Ohio plaster to the acre.

The two acres yielded one hundred and fifty bushels of oats weighing 33½ lbs. per bushel. Had it not been for immense flocks of blackbirds living on them, and a heavy storm a few days before they were cut, I have no doubt I should have had eighty bushels to the acre.

Ann Arbor, Dec. 17, 1851.

BROOM CORN.

GARRET TEN BROOK'S STATEMENT.

J. C. Holmes, *Sec. Mich. State Ag. Society:*

Sir—I herewith hand you a statement of my crop of broom corn, entered at the late State fair for a premium:

I gave my whole field, which measured seventeen rods and four feet, by nine rods and fourteen feet, or one hundred and seventy rods of ground, which yielded seven hundred and ninety-one pounds, or seven hundred and thirty-six pounds to the acre, after being cut, dried, and the seed cleaned off, ready for making up.

Adrian, Dec. 5th, 1851.

We the undersigned, having helped to harvest and take care of the above described broom corn, do certify that the above statement is correct.

WM. TEN BROOK,
URIAL STEWARD.

WINE.

W. A. BUCKLAND'S STATEMENT.

J. C. Holmes, *Sec'y Mich. State Ag. Society:*

Dear Sir—This wine was manufactured by W. A. Buckland, of Howell, Livingston county, Michigan, from the Isabella grape, in the following manner, (in October, 1850:)

The grapes were picked clear from stems and smashed in a wooden vessel and allowed to stand forty-eight hours, when a slight fermentation commenced producing the color, (otherwise it would have been white wine.) The juice was then pressed out and two and a half pounds of loaf sugar added to each gallon. It was then put away in the cellar, and allowed to stand until the first of January, well vented. It passed through the sacharine into the vinous fermentation, was racked off, all that would run clear, and corked tight, where it has since remained, having no trouble in keeping it from passing into the acetic fermentation.

The wine has nothing in it but grape juice and loaf sugar, as above, and is pure as wine can be made. The undersigned is of the opinion that such wine can be made for one dollar per gallon, and contemplates extending his vineyard and the manufacture of wine. The soil is gravel and sand, descending to the south.

Howell, Sept. 23, 1851.

DIAGRAM OF S. G. PATTISON'S FARM, MARENGO, CALHOUN COUNTY.

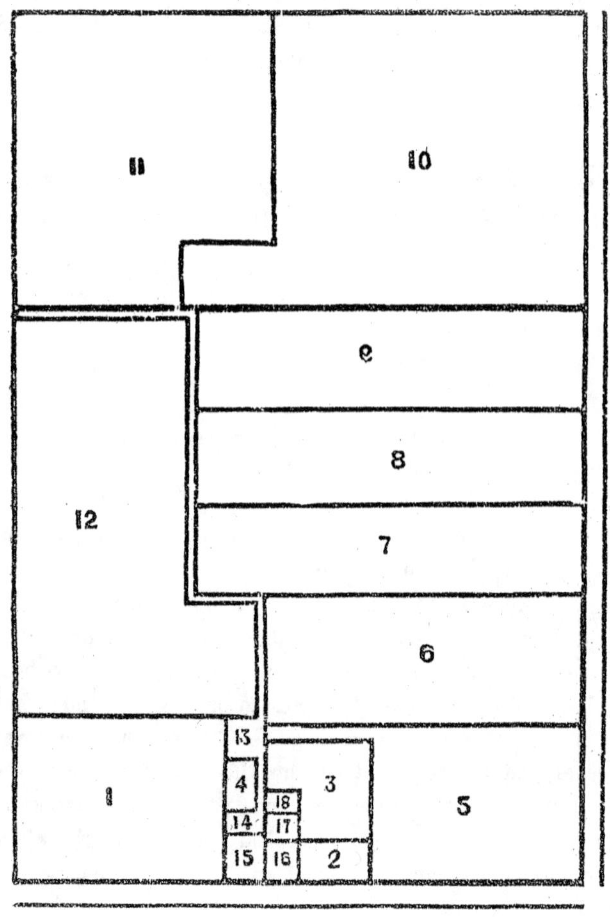

FARMS.

S. G. PATTISON'S FARM.

In the Society's Transactions for 1850, page 143, may be found a statement of S. G. Pattison, Esq., respecting his farm. The diagram furnished by Mr. Pattison, not having been published, I herewith present it with the following references:

This farm is situated in the town of Marengo, Calhoun county; it contains two hundred and forty acres of land, being one hundred and sixty rods north and south, and two hundred and forty rods east and west. The buildings are situated near the north end on a main traveled road, leading from Marengo to Homer. This farm was located of the government, in 1833, by its present occupant. It is one mile south of the Kalamazoo river, from which there is a gentle assent to near the centre of the farm, which is about one hundred and thirty feet above the level of the river. The land is Burr Oak and Hickory plains. Good water is obtained by digging from five to fifteen feet. As the farm is laid out there is living water on nearly every lot.

Lot No. 1 contains 16½ acres, in corn and potatoes.

Lot No. 2 contains 1½ acres, garden and fruit yard.

Lot No. 3 contains 4½ acres, in orchard and meadow.

Lot No. 4 contains 1½ acres, yard for calves and other young stock.

Lot No. 5 contains 17½ acres, meadow.

Lot No. 6 contains 20 acres, oats, rye and meadow.

Lot No. 7 contains 15½ acres of pasture.

Lot No. 8 contains 17½ acres of pasture.

Lot No. 9 contains 22 acres of meadow.

Lot No. 10 contains 48 acres of wheat.

Lot No. 11 contains 30 acres of wheat.

Lot No. 12 contains 33 acres of thrifty young hickory.

Lot No. 13, sheep yard and sheds, barn and stable for cattle.

Lot No. 14, horse barn, hen house, and yard attached to barn.

Lot No. 15, yard in front of horse barn.

Lot No. 16, dwelling house and front yard.

Lot No. 17, wood yard.

Lot No. 18, hog yard.

DIAGRAM OF LINUS CONE'S FARM, TROY, TOWNSHIP OF AVON, OAKLAND COUNTY.

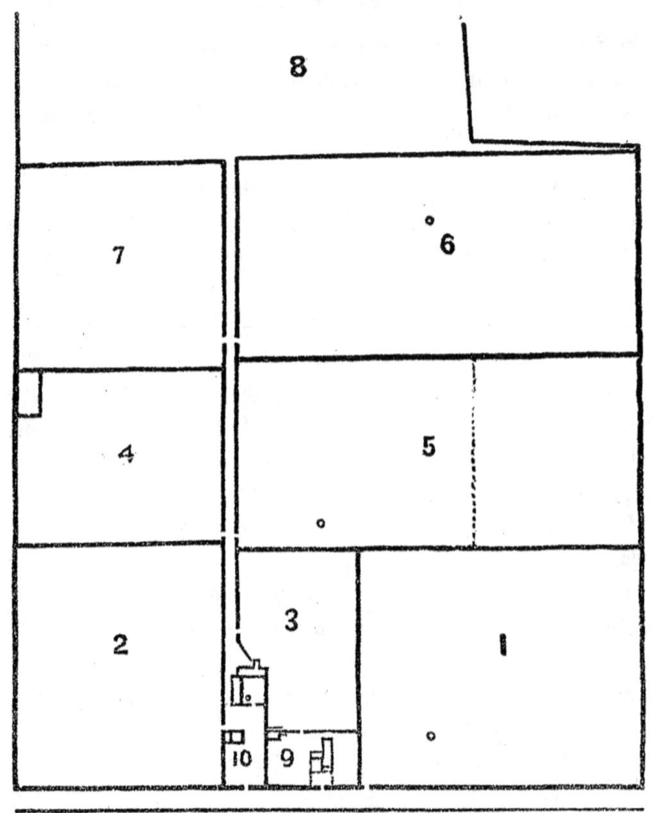

LINUS CONE'S FARM.

[A description of Mr. Cone's farm may be found in the transactions of the society for 1850, page 152. The diagram and statement here presented were furnished but not published.]

State of Michigan, Oakland County, ss:

Hervey Parke, being duly sworn, deposes and says, that he is a resident of Pontiac, in said county; that he is by profession a practical surveyor; that on or about the 16th day of December, 1850, he made a survey of the improved portion of the farm of Linus Cone, situated in the township of Avon, in said county, located principally on the east half of the south west quarter of section thirty-three in said township; that the plat hereto attached and to which reference is made, is a correct plat of said improvements as surveyed by me, excepting the west line of lots number five and six, and the east line of lots number four and seven, which were ascertained by the bearing of the lines, distances being determined by the survey of the outer lines of said lots, and that lot number one as delineated on said plat was actually surveyed by me and contains twelve acres and two hundredths of an acre of plowed land.

HERVEY PARKE.

Sworn and subscribed before me this 27th day of December, A. D. 1850.

CHARLES M. ELDREDGE,
County Judge, Oakland Co.

Lot No. 1 contains 12.46 acres or 12.02 plowed.

Lot No. 2 " 9.60 " 9.21 plowed.

Lot No. 3 is the orchard, and contains 3.42 acres or 3.18 plowed.

Lot No. 4 contains 6.70 acres including the small piece of wet land in the north west corner, or 6.21 plowed.

Lot No. 5 contains 14.09 acres or 13.70 plowed.

Lot No. 6 " 14.50 " 14.01 "

Lot No. 7 " 7.99 " 7.64 "

Lot No. 8 unimproved.

Lot No. 9, dwelling house and garden.

Lot No. 10, yard, barns and sheds. A lane runs from this lot and terminates at the unimproved lot No. 8, with gates opening into the fields upon each side.

There is a well in lot 1, 5, 6 and 10, designated by a dot.

[The reader, in order to understand the following reports on farms, should turn to the premium list, page 42 of this work, where a series of questions will be found, to which the numbers in the reports refer.]

FRED. W. CURTENIUS' FARM.

The committee on farms for the county of Kalamazoo, beg leave respectfully to report the following, as the result of their official labors in said county. Among the farms presented for examination, and which was adjudged by your committee as entitled to the State Premium, was that of Frederick W. Curtenius, lying upon Grand Prairie, in the township of Kalamazoo.

The farm consists of one hundred and sixty acres, one hundred and twenty acres of which is prairie soil, and all under a high state of cultivation, and forty acres of timbered land, covered with a growth of maple, beach, ash and bass wood.

The buildings are all in a good state of repair, and have been erected at an expense of some $2,000. With regard to the various details of soil, management, manures, crops, animals and fruit, we refer you to the annexed answers to the interrogatories propounded by your committee.

SOILS, &C.

1. The farm consists of 160 acres, 120 of which is under cultivation, and 40 heavily timbered. It is furnished with stock water, by a pond of artificial formation, covering perhaps the eighth part of an acre. Water for other purposes is furnished by cement cisterns.

2. The soil is a vegetable loam, having a clay subsoil, with sufficient lime intermixed to answer present agricultural purposes.

3. To improve the soil of my farm, (all of which partakes of the same character,) I consider stocking it down with clover, a moderate use of plaster, and *depasturing* with sheep, as the cheapest, surest and best method,—indeed, I know of no better method to be adopted upon a sandy or gravelly soil.

4. For years I adopted the system of shallow plowing, and for the most part succeeded in obtaining average crops when the soil was comparatively new; but for the last four years I have gradually plowed deeper, until now the average depth is about 9 or 10 inches, and my wheat crops have since raised from crops of 18 and 20 bushels, the former yield, to 28 and 33.

5. I have never failed to secure an ample crop after deep plowing; after shallow plowing, I have sometimes had to contend with partial failures.

6. I have never used a subsoil plow. My method has been to follow with an ordinary plow in the same furrow. The effect on wheat has been to increase the yield of the crop, (in connection with the use of the wheat drill,) at least 33⅓ per cent. In the corn crop I

have never experienced any great benefit in the early part of the season, but when the corn commenced maturing I could discover a benefit. I have never pursued a system of draining. Some portions of my farm is slightly undulating, and during the winter and spring my wheat crop has received some injury by the water standing whereever there was anything in the form of a basin. I have recently rectified this difficulty, by digging in each of these basins a hole perhaps three feet square, and deep enough to penetrate the hard-pan, say six feet deep, and have then filled up the hole loosely with stone, to within a foot of the surface, and levelled it with ordinary soil.

7. My farm is on a prairie; of course no timber on it. I think, however, that the burr oak and the wild cherry were indigenous.

MANURES.

8. I apply about 20 loads of manure to the acre. Upon my meadows I spread about the quantity above mentioned in the fall of the year, and in the spring pass over it with a light harrow or some thorn bushes bound together, in order to level the surface as much as possible. When I apply it to ground intended for corn, I haul it out in the fall, spread it and plow it in immediately after. My manure is principally gathered from my barnyard, and a portion of it exposed to the atmosphere. In May or June, I throw it into heaps or rather winrows in my barnyard, bringing the heaps to a peak as much as possible in order to prevent the rain from drenching it. And after lying in this manner until October or November, I then cause it to be hauled upon the field.

9. I annually manufacture about one hundred and twenty loads of manure. The manure is made from the straw of about thirty acres of wheat, fed to cattle, sheep and horses.

10. The manure is applied partially in a green state and partially in compost. I prefer using my manure on land intended for corn, and choose to have it in a rotten, in preference to a fresh state.

11. I could cheaply and essentially increase my supply of manure by carting muck from hollows on my farm, or by gathering leaves from the timber land, all of which extra labor would pay liberally.

12. I have used plaster frequently upon my lots, and successfully. Plaster I use at the rate of a barrel to five acres. Guano or salt I have never applied—lime only in minute quantities upon wheat. I

have no doubt but that lime sown on land somewhat exhausted, when in wheat, would be very beneficial.

TILLAGE CROPS.

13. I till 80 acres—of which, generally, 30 is devoted to wheat; 20 to corn; 5 to oats and potatoes, and 30 is tilled for summer fallow.

14. Of wheat I drill in from 1 bushel to 5 pecks per acre, and can discover at harvest little if any difference in the yield or in the appearance of the field. I aim to sow from the 7th to the 15th of September; and when there is no prospect of insects, I prefer to sow during the last days of August. My wheat ground (when not a clover sod) I endeavor to turn over in the fall, and the summer following repeatedly drag and cultivate it with a two horse cultivator, and suffer my sheep to feed upon it whenever there is any thing to graze. My yield formerly was about 20 bushels, but since I commenced using the cultivator and drill and have turned my attention to wool growing, I am not satisfied with less than 30. I have never suffered as much as many others from insects, and perhaps one reason may be, that when their appearance was anticipated, I deferred sowing until just before the first frosts were expected, experience informing us that after two or three frosts their ravages are suspended. Late seeding, however, exposes us to hazard from rust, but we run the gauntlet between the many foes to wheat cultivation.

I am not prepared to say positively what amount of fertilizing matter is taken from the land by a crop of wheat, but have been led to believe the value of a bushel of bone dust and perhaps ten bushels of lime and ashes mixed, will supply the loss of phosphorus, lime and alkili, which would result from taking 20 bushels of wheat from an acre. The better way, however, and indeed the more economical one, is to clover down and turn under in order to supply these deficiencies, making use freely of plaster.

Corn I generally plant about the 10th of May—make use of about 4 quarts per acre—cover before plowing as much of the field as possible with barnyard manure. Usually cultivate it twice and hoe it thoroughly once, and if I think it needs it, immediately after harvest, go through it with a shovel plough. If, at the time of cultivating for the second time, I find the weeds pretty well subdued, I seed it to clover, and have always succeeded. My corn crop, one year with another, will come up to about 60 bushels, which, at 2s 6d per bushel, is decidedly the most profitable crop I raise, except in those years when clover does unusually well.

15. The answer to this interrogatory is found in reply to previous questions.

16. Also answered as above.

17. Like most others, I have suffered in a measure from the potato rot, as it is called. I know of no remedy, nor do I believe any has yet been discovered. I have found, however, that potatoes nearest to the surface suffer the least. Potatoes should be placed upon the barn floor, or in some other suitable building, immediately after digging, and be permitted to remain there as long as possible before being taken to the cellar or the pit, if you would materially prevent the further spread of the disease.

Here let me predict, that the time is but a little way in the distance when our corn fields will have to contend with the same malady or one equally destructive and provoking.

GRASS LANDS.

18. I raise timothy, clover and red top. Of clover I make use of 10 quarts to the acre; of timothy half a bushel, and of red top one bushel. I prefer sowing in the spring of the year, and when I sow upon winter wheat I wait until the ground is somewhat settled and drag it in with a light harrow and then roll it. It will answer well to sow with barly, spring wheat or oats just before harrowing these crops for the second time. To soak the seed twelve hours in tepid water and then thoroughly dry them off in plaster, will pay for all trouble attending the operation.

The best meadow for dairy purposes is one which is seeded with clover and timothy, half and half.

In sowing clover seed, I think it to be decidedly the surest and best way to sow it in the chaff. It may not vegitate as quick but it will, I think, stand a drought better.

19. I mow from 25 to 30 acres, and pasture about 10, and when grass does well, as it has done for the last season, the product is about 3 tons per acre. Clover I cut when perhaps one half the heads are dead—timothy just before the seed ripens—red top after timothy.

My mode of making hay is to turn over carefully in the afternoon whatever is cut in the morning, up to 12 o'clock, without scattering it much, and then before sun-down putting it into cocks, and there, if the weather is propitious, suffering it to remain a few days. An hour or two before drawing it in, I open the cocks in order to let the air circulate through them sufficiently to dry out a portion of the sweat, and am careful to salt it in the mow or stack, at the rate of four quarts to the ton.

20. My meadows are all suitable for the plow, and when I wish to sow a piece of grass to wheat, I strive to plow it deeply, (10 inches if possible,) as early in the spring as the frost will permit, roll it, occasionally drag it, and just before harvest cultivate it both ways; and, when I get ready to sow, I cultivate it again a little deeper; then drill it in and am pretty sure of a good crop. If I desire to plant, I pursue a similar course, with a similar result; the only difference being in dragging and cultivating earlier.

21. In regard to irrigating land, I have none requiring that system of husbandry, nor am I particularly anxious to embark in the enterprise.

22. I have no low, bog, or peat land, and consequently can furnish no information touching their reclaimation.

23. I have generally succeeded in eradicating weeds from my farm. I have had some sorrel and yellow dock upon my premises; the former I got rid of by sowing clover, and keeping the field for a pasture for a few years. The dock I exterminate by not allowing one of them to go to seed, and by following this practice a few years, they disappear.

DOMESTIC ANIMALS.

24. My cattle are yet grades, and am getting rid of them as fast as possible, with a view of introducing the Devon and the Durham. I generally plow my land with horses—keep, however, usually one yoke of oxen, 3 or 4 cows, and perhaps a dozen head of young cattle. I keep one span of horses, and perhaps 4 or 5 colts. Horses a portion of them, of the Morgan breed.

25. I have never made any experiments calculated to show the relative value of the various domestic animals.

26. The best and most economical method of wintering cattle is to stable them, turning them out to water twice a day, and feeding cut straw moistened, and sifted over with perhaps two or three quarts of corn and cob meal ground together. This not being within the reach of every farmer in a new State like ours, the next best plan is to have sufficient sheds covered with a thick coating of straw, and well littered beneath to comfortably sh lter whatever stock the farmer may have—let them have access to water whenever nature requires it, and a ra k filled each day with bright wheat, oat, or barley straw, sprinkled with weak brine; an ear of corn apiece, night and morning, and whenever the weather is pleasant, a ramble into a corn field, where the stalks have been left uncut, and in very tedious and boisterous weather an additional daily meal of good hay, or in the absence of that, a few bundles of oats, and cattle will come out in good heart and without a bill of expense against each head sufficiently large to absorb its value.

27. I make no more butter than will supply my own wants, and my reason is, that I do not wish to increase the labor of the female portion of my family, though there is no doubt it would pay well to make butter for market. I can, therefore, give no satisfactory reply to this interrogatory, from actual experiment.

28. I keep, generally, about 200 sheep. They are of the Pauler and Spanish Merino breed, and have yielded for the last two clips, 3¼ pounds per fleece. My last sale was at 41 cents per pound. I only suffer my best and largest ewes to bear lambs, and consequently do not raise more than about 40 or 50 lambs a year. 200 sheep is as many as my farm will keep well, and as I desire to keep such as will produce the heavist fleeces, two-thirds of my fiock are weathers. My weathers are worth to butcher from $2 to $3. My lambs, should I elect to sell them for that purpose, would probably net me half those amounts.

29. In wintering sheep, as much depends upon shelter and water, as upon food. Shelter is not so necessary in clear cold weather as in stormy weather; indeed, I had rather sheep would not cluster under sheds when the weather is fair, never mind how cold it may be, but in storms, nothing can be more essential. So far as water is concerned, it is almost as conducive to success in wool growing. I

prefer to have the watering place from 20 to 50 rods distant, in order to furnish them daily with some little exercise. You may provide every convenience possible, and the very choicest food for sheep, and yet, if permitted to live day after day unwatered, they will not only fall away but actually suffer. Humanity, aside from interest, would direct such a system of treatment.

As to food, I allow them morning and evening, about half as much hay (clover and timothy mixed is best,) as they will eagerly consume, and during the interval I strew the yard with straw, occasionally brined. As often as every other day, I feed them shelled corn in troughs, about a gill each. I choose it shelled because it is more readily eaten, and more evenly. When the weather is tempestuous or intensely cold, I feed extra quantities.

My loss is seldom over two per cent annually. I take care to fatten or trade off any that show symptoms of failure, being perfectly satisfied that some one else should be present at their funeral besides me. Very fine sheep are not as hardy as medium quality.

30. I have at present eight swine—five of them are a mixture of Suffolk and Leicester, one good grade sow, and two pointers. The last two, after feeding some 20 bushels of corn without any improvement perceptible to the naked eye, I turned into the road, hoping never to see their snouts again; in which desire, up to this time I have been fully gratified. I do not know that I have an enemy in the world, but if I have, I really hope he has them. Besides eating up the twenty bushels of corn, they consumed ten ducks belonging to my boy.

31. For fattening purposes I know of no better food than cornmeal, scalded. For increasing the quantity of milk, carrots no doubt take the preference. Turnips are good, but I incline to the belief that it effects the quality of the milk.

FRUIT.

32. I have about seventy apple trees, all grafted fruit; and prominent among these are the Rhode Island Greenings, Esopus Spitzenburgs, Seek-no further, Princes Yellow Harvest, Newtown Pippins, Golden Russets, Northern Spy, Baldwins, &c., &c.

33. I have also Seckel and other varieties of pears, Washington and Golden Drop Plums, and others. Of cherries, black and white,

Tartarian and Ox Heart. Of peaches, some seventy or eighty trees, of good, bad and indifferent qualities.

34. My trees have never received any injury from insects, or from any other source, except during the last autumn, when some six or eight apple trees were girdled by mice. The remedy against this, is by heaping earth around the trees, to the highth of six or eight inches, which will succeed in nine cases out of ten.

35. My method of treating fruit trees is, to plant among them several years after they are transplanted; and whenever I stock down my orchard I do not fail every year or two, to throw barnyard manure around them, and spade it in, to the extent of six feet in diameter. I prune liberally when young, and endeavor to have few, if any limbs, so low as to prevent my team from passing under them when plowing. It is desirable also, to thin out the limbs occasionally in order to admit light and heat.

36. Under this head I think of nothing beyond what is interspersed among previous replies.

FENCES, BUILDINGS, &C.

37. My house consists of an upright part, 24 by 32 feet, cottage form, having an addition 24 by 14 feet, which is used for kitchen, a store room, and large pantry; the main building contains a parlor and dining room, each 16 feet square, and one bed-room below, and a hall; above, three bed-rooms. The house is substantially built, with a good cellar stoned up, having a brick floor; blinds to all the windows, lightning rod, tin eave troughs, and a permanent cement cistern. The house is a very convenient one, and with an occasional coat of paint, will out-last two generations.

My barn is 40 by 50 feet, of the usual internal arrangements; a shed 20 by 45 feet is attached to the barn, sufficiently high to store away above at least ten or twelve tons of hay.

Upon the premises also is a good corn-barn, 18 by 24 feet, which, by the bye, is one of the most convenient buildings upon my farm, being devoted to a great variety of purposes, aside from holding corn. My buildings are all constructed of pine, and that used about the house, perfectly seasoned.

38. My fences for the most part are of oak and ash rails, seven rails high, with stakes and riders. More recently I have turned my

attention to board fences and red cedar posts, the expense of which is about six shillings a rod. This kind of fence in the end is decidedly the best and the most economical. Wire fences I know nothing about experimentally, but incline to the belief that it will never come into use to any extent.

39. With regard to weighing and measuring my various crops, I am somewhat particular. Immediately after harvest I count my sheaves of grain and then thresh out with a flail a sufficient number of bundles to make a bushel, and at once can determine the quantity of my entire crops and the yield per acre, and never yet, I believe, have varried ten bushels on thirty or forty acres. So of corn. I husk out a row and at once have the the amount of that crop; so that I am prepared to make contracts with a degree of certainty as to the number of bushels I have to dispose of. At the same time, I weigh, in order to make assurance doubly sure. I have found it too often to be the case when my brother farmers guess at their crops they are apt to be sadly disappointed.

40. I keep a strict farm account, so that at the end of the year, I can tell with a degree of accuracy the cost of my several crops, their sales, and the profits or loss. This practice should be adopted by every one, as I esteem it highly essential to know what we are doing one year with another; and it will conduce to economy here and there, where it can be practiced, withal fostering business habits and furnishing at the same time food for amusement, as well as a taste for system.

41. The following is a general exhibit of my farm account for the last year:

	FARM	Cr.
By 30 acres of wheat, 28 bu. per acre, 840 bushs., 60 cts,		$504 00
" 18 " corn, 53 " 954 " 31¼ "		298 12
" 200 fleeces of wool, 650 lbs., 41 cents,		266 50
" 50 tons of hay, $7 per ton,		350 00
" 2 acres of oats, 100 bushels, 20 cents,		20 00
" 20 bushels potatoes, 50 cents,		10 00
" 40 lambs, 10s,		50 00
" increase in value of colts and cattle included in corn and hay.		

By profits in swine,		30 00
" profits on 2 cows, 1 yoke of oxen and 2 yoke of steers sold,		93 00
" profits on 18 sheep sold,		12 00
		$1,633 62

DR.

To interest on $7,000, (value of farm,) 7 per cent,	$350 00	
" boy 6 months, $10 per month,	60 00	
" board of boy,	27 00	
" harvest, haying, washing sheep and shearing,	35 00	
" moving manure and threshing,	37 00	
" husking and sundry other jobs,	23 00	
		532 00
		$1,101 62

From this balance of $1,101 62, there is to be deducted a few mechanics' bills, the wear and tear of farm implements, and certain store charges having connection with farm operations, amounting to about $65 00, leaving a net profit of $1,036 62 for my own labor.

I might say much more, but time and the space usually allotted to farm reports forbid—for should every competitor for a premium upon his farm be as voluminous as I have been, it would require two volumes of Transactions to record them all.

F. W. CURTENIUS.

Grand Prairie, Kalamazoo county.

The above is submitted as the report of your committee.

A. Y. MOORE, *Vice Pres. of St. Ag. Soc.*,
N. A. BALCH, *Cor. Sec. of St. Ag. Soc.*,
JOHN MILHAM, *Pres. Kal. Co. Ag. Soc.*,
Committee.

GEO. CLARK'S FARM.

1. My farm consists of 80 acres, with house, barns, sheds, and good yards.

2. Nature of the soil is gravely and loamy, mixed with limestone.

3. I consider that it should be summer fallowed, properly subdued, and then sowed with wheat.

4. I plow six inches deep, and ten inches in width. This I consider the best mode of plowing for crops.

6. I have drained with open and blind ditches, to the amount of one hundred and fifty dollars, and with good success.

7. My up land was oak openings, and the low land was tamarack, elm, black ash, and poplar, with hazle brush.

8. I apply about fifteen loads of manure per acre. I carefully see that my manure is kept in the yard, then taken to the field to rot. I think that all manure should be well rotted before applied to crops.

12. I have not used any lime for it is not easily procured. I use plaster, and like it much.

13. I till 40 acres of land, and occupy it with wheat, corn and grass. I harvested eighty acres of wheat. My wheat last year was 30 bushels per acre, but I think it will not yield quite as much this year. I have four acres of corn. My corn last year was 50 bushels per acre, but will not yield quite as much this year I sow 1½ bushels per acre of wheat, and I think about the 15th of September is the time for sowing wheat. I take my manure from the pile, and spread it before plowing for seed.

18. I use clover and timothy seed. I sow 10 lbs. of clover and 4 lbs. of timothy per acre in the spring; about April.

19. I mowed 14 acres this year, and my crop was about 30 tons. Grass should be cut when in the blossom.

22. I have in progress of reclaiming, about three acres of tamarack. I drained and ditched and then cleared it, and sowed it with timothy seed, and now pasture it.

23. I keep the weeds out of my crops. I think the yellow dock is the worst. I dig them and burn them.

24. I keep one yoke of oxen, two cows, eight head of young cattle, and three brood mares, two 2 years old and one 1 year old colts, They will class with the first class of horses.

30. I have 8 swine, young and old. I kill them at 1½ years old—the weight when dressed is from three to four hundred—the breed is good—they come to maturity quick and are easy kept.

37. I have one grain barn 30 feet by 40, one horse barn 36 by 45; also two sheds, one 20 feet by 70, the other 18 feet by 35; the yard is large and well fenced with boards, so that my buildings are well constructed for convenience and yarding stock.

GEO. CLARK.

Lapeer, Lapeer Co., Sept. 27th, 1851.

W. H. MILLER'S FARM.

The committee on farms, of the county of Hillsdale, respectfully report that they have examined the farm of W. H. Miller, of the town of Moscow, and received the following report on management of his farm, containing two hundred and eighty-six acres:

1. Number of acres improved, one hundred and fifty-six; woodland, one hundred and thirty acres; waste land, none.

2. The nature of the soil is a gravelly sand; sub-soil is the same.

3. Limestone, plenty.

4. I consider the best mode of improving, by plowing ten inches deep or more, on all soils.

5. I have found deep plowing preferable.

6. Never have used the sub-soil plow, nor drained any land.

7. White oak, black oak, hickory and burr oak. The Columbo root grew spontaneous, and other flowers in great abundance.

8. Of manures, not more than eight or ten loads per acre—it is not kept under cover.

9. Manufacture about one hundred and fifty loads.

10. Manure applied is in a rotten state for spring crops.

11. I could very essentially increase the supply.

12. I have used no lime, but have used plaster on clover and corn, with good results.

TILLAGE CROPS.

13. About fifty acres—twenty-five of wheat, eight of buckwheat, and the remainder equally with corn and oats; and one acre of potatoes.

14. The amount of wheat sown, is one bushel and a peck per acre; of corn, four kernels in a hill, four feet apart—bushels per acre, 40. Time of sowing, first of September; about the 10th of May for

planting corn, as the season requires. The fly has been very injurious to wheat—not any remedy.

15. The quantity of manure for wheat, 10 loads to the acre, and applied in August, in a rotten state.

16. Covered from six to ten inches for all crops except potatoes; from three to four inches for those.

17. The first potatoes that grew this year rotted, from being too wet; a few of the second growth survived.

GRASS LAND.

18. Timothy and June grass—of clover seed, 2½ quarts to the acre—seed in March, on wheat—in spring, on oats.

19. Mow about fifteen acres—mow twice—cut in the blow, and cure in the cock.

20. All the mowing ground is suitable for the plow.

21. No irrigating needed.

22. I have no such land.

23. The burdock is the most to be looked to.

DOMESTIC ANIMALS.

24. One yoke of oxen is used, and a span of horses. The Devon cattle appear to be the hardiest. and best adapted for work.

25. I have full blood Devons—a cross with the Natives, for oxen; produces fine forms and good size.

26. For wintering, prefer cornstalks, cut early, and good straw; water twice each day.

28. Sheep—about one hundred good common merino. They yield 2½ to 3 lbs per head—price, 25 to 33 cents—sold from one to two dollars a head.

29. Feed three times in the day of coarse food—either oats in the sheaf or buck wheat, good water and open shelter—lose from one to three by the grub—think the fine needs the most care.

30. Swine—I have from 15 to 16, keep well through the winter, and nothing but clover pastures in the summer—kill at 18 months and weigh 300 lbs.

FRUIT.

31. Number of trees, 120 apple and forty or fifty peach; one-half grafted with the best kind of apples.

32. Have fifteen cherry trees of various kinds.

33. Have no trouble with the insects.

34. Keep well pruned, and the ground loose with chip manure.

35. Good care that everything is in its *place.*

FENCES, BUILDINGS, &C.

36. Barn 30 by 40 feet, cow house and hog pen constructed together, 22 by 24 feet.

37. Fence of rails, 10 feet, 8 high; one dollar per hundred in the fence.

38. Have never kept a correct account.

39. I have no accurate account of expenses, but hope to the following years, and report accordingly. I think I have cleared $500 this year.

LEWIS T. MILLER,
Committee.

Moscow, Dec. 1st, 1851.

SAMUEL RAPPLEGE'S FARM.

SOILS.

1. My farm consists of 160 acres; 100 acres improved; 60 acres timber. About two acres of my cleared land is, I consider, nearly waste, it being wet and springy.

2. My soil consists of about 30 acres light sand; that part of the farm my buildings are on, which makes it convenient to manure from the yard; 60 acres clay, sand and gravel, clay predominates; the balance black sand and muck, the whole laying sloping or gently rolling. On parts of the farm there is lime stone gravel.

3. I consider barnyard manure the best for my sand; clover, salt plaster and ashes for clay.

4. I plow from eight to ten inches deep. Deep plowing does not suffer from extreme wet or dry and consequently produces better crops.

5. My plowing for the last five years has been from eight to ten inches deep for all crops.

6. I have not used the sub-soil plow. I have drained much of my land and almost all varieties of soil, and with good results, adding two-thirds of the value to some of the soil.

7. Oak, basswood, walnut, hickory, ash and elm is the timber.

8. When barnyard manure is used, thirty-five loads; when hog manure, thirty loads per acre. In fattening my hogs I use saw dust for bedding. It makes good manure.

9. I make about three hundred and fifty loads of manure yearly. About fifty is made in my hog pens, and three hundred under and about my barns. I use saw dust in my cattle stables for bedding. It is a cheap and easy way to keep cattle clean and makes valuable manure.

10. I apply all my manure to corn and patotoes, and all is applied in the spring that is made through the winter which is partly rotten.

11 & 12. I have used salt, plaster and ashes with wonderful success for corn and clover. The proportion was one bushel salt, two bushels plaster and four bushels ashes. One half bushel of the compound per acre, sowed on the clover first April, sowed broadcast on corn ground before planting.

TILLAGE CROPS.

18. I plant from fifteen to twenty acres with corn, from two to three acres with potatoes; five to eight with oats, and sow the whole to wheat in the fall. My corn has averaged for the last five years from fifty to eighty bushels per acre. This year nine acres yielded ninety bushels per acre. I manured the nine acres with barnyard manure, at the rate of thirty-five loads per acre—hauled it out in the spring on a clover sod and plowed it ten inches deep; worked the ground well with the cultivator before planting, and harrowed after the corn was up. I use the plow principally.

14. I plant the eight rowed yellow corn, about eight quarts of seed per acre; that plants thick. The ears are smaller, but I get more of them, and the stalks pay the expense of raising the corn. I feed all the corn on my farm. I planted the nine acres the 12th and 13th of June; it ripened well.

17. My potatoes this year yield about one hundred and fifty bushels per acre. They were affected with the rot; about one-quarter of them decayed before digging. The rest keep well.

I do not succeed well with oats. They pay less than any crop I raise.

I sow no wheat on summer-fallow. I sow after corn, principally. My wheat yields from eighteen te twenty-five bushels per acre, one year with another. This year, five acres Poland wheat gave thirty-two bushels per acre, after corn. It was sowed on corn ground the second week in October, cultivated in, one bushel and eight quarts of seed per acre. I think it the best variety of wheat that I know. It stands well, and is a very early variety.

GRASS LANDS.

18. I cut about twelve acres of grass, which yields two tons of hay per acre; and in this way I have more pasture land, and can keep more sheep than I could with the fallow system. I use clover, and clover and timothy seed together—sow six quarts of clover, or four quarts clover and two quarts of timothy seed per acre; sow in March and roll with a heavy roller in April or the first of May following, then a light dressing of salt, plaster and ashes, insures the crop.

19. This season I have mowed about two acres of clover for seed, that I sowed last March, and managed as above, which I think will yield three bushels of clover seed per acre. I think that clover, timothy and red-top, are all good and necessary for successful dairying. I cut grass just before it is fully ripe, let it lay in the swathe half a day, then put in cock, and if the weather is fair, let it remain until cured, which will be about two days.

20. My mowing land is all suitable for the plow, as I have ditched most of my wet land.

21. I have but two or three acres of bog land, and that I have not ditched. My mode of ditching is to cut a ditch in the lowest part of the field, sixteen inches wide on the top, twelve at the bottom, and from three to four feet deep, then fill 16 inches with straight and well-trimmed poles, taking care to lay them as tight and close as possible, of any size, from one to eight inches through, then fill with brush closely packed, sixteen inches more, then put on straw, if the leaves are not on the brush, and the ditch is ready to cover, which is done by throwing on a light covering of earth with the shovel, keeping one foot on the brush to keep it down, and then the rest is done with the plow.

22. The surface should be the highest over the ditch when it is complete, to prevent the surface water from washing and breaking into the ditch. I succeed well with my crop after ditching.

23. Thorough husbandry to all crops is the surest weed eradicator that I am acquainted with.

DOMESTIC ANIMALS.

24. I have three yoke of oxen, fourteen cows, forty-eight head of young cattle, from three years old down to calves. I winter my store cattle on cornstalks and straw, principally, giving them each one ear of corn a day.

25. I have four horses; one of them is a stallion, of the Messenger blood, three years old; the other three are geldings, two of them four, and the other five years old, one Eclipse, one Sampson, and one Magnumbonum stock.

26. I consider cornstalks the best and cheapest fodder for cattle and young horses—they should have water and shelter, but not kept too warm—stabling in a warm stable I do not like.

27. I have made this season twenty-six hundred pounds of cheese, and about six hundred pounds of butter, from eight cows and six heifers.

28. I have fifty sheep and lambs—wintered twenty ewes, and they dropped twenty-seven lambs; twenty-four of the lambs they raised; they were dropped in March, and in September I was offered twenty-four dollars for them. My flock I raised from a promiscuous lot of coarse and middling-wooled ewes and a merino buck. They produced a very even flock of lambs. This season they produced three and a half pounds of wool per head, which was worth forty cents per pound; my weathers would sell for three dollars per head to the butchers.

29. For nearly three years my sheep have picked their living, summer and winter—they have not been under a shelter, nor have they been fed anything for the past two winters, and thus far this winter I have not had one sheep die in three years—have not seen a tick on them in two years—do not know that one of them has been sick. I give my sheep salt and ashes once a week.

30. I kill about twenty hogs and pigs yearly. I think them a good breed, but no particular blood. Last year I killed twenty—two

litters from the same sow. Ten were fifteen months and twelve days old when killed; their average was four hundred pounds. Ten were about nine months old; their average weight was two hundred and ninety-seven pounds. The oldest ten were put in the pen from pasture and shack, the first of November, and fed on assorted corn two weeks, then on corn meal two and a half months. The twenty hogs were fed, while fattening, about twenty-five bushels of charcoal and ashes. I make about sixty hundred weight of pork yearly—have made that amount for the last six years. I have never sold for less than four dollars per hundred.

31. I consider corn worth fifty cents per bushel to make pork, potatoes about one shilling; turnips I have not tried.

FRUIT.

32. I have three hundred and fifty apple trees set on my farm, all except a few are grafted with the choicest kinds of fruit. I have a choice collection of peaches, pears and cherries.

35. My management with young trees is to cultivate the ground with a hoed crop, either corn or potatoes. Manure with barn yard manure, lime and ashes.

FENCES, BUILDINGS, &C.

37. My buildings consist of one old, rotten, rickety, but worthy log house; one barn forty by sixty, eighteen feet posts, with a basement, nine feet for cellar and stabling. I use rails eleven feet long for fencing.

40. I do not keep farm accounts, but I intend to. I think it very essential to good farming. I cannot give a positive accurate account of my farm expenses, or the receipts of the farm.

Adrian, Lenawee Co., January, 1852.

DIAGRAM OF GEORGE CLARK, JR.'S, FARM.

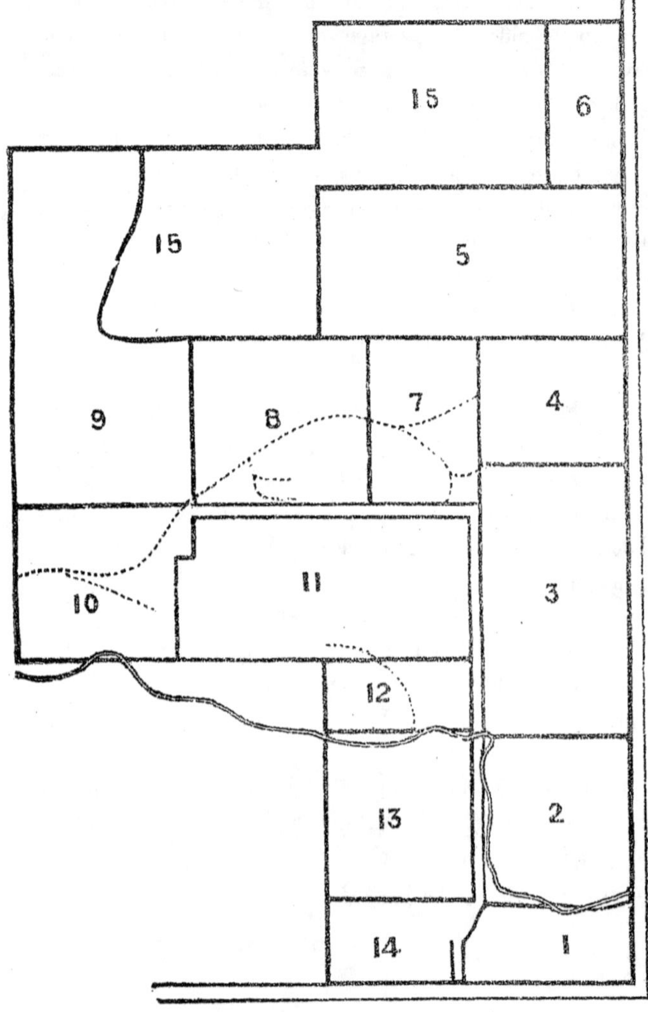

GEORGE CLARK, JR.'S, FARM.

1. I have 180 acres of land, 33 acres of wood land, 3 acres of tamarack swamp, 120 acres of improved land, 20 acres of swale reclaimed land, and 4 acres for roads. It is laid out as follows:

Lot No. 1 contains $3\frac{1}{2}$ acres meadow.

Lot No. 2 contains 8 acres wheat stubble and grass.

Lot No. 3 contains 16 acres pasture.

Lot No. 4 contains 7 acres meadow.

Lot No. 5 contains 16 acres sowed to wheat.

Lot No. 6 contains 5 acres meadow.

Lot No. 7 contains 6 acres sowed to wheat.

Lot No. 8 contains 10 acres oats, corn, potatoes and beans.

Lot No. 9 contains 22 acres sowed to wheat.

Lot No. 10 contains 12 acres marsh and swale.

Lot No. 11 contains 16 acres wheat stubble.

Lot No. 12 contains 4½ acres meadow.

Lot No. 13 contains 9 acres meadow.

Lot No. 14 contains 3 acres, dwelling house, barns, sheds and orchard.

2 acres for lanes.

36 acres woods and tamarack swamp.

4 acres roads.

Total, 180 acres.

A brook crosses the farm on lots No. 1, 2, 10 and 13. The under drains and open ditch are represented by dotted lines.

2. My soil is clay, loam, sand and gravel, subsoil generally clay and gravel, some limestone.

3. I summer fallow clay lands for wheat, and raising grass on sand and gravel soils for spring crops and clover.

4. I plough from 6 to 8 inches deep. I find that deep plowing is best on clay and gravel soils.

5. I have learned by experience that deep plowing does well after shallow plowing.

6. I have not used the subsoil plow. I have drained clay loam and sandy soils with good results.

7. My timber is generally oak, with some hickory, indigo weed tea weed.

8. I think I generally use from fifteen to twenty loads per acre. I keep a part of my manure under cover, the rest in a good yard. I have thought of making some cellars.

9. My means of making manure are from straw, hay and stalks; feed to my stock in different yards, sheds and stables. I have made about one hundred loads of manure this season, and applied it.

10. I have used my manure fresh for corn and potatoes; for wheat, in June, and just before seeding.

11. I think I can increase my supply of manure, and have, by using marsh muck.

12. I have used plaster on grass, wheat and corn; salt on asparagus, with good results.

13. I harvested 20 acres wheat, 6 acres oats, 3 acres corn, 1 acre potatoes, beans and turnips, 32 acres meadow, 18 acres pasture, 40 acres sowed to wheat.

14. I plant about 4 quarts of corn per acre, sow 1½ bushels of wheat per acre, and 2½ bushels of oats per acre. My corn, this season, was not planted until the 9th of June, for the reason of the land being wet and springy. I generally plant about the 20th of May. My oats were not sowed until about the last of May; wheat sowed about the 15th of September. I plow my land once for corn and oats, for wheat twice, and use the cultivator. I use the cradle and rake for harvesting wheat and oats. My corn yielded about 25 bushels to the acre, oats 35 bushels to the acre, wheat 25 bushels to the acre. I had about half an acre of wheat one year injured by the insects.

15. I use about fifteen loads of stable manure per acre for corn, and twenty loads of coarse manure for wheat and corn.

16. I have generally covered my manure from 6 to 8 inches deep.

17. Some of my potatoes were injured by the rot this season. I have not discovered any cause or remedy for it.

18. I sow clover and timothy and some red top seed on low lands. I sow about 6 lbs. clover seed, and 2 quarts of timothy seed to the acre. I sow my grass seed in the months of March, April and May. I sow broadcast.

19. I mowed 32 acres. I think my hay would average about two tons to the acre. I generally cut my grass when in the blow. My mode of making hay is when cut to spread it out, rake with a horse rake, sometimes cock it, other times draw it out of the winrow.

20. I have some land unsuitable for the plow. I have ditched and cut the brush from it, and pastured it.

21. I have not practiced irrigating or watering meadows.

22. I have reclaimed some low land and sowed grass on it.

23. I have eradicated the weeds by summer fallowing and seeding to clover. The most troublesome weeds are pigeon grass, pig weed and sorrel.

24. I have 1 pair of oxen, 3 cows, 15 head of young cattle and 3 horses, good common breeds.

25. I have learnt from experience that it is the best to raise good cattle and horses.

26. I consider the best and cheapest manner to winter my cattle is to feed them straw, hay and corn stalks, and make shelter for them. I have running water near my yards.

27. I make butter and cheese for the use of my family.

28. I keep one hundred and two sheep. They are grade merinos; average three pounds of wool per fleece; sold last June for forty cents per pound. Twenty-seven sheep produced lambs, and all reared. I have butchered a good many of my sheep. They generally weigh from fifty to eighty pounds per head.

29. I generally winter my sheep on hay; they have water and shelter; did not lose any last winter.

30. I have got one hog and ten spring pigs; they are good common breed of hogs. I feed them on Indian corn. I generally kill them at one year and a half old. They generally weigh from three to four hundred pounds.

31. I have not made any experiments relative to roots.

32. I have one hundred apple trees, chiefly grafted fruit, consisting of Greenings, Spitzenburg, Pippin, sweet and sour, Bough, Twenty ounce apple, Fall Pippin, Black apple.

33. Pears, plumbs and cherries.

34. Bark lice and caterpillar; wash them with lye.

35. Pruning and washing.

36. I have ditched low grounds that used to be covered with water, upon which I now raise good wheat and other crops. I sow broadcast with both hands with good results.

37. Three in number; one barn forty and fifty feet for grain and hay; one shed for stabling, granary and hay up above, twenty-four by fourty-four feet; another building for wagons, farming tools, hog pen and corn chamber, twenty-four by forty-four feet.

38. I construct rail fences; some of them stake and rider; some stone wall; height from five to six feet. The different kind of fen-

ces are marked on the plan, generally in good condition. No wire fences.

39. I generally weigh and measure my grain. They are not daily registered.

40. I do not keep regular farm accounts. Think it a good plan, but have not practiced it.

Lapeer, Lapeer Co., Dec., 1852.

FARM ACCOUNTS.

BY REV. CHARLES FOX.

In every sort of mercantile business, a strict and accurate system of book-keeping is considered necessary. Even the smallest store keeper has his day book and ledger; while the manufacturer employs a still more complicated system of accounts, without which it would be impossible for him to carry on his business with any chance of success. Agriculturists, as a body, are perhaps the only class of business men in the country who are generally either ignorant or negligent of strict book-keeping; and, without casting any slur upon them, we think common experience will bear us out in this opinion. But why is this? Are our farmers a more ignorant class of men than the generality of those who live in villages and cities? By no means. It requires a far more extended knowledge, a much more accurate judgment, a greater diversity of ability, more constant forethought and attention, to farm even tolerably well, than to be able to buy and sell again a stock of dry goods or groceries. Is the manufacture of grain and meat, a less complicated operation than the making of cotton or woolen cloth? Certainly not; on the contrary, there are many more elements entering into the profit or loss of the former than of the latter.

We entertain the highest estimate of the science and art of agriculture. With the exception of two or three other professions, we believe that farming affords a field for the exercise of mind, of talent, and of information beyond all other occupations of men; to attain *eminence* therein requires an ability, which exercised in any other direction would place the individual high in the estimation of the world. Yet by wrong management, farming may be rendered ruinous; while on the other hand, with knowledge and prudence, we believe it to be, on the average, quite as lucrative a pursuit as any other in the country, if it be not more so. The only reason, then, that we conceive, for the neglect to which we refer, is that farming proves, as a general thing, *too lucrative.*

In this connection it appears worth our while copying the following statement from *Hunt's Merchant's Magazine,* the best authority on such a subject. It appears that throughout the east, and we believe it to be nearly the same in the west—scarcely a merchant goes through life without *failing,* or in other words, losing the labor and capital of years of toil. But it is not so with farmers. Such a thing as a farmer's failing is rarely or never heard of. It is true that a few years ago it had become the custom with some wheat-growers in this State, to use up their crops a year before they reaped it, by going in debt to the store. Of necessity, they were obliged to pay interest, in either a direct shape or in that of higher prices; for the store keeper could not afford to lend his money and run the risk without this interest, and often the farmer was obliged to sell his crop of wheat at the worst time, at the lowest price, to liquidate the debt. A bad season or two came, the debt was not paid,—a mortgage was required. This but added to the previous recklessness; interest accumulated, and the farm was lost. Such a system we hope is disappearing from among us; for both in its moral and mercantile aspect it is ruinous. The farmer and the minister, of all men in the world, have the least moral right to run in debt. But such persons were the exception, *not* the rule. They were not strictly farmers, but speculators on the chances of a good season and a good market. They traded on borrowed capital—borrowed, to often at an usurious interest; they put themselves in the same category with the merchants, and for the same reasons came to a ruinous end. Let us then compare the profession of the agriculturist and the merchants together, and how infinitely, even in a merely pecuniary point of view, is the former superior. He may not, for a time, make as large profits, but his profits are sure. He may not *seem* as rich, but what he has he keeps, and hands down to his children. In the one case there is certainty of prosperity; in the other case, certainty, sooner or la-er, of entire loss.

"It is asserted that but one eminent merchant (and his death is recent and lamented) has ever continued in active business in the city of N. York, to a close of a long life, without undergoing bankruptcy or a suspension of payments in some of the various crises through which the country has necessarily passed. I have no means of determining the truth of this assertion, but it must have some foundation, and I think it would be difficult to add to the number.

"It is also asserted by reliable authority, from records kept during periods of twenty to forty years, that of every hundred persons who commence business in Boston, ninety-five, at least, die poor; that of the same number in New York, not two ultimately acquire wealth, after passing through the intermediate process of bankruptcy, while in Philadelphia, the proportion is still smaller.

"By statistics of bankruptcy, as collected under the uniform bankrupt law of 1841—

The number of applicants for relief under that law were	38,789
The number of creditors returned	1,049,503
The amount of debts stated	$440,934,615
The valuation of property surrendered	43,597,307

"If this valuation were correct, nearly ten cents would have been paid on every dollar due, but what was the fact?

"In the southern district of New York, one cent was paid on an average, for each dollar due; in the northern district, 13⅔ cents, being by far the largest dividend. In Connecticut, the average dividend was somewhat over half a cent on each dollar.

In Mississippi it was	6 cents to	$1,000
In Maine,	½ "	1,000
In Michigan & Iowa,	¼ "	100
In Massachusetts,	4 "	100
In New Jersey,	1 "	100
In Tennessee,	4½ "	100
In Maryland,	1 dol. to	100
In Kentucky,	8 "	1,000
In Illinois,	1 "	1,500

In Pennsylvania, East Virginia, South Alabama, Washington, nothing."

A manufacturer who kept no accounts would soon become a bankrupt. He would cease to be a manufacturer. But so liberal is the earth; so readily does she respond to industry; so patient is she of reckless mismanagement; so many different resources has the farmer, that even the most ignorant among us, if he has but strengh, health, and heart to labor, can make at least a living; whilst most of us can make more than a living. Farmers do not feel *compelled* to keep a set of books. In some manner or other they can get on without them. Perhaps it has not been a part of their education. The ability requires some pains to acquire; and few men will willingly undergo mental labor—the labor of self-instruction—especially in mature age, unless they feel that their occupation actually compels and obliges them. As there seems to be no absolute necessity, the thing is not done.

But is the mere gaining of a living, or a little beyond a living, all that intelligent men should aspire to? Are we to dig and delve, to bear the scorching sun of summer and the bitter frosts of winter, from youth to hoary age, and never be anything but workers? While every improvement in heart and mind raises us in the scale of beings, and every spiritual acquirement is a deposit for eternity, the mere labor of the body, beyond a certain extent, is in itself an evil—a great evil. It was not till Adam was driven from Eden that the sentence was pronounced, "cursed is the ground for thy sake. In the sweat of thy face thou shalt eat bread." Every thing, therefore, which helps to relieve men from bodily toil, and afford them more time for higher occupations, is a blessing; and if we can show, that, by the keeping of strict and systematic accounts, farmers, in general, would be a more prosperous, and therefore a more intellectual class, we shall not have written in vain.

To illustrate how farmers may be working in the dark, and how they may be actually throwing away their time and losing money, without suspecting it, let us take one example. As we cultivate our farms generally in Michigan, it is difficult to ascertain what any particular crop really costs us; and yet unless we know this we are exactly in the position of a merchant selling his goods before he has received his invoice. He *may* make a profit, but he is just as apt to lose. The owner of the farm does most of the work himself; it is done at various times; the expenses run up without being noted; and when the grain is sold, the ready money looks well, and the receiver feels rich. Few, perhaps, consider whether the amount that they have received is more or less than the aggregate of their wages; the interest on their capital; and what they have paid out for transportation, &c. Yet it is very clear, that if a farmer does not receive all these items in the price of his grain, he is a positive loser. He has either worked for little or no wages, or he has realized a much lower rate of interest on his invested capital than he ought to do. If he does not receive *more* than all these items, he has made no profit. At the end of the year, he is just where he was at the beginning of the year; and supposing that the price he has received is the fair average of the market for a series of years, it is quite apparent that it were foolish to continue raising that crop unless he can hereafter do so more cheaply.

In the *Patent Office Report* for the year 1847, the Hon. Jonathan Shearer, of Plymouth, Wayne county, Michigan, estimates the cost of raising wheat as follows:

1st.	Two years interest on land at $10 per acre, 6 per cent,..	$1 20
2d.	The cost of ploughing, sowing and harrowing,...........	4 50
3d.	The cost of seed, 1½ bushel per acre,.....................	1 40
4th	The cost of harvesting, threshing and cleaning,..........	3 00
		$10 10

In the same publication for the year 1848, the same gentleman estimates the cost a little lower, viz: $9 40 per acre; but we are inclined to believe, from our own experience, that the first estimate is nearest the truth; and it will be observed that nothing is allowed for the expense of marketing, which ought to be added as a part of the cost. If the land is worth more, the interest is higher. Now, this year (1851–2) a great deal of wheat is reported to have been sold in Detroit at sixty cents; the highest at sixty-four. In the interior, the price has been, of course, in many places less, in proportion to the distance from market. The average crop of the whole State is said to be fifteen bushels to the acre. Now fifteen bushels at sixty cents is exactly $9 00, showing a *loss* of $1 00 an acre on all raised. Or allowing the crop to be a large one, and calling it twenty bushels per acre, the profit it just $2 00 an acre; that is, it requires fifty acres of excellent wheat to return a *profit* of $100 00 to the farmer. But fifty acres of wheat is quite as much as the most industrious man, under the most favorable circumstances, can put in and attend to; therefore he receives for his labor, risk, interest, and anxiety about one half of the average wages of a hired man, including board, even when his crop does well. Were it not for the many other items on a farm which have perhaps done better, such an one could not long go on; and hence the extreme danger of cultivating one standard crop.

Now when we speak of a farmer keeping a set of books, we do not mean that his boaks are to be the same as those of an ordinary merchant. The two pursuits are different; the ends in view are different; and the books must be different. When, some ten years ago, the writer first commenced farming, he was as *practically* ignorant of the subject as he could possibly be. He had never lived in the country; and all his knowledge was confined to what he had picked up from books and conversation. He knew, however, that without great care, he might easily lose all he was worth; and unable to find any instructions for keeping farm accounts, he was obliged to invent a plan himself. He accordingly procured a blank book of quarto size. In the beginning of it, he drew the best plat of his farm that he was capable of doing; and had the fields surveyed to ascertain their exact area. Each field was then numbered, and in the succeeding pages, a debtor and creditor account was opened with every field. After the days work, in the evening, whatever work had been done to a field was entered on the debtor side, from ploughing to selling; and when the crop was sold or used, the nnmber of bushels and their value was entered on the credit side. Thus, the book exhibited an exact daily *auto-biography* of the farm; the amount and cost of work done on each field, at the current rate of wages; and when the account was closed, by adding interest on the land and implements, and the value of the work of the teams, there was the exact profit or loss, as the case might be. Adding all these totals together, the result of the operations of the whole year was seen at a glance. It is not fair to judge of a field or crop from one year's experience. A series of five or seven years are necessary to arrive at a *positive* certainty; but the writer soon ascertained that some crops paid better than others and some fields were naturally more grateful for the work bestowed upon them. The time occupied in keeping such a book is not five minutes a day; all that is requisite is a habit of punctuality. It is very probable that others make use of the same plan, but not to the writer's knowledge.

But without keeping such a book as this, or one similar to it, we do not know how it is possible to farm with either judgment or security. Unless we know what our expenses are, we have no inducement to try to farm better; and what, in this country, is of most importance, to farm *cheaper*. We hear, on every side, a great deal said of large crops, but rarely anything of the cost of comparatively producing such crops, while it must be apparent on a moment's consideration that a large crop may be raised at a losing price; and that so far as *profit* is concerned, a small crop, worked with economy, may actually pay the best. The chief practical benefit, however, to be derived from this plan, is that the farmer is impelled to study economy by the use of improved implements, and labor saving machines; and he avoids the probably too common mistake of raising a losing crop, year after year, just because his neighbors do so; while he might grow rich by a more judicious system. The same remarks apply to manure; and we feel sure that were such a system prevalent, the dung heaps would rise not only in height, but everywhere, in general estimation. "He who carts dung, carts gold," says the old proverb.*

For ordinary and extraordinary accounts we make use of a modification of the cash book, journal and ledger, so arranged as to require but little time and attention, and yet sufficiently complete for the business of most farms.

Simplicity is what farmers require in such matters; and after trying two or three different arrangements, we have adhered to the fol-

*Since writing the above, we have met with Mr. Linus Cone's (of Troy, Oakland County,) statement of a premium crop, in the Transactions of the Michigan Agricultural Society, 1850, page 436. It cost to raise 12 2.100 acres of wheat, $128.95. Of this, $7.75 is charged for manure; but on the other hand only $7.56 is charged for interest, while such land cannot be worth less than $25 to $30 an acre; and no charge is made for marketing. These additions would make the cost above $12 per acre. Mr. C. is high authority.

lowing as the best we are able to devise. We have a little, leather covered account book, resembling, but somewhat larger than those used as bank books. In this, every payment is entered as made, whatever that payment may be for. At the end of the month, the entries for that month are added up; and then the items are abstracted under the following heads: 1. *Farm wages.* 2. *Family Expenses,* including all payments used for the household in every way. 3. *Permanent Farm Expenses,* including tools, repairs of buildings, &c., in fact, everything appertaining to the farm, which is of a permanent character. 4. *Yearly Farm Expenses,* such as seeds, taxes, &c.—things used at once, or the purchase of which is apt to recur annually. 5. *Investments,* cattle, horses, sheep, &c., purchased, and everything that may be considered as the profit-giving capital of the Farm. The gross amount of each of these items is then entered in the Ledger, under its proper head, for the month only; the whole month making only one entry of each.

This ledger, ruled as such, with index, is in reality a day book, ledger, and memorandum book; and all items not inserted in the above, are at once entered here. We once used a day book, but found it required too much time, while the present plan served every purpose. The *headings* are as follows: 1. *Improvements on Farm,* viz: buildings, fences, painting, clearing land, &c.—everything that adds permanent value to the property. 2. *Farm wages.* 3. *Family Expenses.* 4. *Capital,* value of farm, stock, implements, furniture, cash, food and grains on hand, debts to be received. This is made up the first of January of each year, and gives a full view of your entire property; the articles being estimated, as nearly as possible, at their *then* value. 5. *Implements* purchased during the year. 6. *Live Stock,* of all kinds, enumerated on the debtor side, and all sold or died entered on the credit side. 7. *Grain and Straw and Hay.* The two latter entered at *market* value as put by, and the grains as threshed, Dr., and the actual cost to you, Cr. 8. *Cash Receipts,* entered monthly only. 9. *Sheep.* 1. One year's interest on value of flock at ten per cent. 2. Feeding, say 50 cents per head a year. 3. Washing and tagging. 4. Shearing. 5. Twine, tar and salt. 6. Loss. 7. Marketing wool, Dr. 1. Wool sold. 2. Lambs. 3. Manure. 4. Sheep sold. 5. Sheep eaten in family, Cr. The balance is the clear profit. 10. *Grain in use,* entered as put in granary, at market price, Dr. As sold and whom to, Cr. 11. *Permanent Farm Expenses.* 12. *Investments.* 13. *Yearly Farm Expenses.* And if necessary, 14. *Cord wood,* cost of cutting and hauling, Dr. Receipts for it, Cr. 15. *Bank account,* and 16. *Accounts current, &c.,* with hired men, and other persons with whom you may have dealings

These various entries are in themselves, we believe, sufficient, but can be altered or modified to suit convenience. We, however, go farther, and keep by itself, a list of *every thing* we get off the farm, at its market value. This gives the gross annual products of the farm; and by deducting, at the end of the year, 1. Interest on the value of real estate. 2. Interest on stock. 3. Interest on implements. 4. Taxes. 5. Wages and board of hands. 6. Food of stock, hay, oats, &c. 7. Miscellaneous expenses. By deducting these from the gross receipts, you find how your farm has paid. To be truly profitable, there ought always to be a balance to remunerate you for your own labor and superintendence.

And here a word may properly be said on a point regarding which we have often met with misunderstanding; that is the leaving out of calculations, the produce of the farm which goes to feed the family. We once heard a farmer, who lived very comfortably, with a large family, bitterly complaining that his farm was a loss to him—"he made nothing at all by it." We were surprised to hear this, for his daughters played on the piano; his wife went to church in a silk dress; they entertained a good deal of company, and had, always, the best of cheer for their friends. So we began to make enquiries of our acquaintance, how he managed to sustain his family as he did, for we knew that he had nothing else than his farm, and he was too honest a man and too good a christian to run in debt, where he was unable to pay. But it soon appeared that all he used at home, went, in his eyes, for nothing, and as the balance was comparatively small, he insisted that he lost instead of making money. We believe that this feeling is common, but a moment's consideration will set it right. If your farm has a house upon it, the farm is worth more than if there was no house. You debit the farm with interest on its full value; you should, therefore, credit it with the rent of the house. If you had not that house, you must either build one or hire one. In either case there is a rent, and why should you not consider it in this case? If you did not raise corn, and wheat, and pork, and beef, and chickens, and vegetables, and fruit, you would be obliged to buy them; you would not starve. And you do buy them with your labor, though you get them cheaper than in market. If you had no wood lot of your own, you must purchase cord wood from your neighbor. But you purchased it growing, you charge the farm interest on that purchase money, and of course must allow the value as you consume it. So that in farm accounts, *everything* should be taken into consideration, whether you sell to Mr. A. or to yourself, for your own use. If you get a comfortable living off the farm you do well; most of or all the rest is clear profit.

In making up farm accounts, another very plausible error, apt to deceive beginners, must also be guarded against. This is *crediting* both the stock and wool; and hay and other winter fodder at the same time. All farmers have a certain quantity of live stock; some young and growing animals, some kept for labor, some giving milk, and others producing wool. In a proper account, all these various articles are carried to the credit of the farm. Every farmer, likewise, cuts so much hay, feeds out so much straw and grain. These have a market value and he credits them also. But in so doing he takes credit twice over for the same thing. For what is a calf, or milk, or wool, or the labor of the horse, but that identical hay and oats in another shape? We all know that if we starve our sheep, we get very little wool; if we feed to them throughout winter not only hay, but corn and oats, we have a clip to boast of. But that extra wool is nothing more than the food converted. The hay is the raw material, the sheep the manufacturer, the wool the article ready for market; and as well might the cotton manufacturer charge the price of the cotton twice over as the farmer credit both articles at once. There is, however, some practical difficulty in deciding this matter. Straw has a merely nominal value; and the actual *cost* to the farmer of a ton of hay varies considerably from the market price. Besides, it is difficult, if not impossible to arrive at any certainty regarding the value of the improvement of a young animal; or the work done by a team. Suppose, however, a yoke of oxen eat two tons of hay a winter; and they can be hired for two shillings a day.

While eating this hay they work ninety days: their food is therefore worth $22 50 or $11 25 a ton, without counting the manure; and so on in proportion with other things. In practice, however, the readiest way is to credit both stock and food; or the increased value of the former; and at the end of the year debit the food they have actually consumed, as nearly as can be ascertained. The Germans, who have reduced the science of agriculture to the utmost exactness, not only weigh the food daily, but have rules for proportioning the fodder to the weight of each animal.

The following table from the German work of J. Burger, on "The Economy of Farming," will be interesting, and may prove useful in this respect:

	Living w't of beast.	Quantity consumed in a day.		
	lbs.	Hay. lbs.	Water in winter. lbs.	Water in summer, lbs.
Working Horse, ...	1050	26	35	50
Working Ox, ...	1000	21	55	70
Cow, ...	700	17	42	60
With some salt in the water, ..	....	....	55	....
Sheep, ...	70	1.8	2.5	3–
In a day when salt was given,	....	....	3–	6–
Hogs, ...	140	in hay, val. 4	16	20

The mode of ascertaining whether any profit is made on fattening hogs and cattle in winter is very easy; but were it more strictly attended to, we believe that it would originate so many experiments, and so many curious facts would come to light, that the principles of this business would be entirely changed in the west, and double the profit now made would be realized. Chemists have paid much attention to the *philosophy* of fattening; and the great *principles* of this change in the animal economy are correctly understood. What now remains to be done is, that farmers themselves should try the fattening qualities of the various grains in use; these grains again in various shapes, whole, ground, or cooked, and in various combinations; and learn besides, the circumstances in which an animal prospers the best, whether tied up tight, at large in pen, or roaming about. If we recollect rightly, Mr. Ellsworth has ascertained that one bushel of corn ground fine, and boiled, produces twice as much pork as one bushel fed in the cob. If so, the west annually loses many thousands of dollars; for if one million of bushels of corn are now required to bring our annual hog crop to market, by more careful feeding, one-half million would suffice, and the remainder be nearly clear profit. Taking things as they exist, and as they *might* exist, we believe a far still greater reduction may be made. As the Indian now carelessly dresses his Buffalo robe for clothing, so do we fatten animals in the most careless and extravagant manner; and as the tanner by his art produces out of the same hide the most beautiful and lasting leather, so may we, by the aid of science, produce fat meat at a comparatively small cost. To accomplish this, however, the first step is for every farmer to keep an accurate account of the cost of making pork or beef; to learn that his *profit* is often very trifling, and then by improved means to begin to reduce the cost. The account, of course, stands thus: market value of hog when put up to fatten; grain, etc., consumed by it; and labor of attending, feeding, killing and marketing, Dr. Value of pork sold, or used; gut-lard, refuse and manure, Cr. Balance, the profit. In England, many farmers are now content if the fat animal merely pays its expenses; the manure made being considered equivalent to a full profit.

We have thus thrown out a few hints on the subject of farm accounts; and we trust that sufficient has been said to prove to our readers that he who carefully keeps such accounts, is more likely to farm profitably than those who neglect accounts. In the latter case, one crop may swallow up all the profit made upon another; the loss in fattening cattle and hogs in winter neutralize all the gain of summer. But the intelligent man who has learnt carefully what figures can teach, rarely runs such a chance. He knows with great accuracy what every thing costs him. If he finds one crop steadily less lucrative than another, he ceases to raise it. If he finds certain expenses stand in the way of profit, he sets mind to work to invent some cheaper substitute for his present management. If he finds the employment of a hired threshing machine cost less than the interest upon one of his own, he continues to hire it; if more, he buys one for himself; and he soon learns that invaluable lesson—a lesson, indeed, necessary for success in all the affairs in life—that if he would progress he must "take care of the little things, and the large ones will take care of themselves." More *profit* is made on a farm from trifles than from the large crops; but even that the latter may be profitable, intelligent care must be bestowed upon all, even the minutest elements that enter into their composition. Farming then will cease to be a mere art, and that a rough one. It becomes a science; including and involving and rising superior to, nearly all the physical sciences in existence; and if properly learnt, including geology, chemistry, astronomy, medicine, botany, &c., the earth the raw material, but the comfort, happiness and blessing of mankind the product.

Grosse Isle, January, 1852.

CATTLE BREEDING.

BY JOHN STARKWEATHER.

Cattle raising in Michigan is beginning to receive that attention it properly deserves, and from present appearance, must, at no distant time become one of the most important, as well as the most profitable branches of industry, to which the farmer can devote his labors' Most of the lands in our State, especially the best of timbered land, adjoining the lakes and rivers of our Peninsula, are well adapted for grazing purposes, affording abundance of rich pasturage, at least seven months of the year. These lands are not in condition suited to wheat culture, nor can they be unless a thorough and systematic course of draining be resorted to, which at the present low price of land, would hardly afford remuneration for the outlay. There is every prospect of ample facilities soon being furnished us for quick and cheap transportation to the markets on the sea board for all the cattle we can raise; and what is of more importance, there is an increasing demand for them at home and abroad. With such inducements before us, we should awake at once, and earnestly enquire after the best method to be pursued for the purpose of improving the quality and increasing the quantity of our neat stock, to supply the home demand, and to compete with eastern graziers in their own markets for a share of the profits; this is no easy task, nor can we expect to succeed until public sentiment accords a standard of merit worthy to be pursued in breeding and rearing our stock, which shall attract the attention of connoisseur and drovers from abroad, who will doubtless become purchasers when the excellence of our animals becomes established, and plain to be seen.

The few can accomplish but little in promoting the work; it ought to become general in practice, and every farmer should make the effort to produce a better grade of cattle—excel, if possible, be his motto. Those in Michigan who have devoted their attention to this branch of improvement in agricultural pursuits, need encouragement; they cannot be expected to succeed in breeding and keeping thorough bred animals unless they are patronized; thus far they have been but poorly rewarded, having had to contend against the jealousies and prejudices of men, as well as in overcoming the defects of animals. It is manifest, however, that such a state of things are fast giving way to progressive principles in breeding. Neither prejudice nor jealousy can stay its onward coarse.

Our cattle are mostly natives; they are active and hardy, possessing the advantage of thorough acclimation, which is about the extent of their good qualities, and may be considered highly important in crossing them with stock recently imported from a climate differing widely from ours, and should form a sufficient guarantee for our success in trying to remedy some of the defects which meet the eye at almost every point.

So far as my experience and observation extends, I am decidedly in favor of the short horn Durham breed of cattle, for the purpose of crossing with our natives; their advantages are early maturity, large size, good milkers, quiet disposition and an aptitude to receive and obey instructions. If properly reared, the heifers are in condition to have a calf at two years of age. Thus early maturity is obtained, which adds much to their value.

Extra care and feed is essential; it is also highly important that heifers early put to breeding should be milked regular, and persevere in milking after the first calf until within six months of the second period of calving, providing she is intended for milk. Habit is thus established, and is easily followed; it forms an excellent trait for a good village cow. Steers of this breed also mature young; they are fit for the shambles or the yoke at two and a half years of age, and weigh alive from eleven to thirteen hundred pounds; their flesh is juicy and tender, with the fat well mingled with the lean. Good size should be regarded highly in neat stock; the butchers stall is their final market, and the drift to that point is no more on a large than on a small bullock, neither is the expense of butchering and dressing any more in the former than in the latter case.

For working cattle size is required, and it is idle to think that light oxen can perform the labor of heavy oxen, especially if draught be the object; and to settle the point of their relative value we will have reference to lumbermen's opinion. Messrs. Smith & Dwight, who are engaged in the lumber trade in St. Clair county, bought fifteen or twenty yoke of working oxen last fall, the prices paid ranged from $85 to $125; and in no case could they be induced to purchase a yoke of cattle whose girth would not exceed seven feet, with a proportionate build for size in all other respects; indeed, they willingly paid prices in ratio to the excess of measurement over seven feet. Thus it will be seen that large size is considered of the first importance with them, and so it may be considered with farmers who believe deep plowing essential to secure good crops. Light oxen have no business with deep plowing.

With regard to the milking qualities of the short horn Durham, much has been said and written on the subject, and where a difference of opinion existed it was doubtless in consequence of the variation of this desirable trait in the same family, which is the case with all breeds of cattle. They, however, most invariably come to the conclusion that a cross with our native stock improves the latter for dairy purposes; and some are inclined to think them better as they approach nearer to full blood. Instances occurring within scope of my practice and observation, lead me to pronounce them excellent for the dairy. The cow of Mr. Fuller, for which he received the first premium for milk at the society's first annual fair, is half short horn and half native. Mr. Bissel's cow, for which he received the second premium in the same class, is entered as full blood. At the society's second annual fair, Mr. De Garmo was awarded the first premium on his cow for milk and butter; she is five-eights short horned Durham, the balance native. She is referred to in another part of this communication.

The cow of Mr. Tibbets, milked in presence of the committee, for which he was awarded a special premium, is half short horned and half native. I own a full blood short horned Durham heifer, three years of age, from the herd of J. W. Van Cleve. She is not easily surpassed by one of her age for milk and butter. I also have one the same age, of my own raising, from the same herd, three-quarters

Durham and a quarter native. She is good for milk and butter. Much more might be added in support of their milking qualities, but it is thought unnecessary.

For working oxen, half Durham and half native have proven to possess the qualities essential for the purpose; they command the highest price in market; they are quiet out of the yoke, and kind and tractable while in it. In fact they cannot be surpassed, all things considered. I esteem oxen equal to horses for farm work, and more economical.

I know by experience that good calves can be raised by taking them from the cows as soon as the milk becomes suitable for use, and learned to drink new milk at first—a day or two after, the milk drawn from the cow should be allowed to stand from morning until night, or from night until morning; the cream that has risen taken off, the milk warmed and fed to the calf; it should in no instance be allowed to sour before feeding, as it is from this cause that so many poor calves are found, having been scoured until they are scarcely able to stand. An egg or two broken into the milk adds materially to the growth and sleek appearance of the animal; they soon learn to seek for the egg first; this shows a fondness for them. When eggs are worth but six cents per dozen, they are cheap food for man or beast. If the breeder thinks it important to force the growth of his calves, the addition of Indian meal finely ground, and given in small quantities at first, with the mess, is good; if the meal is scalded, it is better, and less liable to scour, which is the only objection to feeding it; though in skillful hands there is no danger to be apprehended. The calves should be allowed to run in good pasture, and have access to salt and pure water. If thus reared, they have the advantage in many respects, over those fed wholly by sucking the cow. The latter on being weaned, receive a check, and fall off to a degree from which it is difficult to recover them. This generally occurs at a period when vegetation has lost its tender, and some of its nutritious substance, so at the years end the former have the decided advantage, and less expense and trouble in rearing; besides, cows become much injured by the calves being allowed to suck them; in fact, they are comparatively worthless, except for breeders.

The prize steer exhibited by H. E. De Garmo at the society's second annual fair, and which he sold for one hundred dollars in August after, three years of age, was reared on skimed milk; and so are all the cattle of his raising; and certainly, if we judge by their looks, none of them show a want of proper care and feed while young; indeed it is a rare thing when we find better.

The excellence of his animals is doubtless owing mostly to the stock from which they originated. The dam, his prize cow for milk and butter, was sired by Splendor. Her descendents are from the herd of J. W. Van Cleve; thus are they at least three-fourth blood short horned Durham. Her reputation as a breeder is universal in this place and vicinity. Calves should be brought into winter quarters early and receive the best of care and feed. They do well to run with sheep and consume much of the coarser quality of fodder rejected by them, and are always free from lice. After one year of age, they will winter on cornstalks, straw and chaff. It is economy to provide comfortable shelter against the pelting storms, for all neat stock, and no farmer can furnish a reasonable excuse for neglecting to do so when straw and a few poles are at hand.

Raising neat stock involves but little expense; and from the droping of the calf to the period when it is worth fifteen to twenty dollars, the cost is scarcely perceptible; a little humane attention to its welfare and it instinctively repays the debt. Besides, it is one of those branches of industry in which the labors of the husbandman are divided, and frequently the toil is made more easy by the change. Cultivating the cerial grains affords employment in spring, summer and autumn; raising stock that of winter and fore part of spring, which fills up the period.

Indeed we cannot rely with certainty on the successful culture of wheat for any great length of time, unless we practice the rotation system. Thus will beef, butter and cheese, as well as wheat, wool and mutton compose a portion of the products in perfecting the rotation. They are now telling on the right side for the producer, with a prospect of increasing profits. It may be safely asserted that raising stock aids and extends fertility of soil to any desirable period, which is of the first importance; in fact it may be termed the philosopher's stone. Fertility is the grand secret of the farmer's success, and those who neglect it must expect a wide difference in circumstances between them and those who observe and practice it. There is no getting around this, and there are no short roads to avoid it.

In comparing the relative value of the different breeds of cattle, I am inclined to give those the preference which weigh the most, all other prominent points of excellence being equal; and the short horned Durham variety of cattle appear at present to possess the most good qualities to recommend them to the Michigan farmer, allowing the foregoing to be a correct standard in judging.

The Devon are a beautiful race of cattle; they excel in color and are doubtless destined to aid in improving our native stock. The friends of this variety are evidently permitting color to usurp the more valuable trait, large size. Still they appear determined to press forward in the enterprise with becoming zeal.

THE SHEPHERD'S DOG.

BY D. D. GILLET.

J. C. Holmes, Esq., *Sec'y Mich. State Ag. Society:*

Dear Sir—I improve the earliest possible moment, to gratify your renewed request, for an article relative to "Shepherd's dogs, and the method of training them." I am the more sorry that this subject had not been committed to more experienced hands, from the fact that upon the last and most important branch of it, little has been written—and also, the great want of information, in order to induce any considerable number of Michigan flock-masters to undertake what they have hitherto considered the hopeless task of rendering a dog a valuable assistant in the care of their flocks—a task as largely over estimated, as the sagacity and capability of the genuine Shepherd is under estimated by those to whom his usefulness, when broken, or the ease with which he is broken, is alike unknown.

No animal possesses, in its capability of developement or in its susceptibility, so diversified characteristics as the dog. He may be made the ornament of the plantation, or the nuisance of the neighborhood—affords, with proper tuition and care, protection to the domicil and its inmates, to the farm-yard and counting-house, "when all around is still," always rendering the strictest allegiance to his master, and obedient to his every wish. When neglected and indifferently disciplined, he skulks around his neighbors out-houses, peering wistfully into half closed doors, windows, and unsecured repositories of refuse kitchen articles, and after graduating in this class of lessons, pays "Tray" a visit, secures his company in his noctural rounds, and commences (and prosecutes too) an indiscriminate slaughter of the flocks of the neighborhood. Baying at travelers by day, and prowling about out-houses, poultry yards, and sheep folds by night, if not the sum of the duties of the great majority of dogs of the country, are pastimes in which they are frequently engaged; and the man who shall restrain so as to prevent these vices, either by positive enactment, or other effectual means, will deserve as well of his fellows, and be doomed to as enviable a notoriety, as him who of old purged the Augean stables of their defilements.

These intolerable annoyances, and a thousand other nameless ones, may be entirely prevented by proper discipline and care—such care as provides properly for his comfort—such diecipline as cultivates every susceptibility, subjects every faculty, lops off his vicious habits as they are developed, and in short, fits him for use. Indulge an example of a dog with which I am well acquainted, bounding away to the distant pasture with the speed of the hound, and bringing thence the flock as carefully as if he knew all their weaknesses, holding them in the near by corner for the purposes of selection, as for slaughter, show or sale—keeping flocks in adjoining yards from passing through the open gate—yarding them for various purposes, as tagging, washing, shearing, marking—tracking the breachy pig to his distant retreat in the adjoining wheat or corn field, and bringing him out with a snap to him—fetching cows or oxen more quickly, and with as much care as the cow-herd; doing all so willingly as seeming to ask the privilege of the next performance, with such a knowing manner as often to extort from intelligent but indifferent beholders, the exclamation—"Can such performances be thr result of instinct."

It is, perhaps, needless, to say that the animal referred to is a full blood Scotch "Colley," or English "Curr," or Shepherd's dog, (by each of which synonyms he is known,) whose pedigree is well known and easily traced to recent importations. He has been owned by me since June, 1848, and was, when I obtained him, six weeks old. His training commenced with my ownership, has been continued to this time, and is still far from complete—but during it, on more than one occasion, has the falsity of an old apothegm been demonstrated, to wit: "you can't learn an old dog new tricks."

To train a dog well, he must be handled right at the commencement, and must be commenced with at the right time. He should be broken of his "puppy tricks," such as chasing chickens, pigs, and (if near the sheep walk,) lambs, lugging about small articles of clothing, barking at travelers, entering the highway to play or quarrel with other dogs or animals—as he grows up and such inclinations develop themselves, so broken and so effectually, as by no means to allow them to become "*habits.*" Before he is one year old he should be taught implicitly to obey his masters commands, either of voice or motion—of voice, as "come here," or "away," or "over," or "steady," or "down"—of motion, as "right," or "left," or "down," as the case may be, or circumstances require. Especially should obedience to this last command be insisted upon, as more at times depends upon its instant performance, than upon any other of his duties. He may be learned to execute mechanically many, possibly all of these commands, before he is taken to the sheep walk, but can never be learned to manage a flock of sheep, in a proper manner, without many careful lessons in company with his future charge. Care should therefore be taken by his master to have his dog with him, upon all occasions of visiting the flock, so as to make him and the flock familiar.

So soon as you know you can make him "down" at the word, or by the proper motion, get your flock (a small one) between yourself and dog in the highway, or what is abundantly better, a narrow lane, oblige him to keep at some distance behind the flock, while you leisurely advance and keep it behind you. If the flock are as gentle as sheep always should be, (I may add when properly treated, always are,) an effort or two will have proved successful, your dog will have learned that he can drive sheep, and your end is gained.

Suppose the same gentle flock, one that has never been frightened by dogs assaulting them, is in a small field, and you want your dog to fetch them up. Give him his direction by a motion of the hand and the word "away," at the same time calling your sheep in your most familiar manner. Off he starts at full speed, and ten to one, if the flock is in sight and he is left to take his own course, makes for its centre, and scatters it in every direction but the right one. "Down" him at once, and oblige him to take a course to the right or left of the flock, and some distance from it. Having gained its rear, keep him in such a place, either by motion or word, as will give the flock the right direction.

You now wish to hold them in a corner, or to yard them at a small door—he soon learns your wishes, from your words or motion, or both, and holds them in the field's corner, or forces them through the open door as desired.

Should one or more sheep break from the flock, as they will be very likely to do, "by" is the word, which means in the parlance of the sheep-walk, go around, and never after. If the stragglers have gained the road, or a lane, "over" "by" sends him over the nearest fence, and along it, until they are headed and brought back. Continue to practice him at all convenient times, when busied with your flock, for you may feel assured that the more you exercise him the more speedily and perfectly will he be trained.

His disposition must be studied at all times with assiduity and care, and although mild and gentle treatment (if tempered with stability) is generally sufficient to keep him under, yet when wilful transgressions of your requirements are persisted in, discipline him to his duty. He will *acknowledge* but one master, and none but his master should ever allow the young dog to follow him to the sheep-walk, or attempt to train him. After he is broken he may assist any person, whether stranger or domestic, in managing the flock, provided always that his master is not present, and the person in charge conveys his commands in intelligible language.

The question is often asked, "can an inexperienced person thoroughly train a shepherd's dog?" I answer most unhesitatingly, yes. When I purchased mine I had never seen a dog manage a flock of sheep more than ten minutes in my life; and I have received no instruction either oral or written, since, in the matter of his exercise. Notwithstanding such decided disadvantages, his entire training has not cost me a tithe of the time required to prepare this article, having worked him at first, at my own convenience, as I would a wlid steer or colt, simply to discipline him; he soon, however, made himself a useful, an indispensable companion upon all occasions of visiting the sheep-walk.

"Can he be broken without a well-trained dog for a companion?" Mine has; and if I were to break another, my present experience would dictate that he be trained wholly by himself. Stephens (Book of the Farm, vol. 2 page 174) gives contrary advice; but his opinion does not accord with that of any practical man, whom it has been my privilege personally to consult.

Allow me before closing this article to caution those who have Shepherd's dogs in charge, against crossing them with the common dogs of the country. Such crosses almost invariably produce instead of half blood Shepherds, full blood sheep killers. An Englishman informed me at the Ann Arbor State Fair, that he brought from the old country, a full blood Colley bitch, and having settled where he could not not procure a full blood dog, although warned of the consequence, he tried a cross, which resulted as above. Determined to persevere, as John Bull generally is, he tried again and again, until over forty whelps had been reared, all of which, before they were two years old, were caught in the act, and killed for killing sheep.

Sharon, Washtenaw Co., March 24, 1852.

INFLUENCE OF THE WEATHER ON CROPS.

BY LINUS CONE.

By reading the fifth annual report of the Ohio Board of Agriculture, I have been most sensibly reminded of the strong hold of what I believe to be a fundamental error, has upon the minds of a large majority of farmers, western farmers in particular. We allude to charging the failure of crops to the fickleness of the seasons—to the weather. One writer in the work above alluded to, W. D. Emerson, speaking of the heavy taxes that are paid by the laboring man, says: "We might also add the immense losses occasioned by the fickleness of the seasons." He then adds: "This is a cause, to a great degree, beyond human control." Again, in the report on crops from Sandusky county, by Ebenezer Wilson, we find the following:

"OATS.—Usual average thirty-five bushels per acre; less than usual this season. Extreme wet or dry weather immediately after sowing were the causes of injury. Remedy—regulate the weather."

It is presumed that no one who is at all familiar with the reports of the different boards of agriculture, the agricultural journals of the day, or the views entertained by agriculturists generally, that will deny but that these writers have correctly expressed the views of nearly all the cultivators of the soil. Yet there are a few who have so far departed from the tradition of the fathers that they do not hold to these opinions, but on the contrary, believe the seasons are always right, and if the necessary elements of fertility for the growth and perfection of the plants we cultivate are contained in the soil, that we can, by becoming acqainted with the wants and habits of those plants, supply all their wants, or to that extent that the product shall be measurably sure and abundant. We believe that it would be much less difficult to adopt a method of cultivation that would insure a steady, vigorous growth to the plants, than it would be to regulate the weather, even if we had the power, so as to have it conform to the common slovenly, shallow, and exhausting system of cultivation. Hence we believe that the evil effects that too wet or too dry weather has upon our growing crops, may be avoided by adopting a methed of cultivation that will place the elements of fertility, that are necessary to bring them to full maturity, within their reach.

We are aware that the above views will be regarded by most of our agriculturists as visionary; but after many years of experience, wherein this subject has received particular attention, and has been duly considered, and facts as they have occurred been carefully noted, we have become convinced that most of the evils that annually affect our crops, which are generally charged to the fickleness of the seasons, should be charged to our own ignorance or neglect.

Our ignorance of the wants and habits of the plants we cultivate, and a rigid adherence to long established but unsound theories in regard thereto, are the principal reasons why a system of cultivation is persisted in which renders our crops liable to be injuriously affected by every change of weather; also, to the attacks of their enemies and diseases. That this point may be better understood let us examine the common theory of the habits of the wheat plant, which is, that the roots of that plant do not penetrate the soil to any considerable depth, or at least not until in the advanced stages of its growth, and hence it is liable to be injuriously affected by too much rain or too much sunshine, by the snow of winter and the frosts of winter and spring. It is true its roots will not penetrate a soil that is destitute of the elements of fertility, nor a compact, unbroken sub-soil, or a soil that is saturated with water, but remove these obstructions to their downward progress, and they will penetrate the soil in search of food, to almost an unlimited extent.

Thus, deep and thorough tillage is necessary, in order that the plants should continue in a healthy, vigorous state, during all stages of their growth in order to enable them to withstand the enemies and diseases to which they are liable. This will be made apparent to any one that will examine the subject closely. Whenever a spot in our fields of grain has manifested an unthrifty appearance, and the cause was not plainly visible without charging it to the weather, has been sought for by an examination of the roots of the plants. This examination has in most cases revealed to us the cause of unthriftiness and clearly shown it to be occasioned by the circumscribed limits in which the roots have been compelled to travel in search of food. It has been ascertained that when there is no barrier to their progress, they will, in less than three months from the time of sowing, occupy the whole of the soil to the depth of one foot and seven inches, and probably they would have penetrated deeper than that, had they not come in contact with an impenetrable subsoil.

Many other erroneous opinions are entertained in regard to the habits and wants of, not only the wheat plant, but the other plants we cultivate, which have a tendency to mislead the cultivator of the soil, and to prevent the adoption of a method of cultivation that would protect them in a great measure from the evils resulting from the changeableness of the seasons, and from the attacks of their enemies.

A few examples from the many that have been noted during many years of experience, will now be presented to show that the failure of crops, which was supposed to be caused by the fickleness of the seasons, was caused by some important omission on the part of the cultivator.

In a season when it was said that the corn crop was greatly injured by the cool dry weather, at and after the time of planting, as much of the seed did not come up, and what did vegetate was so backward that it was greatly injured by the frost in the fall. A field was planted that season by two men; one trod upon the hills after planting, the other did not. That trod upon came up well, grew vigorously, ripened early, and produced a good crop. That not trod upon was several weeks in coming up, and did not come well; it was backward during the season, and was injured by the frosts in the fall, producing less than half a crop. Again, in a season when the crops of spring grain were extremely light throughout this part of the State, caused, it was said, by the extreme drouth. A field deeply tilled, and in excellent order, was sown to oats and barley, but not rolled after sowing; the crop was not more than one half the usual average, but where a stone boat loaded with stone had been drawn across the field, there the crop was good.

Thus it was clearly ascertained that both failures were caused by not pressing the soil after putting in the seed, which every farmer knows to be necessary, but which nearly all neglect to do, and not to the dry weather.

Several years ago, when the wheat crop was nearly cut off by wet weather, Hessian fly, winter and rust, two adjoining fields were put into wheat. The soil, situation and seed were alike and sown at the same time. One was plowed deep, made mellow with the harrow and cultivator, and well surface drained; the other plowed shallow, made mellow by cross plowing, and drained in the common way by plowing a few furrows here and there, but not clearing out sufficiently to let off the water. The wheat on the shallow tilled field was so injured by wet weather, Hessian fly, winter and rust, that the product was less than five bushels per acre, whilst that deeply tilled suffered from neither of the above causes and produced thirty-eight bushels per acre. Again, in a season when much wheat was injured by rust, it was ascertained that draining prevented rust, for it was found that when surface drains were not over three rods apart the wheat was good, but where they were farther apart a strip in the centre between the drains lodged early, rusted and shrunk, producing not more than half as much as the other parts of the field.

Here are facts showing conclusively that the main causes of the failure of the wheat crop are chargeable to ourselves and not to the weather.

Of the long list to select from to establish the position we have assumed, one more only will be taken.

The last was one of the wettest seasons ever known since the first settlement of this part of the State, and the crops, especially those of spring grain, were extremely light on all clayey soils where the common method of cultivation prevailed. Two fields in this vicinity, containing twelve acres each, which were nearly alike in every respect, both containing considerable land that is said to be "dry enough in a dry season, but rather too wet in a wet one." One was underdrained in part and the rest surface drained, after putting in the crop, by plowing and cleaning out the furrows. The other field was not drained at all. The drained field produced eight hundred bushels of ears of corn from five acres, three hundred and ninety-two bushels of oats from five acres, and eighty-nine bushels of barley from two acres. The undrained field was well tilled and planted to corn and received first rate care during the season; yet the man that tilled the land and harvested the corn estimated the loss, which he attributed to the water, at two hundred and forty or fifty bushels of corn. Here was a loss in a single season of more than enough to have permanently drained the wet places and thus directed the surplus water so that it would have promoted the growth of the crop instead of destroying it.

This is an important subject; one that demands the careful attention of every practical agriculturist—one that opens a wide field for study and investigation—one which we believe when fairly and fully investigated will be the means of changing the views of the great majority of the cultivators of the soil, and lead to a better and more thorough system of agriculture; a system, the very foundation of which will be to prepare the soil for a crop in such a manner that it will retain sufficient moisture in a dry season, and so dispose of the surplus water in a wet one, that instead of its being injurious it will be highly beneficial to the crop.

Troy, Oakland Co., Feb., 1852.

FISH FOUND IN THE RIVERS AND LAKES IN AND AROUND THE STATE OF MICHIGAN.

BY GEORGE CLARK.

HON. J. C. HOLMES, *Sec'y of the Mich. State Ag. Society.*

DEAR SIR—In compliance with your request, I herewith submit a brief description of the several varieties of fish that are found in the Rivers and Lakes in about the State of Michigan. I give the weight, length and probable age to which they live.

I have enumerated twenty-seven varieties that have scales, and four that have no scales, but a tough, smooth, slimy skin.

I will commence with a description of the smallest, and continue in rotation according to their weight, length and age, making thirty-one varieties in all.

CHUB.

These little fish are sometimes used as pan-fish, but generally for bait to catch larger fish. They are found in creeks and along the shores of rivers and lakes. They spawn in the spring, and usually in creeks. Average weight, quarter of a pound; large weight, one lb.; average length, three inches; age, twelve years.

SHINER.

These harmless little fish are used as bait, and sometimes as pan-fish; they are of but little value for food. They are found in the rivers and lakes in large schools, and many other kinds of fish feed upon them. In fact they seem to have been created to serve as food for other fish. They are often caught in seines with other fish, and generally thrown back into the water. Average weight, half a pound; large weight, one and a half pounds; length, six inches; age, twelve years.

SUNFISH.

These pretty little fish are found in all the creeks, rivers and lakes; they are the best of pan-fish, but unfortunately, not very numerous. They spawn in the spring. Average weight, ½ ℔.; large weight, 1½ ℔s; common length, 6 inches; age, 12 years.

SPOTTED OR BROOK TROUT.

These beautiful fish are a favorite with all who know them; they surpass all other fresh water fish as pan-fish. They are found in all the clear creeks, rivers and lakes from Mackinaw to the head of Lake

Superior. It is fine sport to catch them with a hook. They spawn in the fall in vary rapid water. They are not very numerous, but are sometimes caught at the Sault with seines and salted. Average weight, ½ lb; large, 2 lbs; average length, 6 inches; age, 25 years.

ROCK BASS.

These are very fine pan-fish, found in all the creeks, rivers and lakes, mostly in pairs, but are not very numerous. They spawn in the spring. Average weight, ¾ lb; large, 1½ lbs; common length, 6 inches; age, 15 years.

YELLOW PERCH.

These are the best summer pan-fish that are caught in the creeks, rivers and lakes. They are usually found in pairs, but not very numerous. Average weight, ¾ lb; large, 1½ lbs; common length, 6 inches; age, 15 years.

ROACH.

These are very good fish and are found mostly in pairs in most of the rivers and lakes; they spawn in the spring. They are not very plenty. Average weight, 1 lb; large weight, 2 lbs; common length, 7 inches; age, 15 years.

SMALL HERRING.

Are harmless little fish; almost all larger fish prey upon them. They are found in the straits and large lakes, and are taken in considerable quantities for packing; they rank fourth rate in value for that purpose. They spawn in the fall. Average weight, 1¼ lbs; large weight, 2 lbs; common length, 7 inches; age, 20 years.

MINNOWS.

These fish bear some resemblance to both the herring and white-fish, and appear to be a cross of those varieties. They are good fish to salt, but are not very plenty; they are taken in some of the rivers and lakes. They spawn in the fall. Average weight, 1 lb; large weight, 2 lbs; common length, 10 inches; age, 25 years.

WHITE BASS.

These may be called a bunch of fins, scales and bones; it is almost impossible to take them up without wounding the hands with some of their prickly fins; they spawn in the spring and are found in large numbers in most of the rivers and lakes. They are considered about seventh in value for salting. Average weight, 1½ lbs; large, 2 lbs; common length, 9 inches; age, 30 years.

CARP OR RED SUCKER.

These fish are found in large quantities in Lakes Huron, Michigan and Superior; also in many of the creeks and rivers. They run up the creeks and rivers in such large numbers in the spring, to spawn, that where the water is shoal they can be caught with but little trouble. They are very good pan-fish, and rank sixth in value for packing. Average weight 1½ lbs, large weight 2 lbs. Common length, 11 inches. Age, 30 years.

SUCKERS.

Are found in considerable numbers in the creeks and rivers where they run up in the spring to spawn; they are sometimes used as pan-fish, but are not of much value; average weight, 1½ lbs, large weight 2 lbs; common length, 11 inches; age, 30 years.

STONE CARRIER.

These are knowing little fish, and are generally found in pairs. They swim near the bottom, and spawn in the spring, where there is gravel and small stone. They make a pile of their spawn, then throw up the gravel and stone about the eggs to protect them until they hatch; one of them stations itself above and near the little pile to protect it, and to take charge of the young when hatched. They are of but little value, and not very numerous. Weight 1½ lbs; large weight, 2 lbs; common length, 9 inches; age, 30 years.

BLACK BASS,

Are well known and are very good fish; they are found in all the rivers and lakes, spawn in the spring and are mostly found in pairs, but are not very numerous; average weight, 1¾ lbs; large weight, 7 lbs; common length, 10 inches; age 50 years.

LARGE HERRING,

Are very good fish, found only in the straits and large lakes; they spawn in the fall, but few are caught; average weight, 1¾ lbs; large weight, 4 lbs; common length, 10 inches; large, 20 inches; age 50 years.

DOG FISH,

Are found in almost every place where there is water enough for them to swim. They spawn in the spring. They are ugly, disgusting looking fish, having a mouth that opens something like a dog's; they are not used for food, and are of no value; average weight, 2 lbs; large, 4 lbs; common length, 13 inches; large, 20 inches; age, 100 years.

SHEEP HEAD.

These are found in large numbers in the rivers and lakes; they spawn in the spring; they are a lazy, ugly looking fish, and are of no value; average weight, 2 lbs; large, 10 lbs; common length, 14 inches; large, 24 inches; age, 50 years.

PICKEREL.

These are ranked among the best fish found in the lakes; they are the fresh water cod fish, are very good, fresh or salted and dried, and rank second in value for packing. They generally run up the rivers and lakes in the spring to spawn, where they are caught in considerable numbers. Average weight, 2 lbs; large, 20 lbs; common length, 15 inches; large, 30 inches; age, 200 years.

GAR, OR BILL FISH.

These are very singular looking fish; the bill, or mouth, is about one-fifth of its whole length; the body is about half an inch in diameter and almost round. They spawn in the spring, are not very numerous, and of no value for food. Average weight, 2 lbs; large, 10 lbs; common length, 20 inches; large, 36 inches; age, 200 years.

WHITE FISH.

The white fish are more numerous and of more value than all the other fish that are caught in these waters. They are found in the straits and all the large lakes. They spawn in the fall in the straits, and on reefs and shoals about the lakes; they are caught in seines, gill nets, and with spears; rarely with hooks. The Detroit River white fish come up from Lake Erie; they are more juicy and better flavored than those caught in the upper lakes, probably from the fact that they feed on more delicate food, and the water in Lake Erie being softer and warmer than the water in the upper lakes; but the largest fish are found in Lake Superior.

The white fish are the favorite fish of the Indians. They will give, many times, the weight in trout or any other fish, in exchange for them. A person can live longer on white fish alone, than any other kind of fish found in the fresh waters. They are the bread or staff of life of the lakes.

They are very beautiful fish, when they are pulled from the water and thrown upon the shore—as they flounder and twist in the sun, they show all the colors of the rainbow; they may be handled without fear, for they have no fins, thorns or gills, that can wound the hand. Their average weight is 2¾ lbs; large, 18 lbs; common length, 15 inches; large, 24 inches; age, 50 years.

MULLET.

These are not considered as first rate, but are sometimes used as pan-fish, and are packed to a limited extent; they spawn in the spring and are found in considerable numbers in the rivers and lakes. Average weight, 3 lbs; large, 8 lbs; common length, 15 inches; large size, 22 inches; age, 50 years.

BRAM OR RED HORSE.

These are good fish but not very numerous, they are found in most of the rivers and lakes. They spawn in the spring. Average weight, 3 lbs; large, 6 lbs; common length, 14 inches; large, 20 inches; age, 50 years.

EELPOUT.

These are not used for food, and are of no value. They spawn in the spring and are not very numerous. Average weight, 3 lbs; large, 5 lbs; common length, 15 inches; large, 24 inches; age, 15 years.

SISKOWIT.

These are mostly found in Lake Superior, and are thought by some to be the best fish in the lake. They are of the trout family, and are probably the fatest fish known; they are usually taken in gill-nets. They spawn in the fall, and rank as fifth rate for packing; they are also of some value for their oil. Common weight, 4 lbs; large, 8 lbs; common length, 16 inches; large, 24 inches; age, 150 years.

PIKE.

These are exceedingly voracious fish, and appear to have no choice in their food, but will devour anything they can find, and will swallow a fish half as large as themselves. Three pike have been caught at a time, upon one hook, the first small, the next larger, and the third still larger. They spawn in the spring, and are generally found in pairs. They are good in the spring, while the water is cold—they are not very numerous. Common weight, 4 lbs; large, 60 lbs; common length, 24 inches; large, 60 inches; age, 200 years.

LAKE OR MACKINAW TROUT.

These trout are as voracious as pike; three are sometimes caught on one hook; they will eat any fish they can master; they have sometimes been caught with five or six whitefish in them; they are mostly caught in Lake Huron with gill nets and hooks. Saginaw Bay appears to be a favorite place of resort for them. Some winters large quantities of them are caught in the Bay, through the ice, with a decoy fish and spear. They are the most numerous of any of the large fish in the lakes. They spawn in the fall about the bays and lakes. They rank third in value for packing. Average weight, 5 lbs; large, 75 lbs; common length, 24 inches; large, 60 inches; age, 200 years.

MUSKELONGE.

These are the king-fish of the lakes; they are the quickest, smartest and strongest fish in the lakes except the sturgeon. If you strike a large one with a spear, you will soon find out his strength; and if you or the spear, or both, do not go overboard you may think yourself lucky. They pass through the water with great speed. Frequently, when they find themselves enclosed in a net, they will make a break in the water near the shore with their tail in order to get good distance and headway that they may break through the seine, and they are almost certain to go through if they have distance enough and are under full way; if they do not succeed the first time they will continue to try. They are mostly found in pairs in and about Lake Erie. They are the best of the large fish for food. Common weight, 8 lbs; large, 75 lbs; common length, 40 inches; large, 72 inches; age, 200 years.

Fish that have no scales, but a smooth, slimy, tough skin:

LAMPREY.

These fish are unlike any other fish in the lakes—they possess the blood-suching qualities of the leech. They attach themselves to the large fish, generally the sturgeon—sometimes six or eight to one fish, and remain on him until he becomes very poor, from loss of blood and his exertions to relieve himself from his tormentors, or till they are obliged to let go to spawn. It is well that they are not very numerous. Common weight, ½ lb.; large, 1 lb.; common length, 6 inches; age, 50 years.

BULLHEAD.

These fish are mostly found in pairs, on muddy bottoms, in all the rivers and lakes; they are pretty good pan-fish. They spawn in the spring, but are not very numerous. Common weight, 1½ lbs.; large, 2 lbs.; common length, 6 inches; age, 100 years.

CATFISH

Are found in most of the rivers and lakes, and are somewhat esteemed for food. They spawn in the spring. Common weight, 6 lbs.; large, 20 lbs.; common length, 24 inches; large, 36 inches; age, 300 years.

STURGEON.

These fish live to a good old age, and I see no reason why they should not, for they have no teeth to wear out, no bones to become hard and brittle; they live by suction, and their digestive organs are so constructed that they may wear for centuries. They spawn in most of the lakes and rivers, in the spring. The youny fish are good for cooking or packing. The Indians are very fond of them—they jerk them as they do venison. Common weight, 30 lbs.; large, 150 lbs.; common length, 4 feet; large, 8 feet; age, 500 years.

The fisheries of this State are of great importance, when we, take into account the amount of fish caught, the men that are employed, the salt, barrels, boats, seines, and rigging, with all the fixtures required in the business. The annual value of the fish caught is about three hundred thousand dollars. I will venture to say that Michigan receives more income from her fisheries than all the other States in the Union except those bordering on the sea coast. I think we are and always have been highly favored in this regard; for many of the first settlers would have been obliged to abandon their beginnings in this then wilderness, had it not been for the abundance of fish found in all the creeks, rivers and lakes of our Peninsula. I think that some of the inland lakes of this State could be stocked with fish and made to produce more in value than many times the same area of the adjoining land. Should this be done it would prove a great luxury as well as income. I have a pond or a part of the Detroit river picketed in, about ten miles below Detroit, where I have kept white fish several years; and, from what I know of their habits, I think that some of the small lakes in the interior could be stocked with them and they would do well with them. I hope the experiment will be tried by some of the owners of the small lakes.

Detroit, January, 1852.

AN ESSAY

On Agricultural Fences and Enclosures.

BY N. DAVIDSON REDPATH.

The farmer's first care, after having reclaimed a portion of land, either from the forest or the prairie, is to enclose it with a fence of some kind. This is necessary, both for the protection of his crops or the enclosing of his cattle and other animals that may be placed in it. Good fences, besides being convenient and useful, save a vast deal of trouble and annoyance to the husbandman. The construction of fences, in the first instance, must necessarily be rude, and the materials such as are most convenient or abundant. Hence, we find in all new settlements, where timber is abundant, the heavy zig-zag rail fence, or the still more primitive brush and log fence. Stumps, especially of the pine, when properly constructed, make a good and durable enclosure. In time, however, as timber disappears before the merciless axe of the woodman, these rude constructions must give place to others, if not of a more durable, at least of a more sheltering nature. Proper fences, it ought to be kept in mind, are not *only* useful for *protection* but for *shelter* from raking winds and stormy blasts. The best fences, therefore, are those which accomplish this two-fold object.

Various materials, in different parts of the world, have been used for the construction of agricultural fences and enclosures. In France, the willow is extensively used; in Scotland, where stone is abundant, stone walls have been introduced,* while in Wales the slate is used for the same purpose.

*Hedge rows are extensively used in the south of Scotland, but in the more northerly regions, where the thorn will not grow, stone walls, furze or whin hedges, and *feal dykes*, (composed of alternate layers of stone and sod,) are the common enclosures. In the extreme north, flag stones being abundant, are erected perpendicularly side by side; though extremely grotesque in appearance, make an unperishable enclosure.

In England the more prevalent kind of fences are hedge-rows, with trees planted in them at stated intervals. There can be no question that hawthorn hedges, if properly constructed and kept in good condition, besides being beautiful to the eye, make the most durable and proper kind of fence for most agricultural purposes; but when allowed to remain (as is too often the case) untrimmed, and to hold a doubtful sway with weeds and all klnds of dirt, we have no hesitation in saying that they are a perfect nuisance, not only to the farm on which they grow. but to all the surrounding neighborhood.

As much interest is now happily excited amongst the agricultural population of the United States, on the subject of the best materials for the construction of fences and enclosures, it is proposed to devote the sequel of the present essay to this important subject.

THORN HEDGES.

Care being taken in planting and cultivating them, hawthorn hedges will grow thriftily. except in very bleak and barren situations. Any want of success in rearing the thorn is attributable, in most cases, to improper treatment. The thorn, as is well known, is a hardy plant, and if properly planted, regularly cut, and thoroughly cleaned, especially when young, there is little difficulty in soon forming a beautiful and permanent fence. A thorough preparation of the soil on which the proposed fence is to be run, is essentially necessary, for if weeds are allowed to remain unexterminated before planting, it will be found exceedingly difficult, if not impossible to do it afterwards. It has been recommended to fallow foul land, and where poor, to manure the soil previous to planting; but in most cases the manuring may be dispensed with. A good practice is to trench the ground for the base of the hedge, four feet in breadth and about eighteen inches in depth, the surface soil being placed at the bottom of the trench, so that all kinds of weeds will be effectually destroyed by being placed below " vegetation point." Any expense incurred in the thorough preparation of the soil will be amply repaid by the clean and well-growing hedge which will be the result of such well directed labor.

Every farmer, desirous of introducing hawthorn hedges, will be necessitated in the meantime to become his own nurseryman and rear the plant from the pit or stone. This, though somewhat tedious, is a very simple process. In the fall of the year let the *haws* be collected in the desired quantity and put into a rot-heap, in order to destroy the pulp from the stone. In early spring sow them thickly in rows, after the manner of peas, the drills being about twelve inches apart, or they may be sown broadcast in beds, after the manner of onions. In these drills, or beds, they may remain until the following fall or spring, when they must be transplanted into rows about fifteen or eighteen inches asunder, and each plant about three inches apart. In two years they are fit to be conveyed to the hedgerows.

The season most proper for planting the thorn, is the spring, in order that the plants may have acquired sufficient root to resist the heat of summer. There are various ways of planting hedges; some hedgers preferring to plant the thorn on the even ground, while others plant them on a mound, or raised bank, with a ditch at one side for carrying off the surplus water. Ditch and hedge, as this latter mode is called, is that most commonly practiced on wet soils; but where the land is good and dry the ditch may be dispensed with altogether.*

The method generally pursued in England is to plant them perpendicularly into the ground, but this is a plan not to be recommended, as the slightest observation may convince any one that the hawthorn does not naturally sink into the ground, but spreads its roots near the surface. The result of this method, combined with overloading the roots with earth, to use the language of Goldsmith, in his "Deserted Village," will be a stunted,

> ——— "Struggling fence that skirts the way,
> With blossom'd weeds unprofitably gay!"

A more natural method is pursued in Scotland. The plants generally preferred are those which have stood one year in the seed-beds and two years in the nursery rows. In lifting the plant either from the seed-beds or the nursery-rows, great care is exercised that the tender roots be as little injured as possible and no delay permitted between lifting and transplanting.

The thorns are planted from five to six inches apart, in the hedgerows, and at the same depth as they stood in the nursery grounds. The tops are cut off before planting, in order to secure regularity.

*Where the soil is very wet, a ditch is frequently run on both sides of the fence. In very dry soils, however, the ditch is injurious, as it carries off that moisture which ought to nourish the plant. These double ditches and hedge frequently occupy a space from 15 to 18 feet wide.

An inch or an inch and a half being allowed to project above the ground.

If the hedge is to be planted after the Scotch system, but dispensing with the ditch, (which perhaps is the best method for our American climate,) the Scotch hedger nevertheless invariably raises a mound or raised back into which the thorns may easily insinuate themselves and find nourishment. This mound is never made less than four feet broad at the base, and is carried in an oblique direction to the height of from eighteen to twenty-four inches. Opinions differ as to whether the thorns ought to be planted on the top or at the base of this mound; the more general practice, (and in our opinion the best,) however, is to plant them at the base, by which arrangement the roots have ample scope to run in their natural horizontal direction, while the mound behind affords shelter to the tender shoots during the inclemency of winter.

The line of the fence being marked by a cord stretched out in the proposed direction, the first thing that the Scotch hedger does is to form what in Scotland is called the "thorn bed."* The plants being prepared, as before directed, are placed upon it, and if properly done they will lie in an oblique or sloping direction. Fine mould is then collected and laid over the young thorns, to afford nourishment to the tender roots. In forming the mound or back above the plant, where the ditch is used, the material thrown out of the one, is used for the formation of the other; but where the ditch is dispensed with, the materials for the formation of the mound, will be taken where most handy and convenient. A considerable portion of earth, however, may be taken with advantage from the front of the cord or line, (where the ditch ought to have been dug,) and by this means a water course is formed for carrying away the surface water, thereby

*The "thorn bed" is made by simply lifting a thin spadeful of the fine surface soil, and placing it up side down, directly behind the cord, care being taken in forming this "thorn bed," for the plants to rest upon, that it is formed in the shape of a wedge, the front part, next to the cord or line, being raised to the height of 6 inches, gently sloping downwards.

SECTION OF THE "THORN BED."

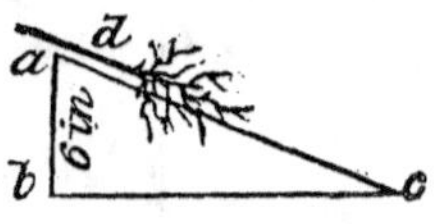

Thus, *a*, *b*, *c*, may represent a section of the "thorn bed"—*a*, *b*, the front, is raised 6 inches, while *a*, *c*, gently tapers down. *d*, represents the thorn lying in an oblique direction on the "bed," with its point projecting out. The mound is thrown directly above this "bed."

preventing the stagnation of water, so very injurious to the young thorn.

As before stated, the mound in Scotland, is never made with a less base than four feet, and is carried in an oblique direction in front, to the height of from 18 to 24 inches. The back part of the mound is frequently carried perpendicularly, by placing alternately layers of earth and sod; by this method, with a rail or two placed above, a good fence is formed for preventing cattle approaching so as to damage the young hedge. The following is a latteral section of the mound.

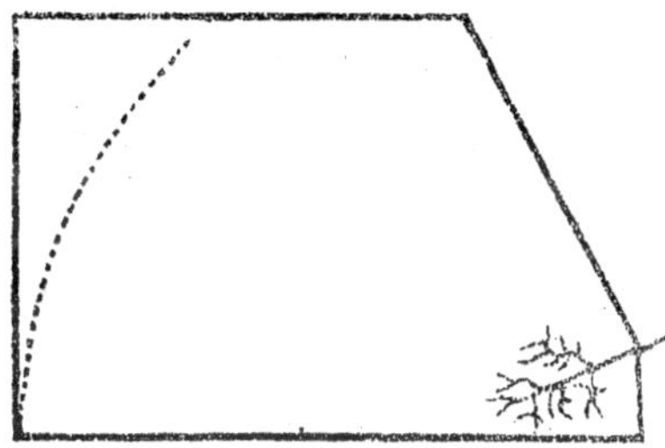

Where cattle or other animals are not allowed to depasture near the newly run fence, the perpendicular back will be unnecessary; the earth in that case will be thrown loosely over the back, and will form an easy round, somewhat as indicated by the dotted line in the above diagram.

Hedges planted out in the spring, ought to be cleaned in the fall of the same year; and if they have attained the height of two feet, or two and a half feet, by that period, then they must be reduced, by pruning the tops, down to 15 inches. The breasts of the fence must be pruned at the same time, in a gentle, sloping direction. If properly done, they will be fitted to withstand the heaviest falls of snow, during the winter, with impunity. Repeat the above operation in the fall of the second, third and fourth years; after which period, a good and permanent fence will be formed, and no further care required as to clearing and pruning,* unless you are desirous to have all things very neat and trim—a laudable ambition, which ought to be encouraged.

*During the first year, the thorns will attain the height of 30 inches. Reduce them to 15 inches. At the end of the second year, they may stand 45 inches. Cut off 15 inches, breast up the hedge and leave it standing at 30 inches. Third year, it will average at least five feet—cut off 15 inches this year, also—breast up again, and leave for the winter, 45 inches, or so. In the fourth and last year, cut off 15 inches again, and you have a hedge five feet high.

The operation of cleaning is performed in Scotland with an instrument, there called a "weeding-iron." It somewhat resembles the "Dutch hoe," but the mouth is only made about three and a half or four inches broad, in order that the tool may move freely between the roots of the plants. It could be made by any blacksmith.

SCOTTISH WEEDING IRON.

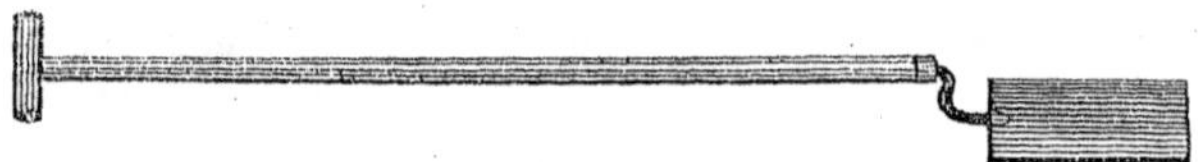

Thorn, in a good soil, will grow at the rate of from 2 feet to 3 feet in a year, but care must be taken not to allow them to run to top; therefore, a free but judicious use of the cutting knife must be exercised. The instruments used for pruning and cutting hedges, are called "hedge knifes" or "hedge bills." They are made of various degrees of strength, according to the kind of work to be done. Pruning knives are made light and slender; those for cutting over strong hedges or old ones, are made thick and strong.

PRUNING KNIFE.

When hedges get thin and bare at the bottom, the general practice is to cut them over by the roots, (about 6 inches from the ground or so;) by this means young and healthy shoots start out in great abundance, and in a short time a good and impervious hedge is again formed. Where blanks in old hedges occur, either from decayed or decaying stock, the intervals, if not too great, are frequently filled up by folding down an old stem or branch close to the ground, which speedily throws out abundance of shoots to fill the vacancy. But if this plan is not practicable, then the vacancy must be supplied by plants which have stood in the nursery grounds for the last three years. They should be planted with the tops entire, and have fresh soil supplied to their roots. Sometimes the vacancies in hedge rows are filled up with crab, beech, privet, &c. When these substitutes are used, it will not be necessary to remove the soil, but only to loosen and pulverize it.

CUTTING KNIFE.

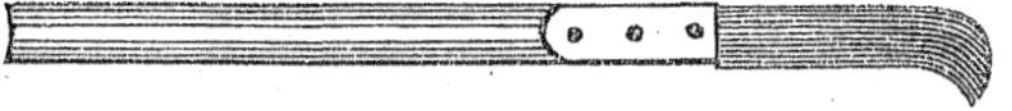

Various substances have been tried as a substitute for the thorn, but none have been found so well adapted for the purpose of the beech. In some of the moorish soils of Scotland, where thorn will not grow alone, the thorn and beech have been planted alternately, or two thorns and one beech alternately. In such soils, the thorn universally dies out, while the beech flourishes. Although beech can never present such a barrier to the inroads of cattle as thorns, yet as they stand much more pruning than the latter, they would form a pretty enclosure around our gardens, door-yards and orchards.*

*In pruning, although a free use may be made of the cutting knife, yet a sufficiency of leaves is necessary to every plant, to obtain nourishment from the atmosphere. Some hedgers round the tops in the shape of a semi-circle, but in pruning, this ought to be avoided, as t e great fall of snow in this country would prove very injurious. The better plan is to prune from the base of the hedge to the top, (on one or both sides,) in a gentle sloping direction, by this means the snow can make no lodgement on the top. Clipping hedges with shears is injurious, as the thorn is only bruised off, not cut.

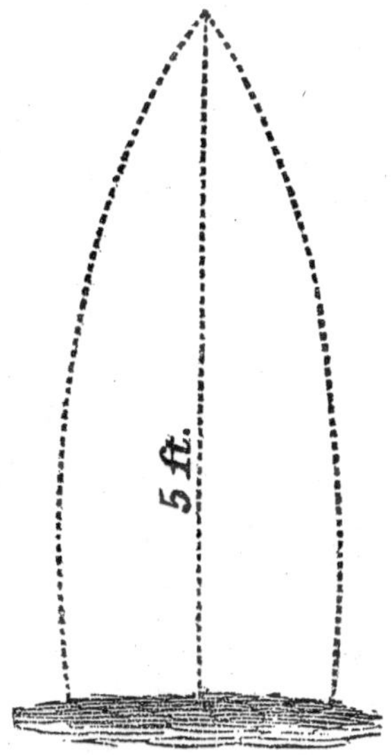

The above represents the most approved shape in which hedges are now cut.

RUSTIC FENCES.

These fences are made from the young saplings or the small limbs of trees. In most cases the bark is allowed to remain, but in constructing oak rustic fence, the general practice is to strip off the bark, as the wood is thereby rendered less subject to rot, besides having a much finer appearance.

By the exercise of a little ingenuity, and at a trifling expense, rustic fences may be constructed around the homestead, by the farmer himself or his assistants. The most simple construction of this kind is made by driving the saplings into the ground, about eight or nine inches apart, (but where pigs are allowed to run at large, four or five asunder,) having the tops fastened together by a long sapling nailed into the uprights.*

SIMPLE PLAN FOR RUSTIC FENCE—STAKES 4 INCHES APART.

No. 1.

No. 2.

*Greater durability is imparted to all kinds of wood driven into the ground, whether barked or unbarked, by placing the timber the reverse way that it stood while growing. The reason is obvious.

No. 3.

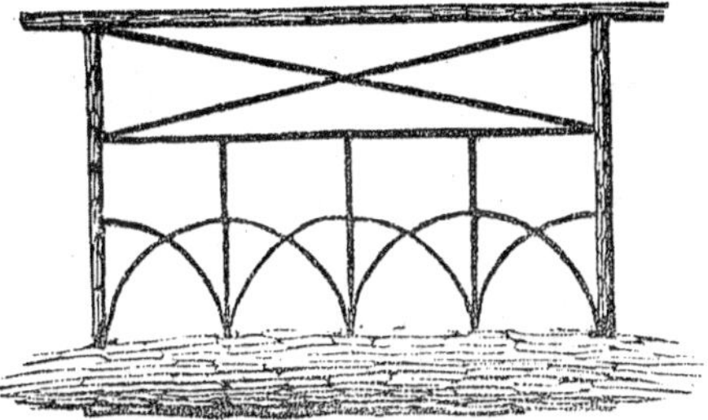

These kind of fences may be made according to the taste or the ingenuity of an individual, more or less ornamental, in all cases, though course, they look remarkably well around cottage or villa enclosures, and would form a great ornament to the farmer's homestead. We present as a specimen a few

PLANS FOR ORNAMENTAL RUSTIC FENCE.

No. 1.

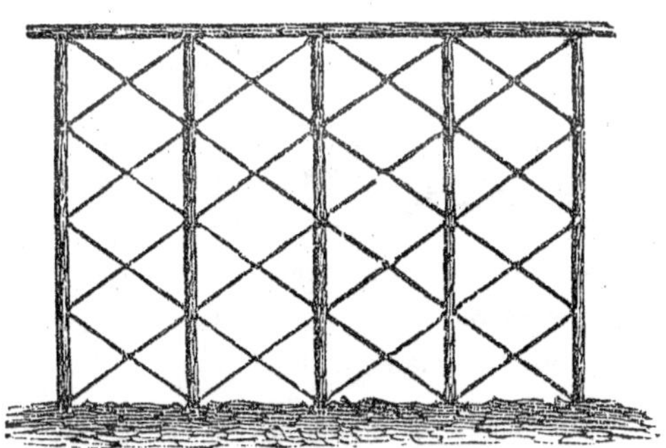

No. 2.

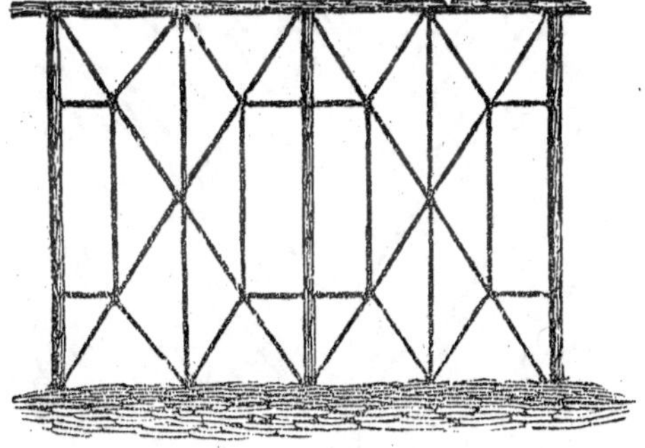

No. 3.

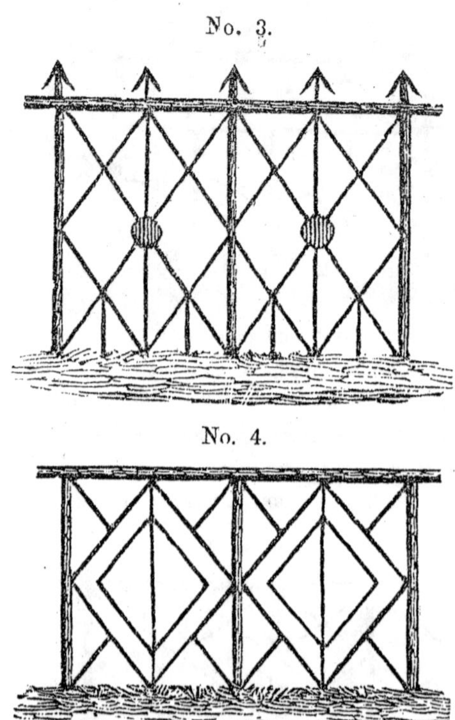

No. 4.

Small gates may be made to correspond with the rustic fence; but where large entrances are required for carriages or waggons, a strong wooden double door may be made, and overlaid with fine small saplings.

PLAN OF DOUBLE DOOR, OVERLAID WITH SAPLINGS, TO BE FASTENED INSIDE.

The saplings are fastened on to the wooden frame with long headless tacks. In the same manner, doors to summer houses, cottages, &c., &c., may be formed.

STONE FENCES, RAISED AND SUNK.

In districts where stone is abundant, cheap and durable fences may be constructed by dry stone alone. If not too round, the stone found on the surface of the land will answer the purpose; but flat stones, when easily obtained from a quarry, are to be preferred. Stone walls are frequently built dry, and pointed with lime, which gives a good finish, and adds much to their durability.

As the method of building these walls is so well known, especially in the eastern States, we deem it superfluous to go into detail.

When it is desired to have fences without obstructing the view, sunk stone walls are excellently adapted for the purpose. A ditch is sunk to the depth of four or four and a half feet, and about double the width at top, tapering to the bottom. The stone wall is built from the bottom of the ditch to the surface of the gronnd. The face of the ditch is sown off with grass. Few animals will attempt to leap down, and from the great slope of the ditch none will be able to leap up.*

CLAY AND MUD FENCE.

In those districts where stones are not to be procured, a clay wall is found to be an excellent substitute. In some regions of our country neither stone nor wood is to be obtained, but where clay is abundant, little difficulty can be experienced in obtaining a substantial and durable fence. Fence walls of clay and mud are very common

* In front of houses where an extended view is desired, or the prospect of some particular object unobstructed, such as a river, these fences will answer the purpose admirably. In the following diagram a, b, c; a, b represent the sunk fence or stone wall, (4 feet high,) b, c, the

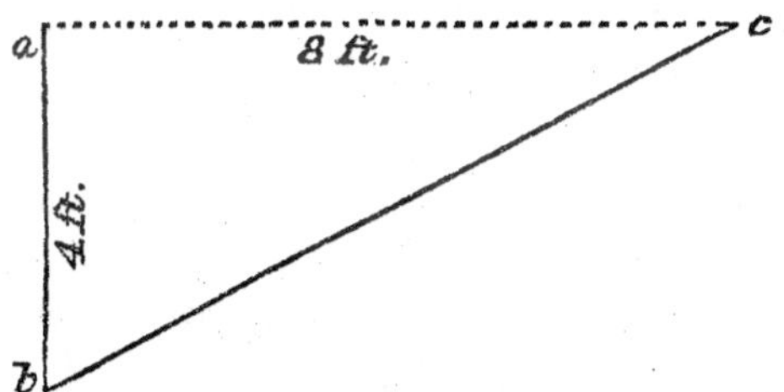

face of the ditch, and the dotted line a, c, (the surface of the ground,) nine feet wide and sloping very gently to the bottom of the wall. Where access is wanted from one field to another a wooden bridge may be thrown across the ditch.

in many parts of England, particularly in those counties where there is a destitution of stone. In building these fences a little straw is mixed with the clay in the same way that hair is mixed with mortar or plaster. In making the mortar, some mix cut straw, two or three inches long, and incorporate the two thoroughly together, while others make the mortar without the straw, and when laid on the walls in layers, the straw to its full length is put on the top of each layer, lapping over and down the sides. The straw is then well beat into the wall and acts as a cordage, binding the whole together. These walls should be built early in the season so as to have the full benefit of the summer's sun in effectually drying them. They should be built a foot or two at a time and allowed to dry. A coating of lime mortar must be added, and the walls carefully examined until thoroughly dry, when they will become as hard and durable as stone. Where the coating of lime falls off, it must be replaced. A cope of some material must be added, or the wall be pointed over the top with lime mortar, the same as on the sides. A ditch may be cut on one side (out of which the clay may be procured) and a furrow on the other, in order that the wall may be kept dry at the bottom. These kind of fences are in some respects superior to hedges as they are complete at once, and when properly finished are neat, substantial and durable, and easily repaired. They are well calculated for those meridians where pigs and hogs are abundant and allowed to go at large.

Cottages may be built in the same manner as the fence wall, and overlaid with lime mortar, intermixed with small gravel and whitewashed. Houses built with clay would neither be so warm in summer nor so cold in winter as those erected with wood, brick or stone. Although this is a subject well worthy the attention of American farmers, yet it is mentioned only incidentally, as the subject of the present essay precludes our enlarging farther.

HURDLE FENCE.*

The hurdle or wattle fence is moveable. They are constructed somewhat after the same manner as gates, but of greater length and lighter construction. The hurdle is generally made 12 feet long and

*Hurdles are, in some parts of England, called "*wattles*," and in others "*trays*." In Scotland they are called "*flakes*."

4 feet high, with four bars, about 4 or $4\frac{1}{2}$ inches broad, and about 1 inch thick, groved into scantlings 3 inches by 3. A bar crosses the middle.

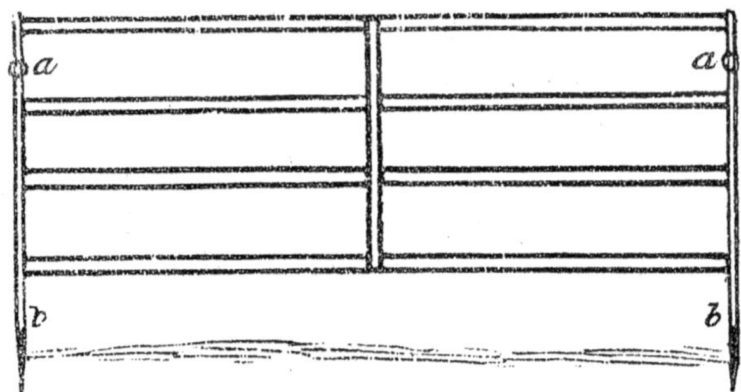

The scantlings are made about 9 inches deeper (*b. b.* in the diagram) than the hurdle, for the purpose of fixing in the ground. The hurdles are connected together and supported by a scantling placed behind and fastened into the ground at bottom, and pinned together at about 12 inches (*a. a.* in the diagram) from the top. The hurdles may be made by notching in the bars, after the manner of a rough gate, and then the farmer may make them himself. These useful appendages to a farm could be cheaply made, were a demand to be created for them, and tradesmen to enter into the business as an article of trade.

The hurdle fence is now extensively used in Yorkshire, in England. In that county, it was calculated that one-fifth of the land was occupied by thorn hedges! The fields were small, generally from three to five acres. The hedges were in bad condition, and formed a nursery for all kinds of weeds. On many farms the interior fences are now thrown down and the hurdle fence introduced with the greatest success. It may be necessary to premise that it is not contemplated to run hurdles around all the lots where fences formerly existed, by any means. No fences are necessary for the separation of grain crops, but only around pasture allotments, or where cattle may be allowed temporarily to graze. Permanent, and if possible, live fences, such as thorn, should be run round the exterior of every farm, and these, were it only for good neighborhood, ought to be kept in good condition. Around the homestead, permanent fences ought to be constructed, enclosing door-yard, garden, orchard and barn-yard. With regard to the farm, great division lines ought to be laid permanently down, as always needed, either for shelter or division. Suppose the following to represent a small farm of 80 acres:

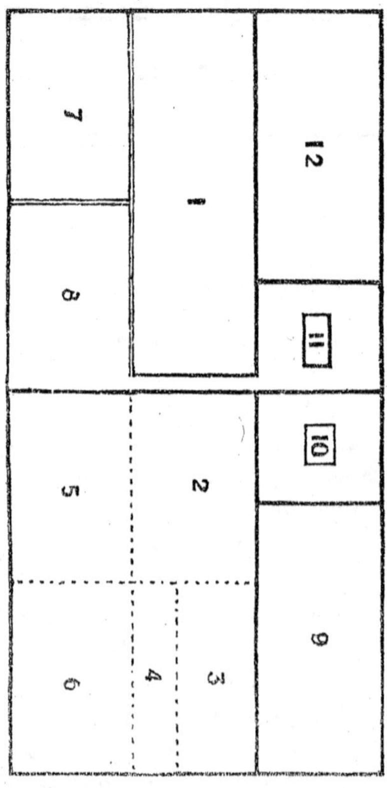

The black lines represent the permanent fences, exterior and interior—the dotted lines the division of crops without any fences—the parallel lines the hurdle fence. A glance at the above plan will show that the farm is divided in the middle by a permanent fence, and that the land to the right of this line is occupied by oats, potatoes, turnips, barley and corn. Now, it is evident that no division is required between these patches, and that consequently a great saving of land is effected, not to speak of the eradication of all kinds of weeds that would otherwise have grown, had fences been laid down. To the left of the line grow wheat, pasture and clover—the wheat and the clover must be protected from the inroads of the cattle, grazing in (No. 8,) the pasture lot. After (No. 7,) the clover lot is cleared, the hurdle fence may be removed in whole or in part and give the cattle free scope over both lots. In the same way divisions may be run at the convenience or pleasure of the farmer, with only about one hundred and twenty rods of hurdle fence, which would be sufficient for a farm of eighty acres. How much more rational such a mode of enclosing would be to the cumbrous and lumbrous system which now prevails. On some eighty acre lots there is more oak used in fences than would build several ships of the line!

SOD FENCES.

In Ireland, sod fence is the common enclosure of the country. Sod fences might be introduced with advantage into prairies, where there is a deficiency of wood. These fences are made in the form of a wall, three feet broad at the bottom, and tapering to about 18 or 20 inches at top. They are raised to a height of about five feet. Sods as thick as they can be raised are placed on both sides, and the middle filled up with earth. Both sides are carried up in the same way, layer upon layer, until the wall attains to the desired height. Furze or whin is sown on the top.* In a short time these sod will become one solid mass. Like the clay wall they are easily constructed, and though not quite so durable, they can be built at less expense and trouble, and where clay cannot be had.

WIRE FENCE.

So much has been said and written, and that to so good a purpose, upon wire fence, that little new can be elicited. Some maintain that wire fence must become *the fence* of the country.

To this we demur. These fences have none of the essentials for good enclosures—they afford no shelter—the materials used are so light and attenuated as to form an "airy nothing" in the eyes of cattle. They might be used, however, for interior fences for guarding in sheep with great advantage, and like the hurdle fence, they can be made moveable. The wire fence would look very pretty around the homestead of our farmers.

*It is doubtful if the British furze would grow in this country, but there is little doubt that the French furze, which is even superior to the British, would luxuriate here. Seed could easily be procured.

Wire fences, where they have been introduced into Europe, have been found unsatisfactory. Whether they may answer a good purpose in this country is yet to be proved. Let them, however, have a fair trial, a fair field, and no favor.

Fogo, Allegan Co., September, 1851.

ADDRESS.

Delivered before the Michigan State Agricultural Society, at its Third Annual Fair, held at Detroit, Sept. 24th, 25th aud 26th, 1851.

BY HON. LEWIS CASS.

That the lines had fallen to him in pleasant places, and that he had a goodly heritage, were subjects of congratulation with the Royal Psalmist of Israel, gratefully acknowledged and beautifully expressed in one of those out-pourings of the imagination, which have come down to us in the pages of scripture, and whose hold upon the human affections is as powerful now as in the brightest days of the chosen people of God. And where is the American who does not feel tha his country offers goodly heritages and pleasant places, and equally, too, to all who seek them, from the shores where our fathers first commenced their stern contest with nature, to that distant ocean, which their descendants have reached by like trials and exertions, and which separate us from the islands and continent of Asia? And nowhere are higher rewards offered to human industry and enterprise, as well by the physical as by the moral and political advantages within our reach, than meet us at every step of our progress upon our own beautiful Peninsula. Around us is one of those pleasant places, one of those *goodly heritages.* Few positions present more objects of agreeable contemplation, than does the very spot where this assemblage is now convened, in peace and prosperity, not to celebrate a victory nor to extol a conqueror, but to commune together upon one of the most interesting subjects that can engage the human attention. Above us and below us are vast internal seas, uneqalled on the face of the globe, stretching away to other regions, distributing their bounties far and wide, and already bearing rich products, "corn as the sand of the sea," from the great granery of the North-West, to supply the wants of other portions of the world, less favored by nature, or more harshly ruled by man.

The solitude of the forest, but yesterday so appalling from its very intensity, is now broken by the busy hum of human ndustry; and wind and steam have united their power, and have penetrated the most secluded recesses, in search of the objects and the rewards of energy andperseverance. But who so bold as to venture to set bounds to the productive extent of this great theatre of human exertion, or to foretell how far and how fast will be its progress? The future baffles all calculation, as the past has out-stripped all anticipation.

But our business is with the present, and our noble river is too prominent an object in the landscape of our "good y heritage," to be passed by without remark. Uniting the Upper Lakes with Lake Erie, it flows between them in a course of many miles, forming by its extension, the small Lake St. Clair, and presenting marked features of beauty and of magnitude, which arrest the attention of the most careless observer, and ministering, also, to the health and wants of our city, while ministering to the wants of a commerce already great, though born of yesterday, and destined to take its place, and ere long, too, among the proudest monuments of human enterprise. With a current deep, broad and gentle, this great stream recalls the tribute of Denham to the Thames, far more appropriate to the American than to the English river—

> "Though deep yet clear, though gentle yet not dull,
> Strong without rage, without o'erflowing full."

Its depth is sufficient for all the vessels that navigate the lakes, while its broad expanse and the mighty volume of water it receives and pours out, announce it as one of the great arteries of the earth, formed to maintain that circulation, which is essential, as well to the order of nature, as to the communication and with it the improvement and civilization of the human family. It possesses one peculiar and invaluable feature—it is subject to no periodical elevation nor depression. As it is to-day, so is it, with slight variations, at all times. The falling of the rain and the melting of the snows, which elsewhere cause the streams to overflow their banks and to carry terror and distress in their impetuous course, are rendered harmless by the great reservoirs above us, which equalize the supply and the demand, and thus prevent some of the worst effects of the extremes of wet and dry seasons.

There is no need that I should recapitulate the rewards that our Peninsula has already secured to those *to whom the lines have fallen* here, and that it continues to offer to all who seek comfort and independence from the life of the husbandmen, under the most favorable circumstances of soil, of climate, and of position, that can excite the hopes of the emigrant or stimulate his exertions. And the progress and prosperity that mark our whole district of country, furnish the best commentary on its adaptation to the wants of society, and its capacity to provide for them easily and abundantly, and with pride and gratification may we point to our own city, spread out before us, and which is becoming every day more worthy of its splendid natural situation, and of its claims to be one of the principal marts of business in all this extensive region. Its noble and capacious harbor, its intermediary position between the upper and the lower Lakes, commanding the communication between them, its relation to the great wheat-growing portions of the State, of much of which it is the natural commercial depot; and the enterprise, capital and industry already accumulated and in active operation within its limits, sufficiently explains the causes of its present prosperity, and justify the most sanguine expectation of its future progress. All the signs of a prosperous community are around us, and yet how short the period, since this very place was the advanced post of civilization, with scarcely a white man between here and the Pacific ocean—a secluded nook in the great primeval forest which extended across the continent, apparently impervious to human civilivation, and sheltering in its recesses a fierce and intractable foe, who defended his heritage with constancy and courage, adding the cruelty of man to the obstacles of nature. But a few brief years have passed away—briefer and briefer they appear as time rolls on—since a war party of Indians issued from the forest, and placing themselves between the spot where this assemblage is now listening to the reminiscence of a bygone incident—removed from us rather by the march of events than by the progress of time—and the inhabited portion of the town, marked their irruption by one of those deeds cf blood, which have made the history of our frontier a record of trials and sufferings, without a parallel in the progress of society. It is difficult for him who addresses you under circumstances so pleasant and prosperous, to realize that he heard the silence of the little village broken by the discharge of the Indian rifles, and to describe the alarm which prevailed, as the strength of the party was unknown, and the country was left defenceless, the troops having been sent away to take part in the operations on the Niagara frontier. But the energy of the inhabitants, few and exposed as they were—most of them the descendants of the original French settlers—supplied the place of numbers and experience, and the foe was driven from the settlement to his native haunts in the forest, but not without a conflict with one of the ranging-parties, almost within hearing of the town; and there is a small remnant of those brave and faithful citizens—faithful at all times—soldiers in danger—whom I now see near me, making part of this great assemblage, and whom I then saw still nearer, making part of the patriotic band, who freely offered their services in that hour of peril, and who will recollect the terror inspired by our return, as the scalp halloo was raised by some of our Indian hunters to indicate the success of the party, and broke the silence of the twilight with that terrific sound, which once heard is never forgotten, and which tells the tale of blood before the bleeding trophies and the victorious party present themselves. Whether the signal was from friend or foe, the helpless women and children—whose husbands and fathers had gone out to defend them—had no means of knowing; and, in the terrible uncertainty, many of them embarked in canoes and fled for safety to the Canadian shore. Human nature is seldom destined to a more severe trial than such a state of doubt brings with it. Happily, the return of their friends removed their apprehensions and secured their safety. And such incidents are characteristic of frontier life; and when they shall have been hallowed by time and traditional associations, they will constitute the romance of Indian history, showing the struggles and sufferings which our settlements have encountered in their long and painful march from the rock of Plymouth and the beach of Jamestown to the shores of that distant ocean, where our fathers' flag and our own is now planted, and where American institutions and American enterprises are doing that deed of power and progress whose consequences no man can contemplate without admiration.

Those who are curious in the statistics of our territorial progress, have given us calculations of its annual average rate; but without seeking precision in a subject scarcely admitting it, we know that though sometimes retarded, and sometimes accelerated, by peculiar circumstances, our march has been onward from the landing of the forlorn hope of a nation, when they exchanged the perils of the sea for the greater perils that awaited them in the new world they were entering, down to our own day. *But old things are passing away*, and that moral energy which seems to be operating with renewed vigor in the great departments of life, intellectual and physical, and which has made our age emphatically the age of progress, is exerting its power on our capacity for expansion, and is now urging us forward in our great mission of *replenishing the earth and subduing it*, with a rapidity which finds no example in the history of our own or of any other country. From the steps of an infant we have assumed the pace of a giant, and are *walking through the land, in the length of it and in the breadth of it*, with a rate of progress scarcely less unexpected to ourselves than astonishing to the rest of the world, who are regarding these new efforts with an interest rarely felt in the concerns of other nations, and which shows the mighty effects they may produce upon the destiny of mankind.

Already have we crossed the continent and established an empire upon the great shores of the ocean of the west, and have laid the foundation—and broadly and deeply too—of freedom, religion and civilization, in regions acquired but yesterday, but which, till then, had been shut out from all improvement—held and governed as they were by indolence and ignorance. The impress of our race is now marked in ineffaceable characters upon a vast outline, and the intermediate space will be filled up with an energy which foreshadows what we shall do, by what we have done. The mountain, the forest, the desert, present themselves to the indomitable emigrant, not as barriers which say to him, "*Thus far shalt thou go, and no farther*," but as obstacles to be overcome by efforts more vigorously renewed as the difficulties are augmented. And along the line of a nation's march, amid fierce and hostile savages, over arid and naked, herbless and treeless plains, and rugged mountainous cliffs and defiles, man goes not alone—but woman, with her power to endure, and her will to exert it, where her affections are engaged, accompanies him in his toilsome journey, to augment his joys, and by taking part in his sorrows, to lighten them. Broad—boundless, almost—is the theatre offered to our exertions—gigantic in its features, like the continent we inhabit—far different from those regions of the old world, familiar to us from infancy, and forever memorable as the scenes of mighty occurrences, connected with the origin and early progress of the human race. In those countries of the east—formerly the cradle, and now the tomb of civilization—events were compressed within a narrow compass, and the limited space within which they passed seems utterly disproportioned to the extent they occupy in the history of the world, and to the influence they have exerted upon its destiny. The breadth of our own Peninsula is greater than the direct distance from the land of Goshen, the residence of the children of Israel upon the Nile, to Palestine, the land of promise, and the object of their migration; and through their journey, by Divine command, was circuitous, still the desert where they wandered is less in extent than several States of our Union, and ours among the number.

If the little bands of self-devoted exiles have become a great people, not less wonderful has been the increase of the means of human subsistence, by which our own augmenting numbers have been supplied, and which is pouring its contributions into other and less favored regions of the globe. History tells us that during the *starving time*—a period of famine, marked by that terrible epithet—the colonists of Jamestown were supplied with food from the bodies of their friends who had sunk under the calamities to which they were exposed; and we are assured by tradition, that the granary of the Puritan fathers had at one time but five kernels of corn for each individual composing the expatriated colony, now become the millions of New England. But, like the widow's meal, the scanty store, by the blessing of God, failed not. I have said upon another occasion—but the circumstance is so striking and characteristic that I must repeat it here—that I have often conversed with a venerable relative, who was a cotemporary of Peregrine White, the first child born to the Pilgrims after their arrival on this continent. What an almost appaling idea does this simple fact present of the progress and prospects of this vast Republican Empire! But one life passed away between the first and latest born of one of its greatest communities—between its infancy and its maturity—between its weakness, almost without hope, and its power, almost without limits—between its granary, holding a few kernels of corn, and all its vast "store-houses," whose contents, like those of Pharoah's, we may *leave numbering, for they are without number*, but which supply kingdoms with abundance, and send plenty where the harvest follows not the seed-time, and where famine seeks to establish an empire, displaying its power by its victims, and not by its subjects—by the dead, and not by the living.

The first want of man is food, and his first care is to supply it. The rude modes of agriculture, originally prevailing, have gradually given way to improved systems, combining sound theoretical principles with enlarged experience and observation. In the dispensation of nature, an increasing population always presses upon the means of subsistence, and if the supply does not augment with the demand, famine, and the terrible evils it brings in its train, will soon come to teach an ignorant, or oppressed, or improvident community, that it is the province of nature to minister to the wants of man by aiding his exertions, and not to minister to his indolence by rendering them unnecessary. For wise purposes, no doubt, there are regions of exhuberant fertility, where articles of food may be obtained without culture, as there are others of irreclaimable sterility, where no industry can conquer the obstacles of soil and climate, and render these stricken portions of the earth desirable residences for the human family. But in the natural, as in the moral world, the system of creation is one of compensations—good and evil go hand-in-hand together. Where man lives without exertion or industry, he lives without virtue or intelligence, and dies as indifferent to the future as he has been to the past. But where necessity—his real friend, though sometimes apparently a stern one—requires him to labor, he attains his true position, and fulfills his true destiny by the proper employment of his faculties, physical and moral, and by their nobler development, which is sure to follow. Experience has shown that in the temperate zones of the earth, where industry is essential to subsistence, and where its rewards are ample, there only the human race have attained their true position, and have advanced in that career of intelligence and improvement marked out by the Creator, and the limits of which are known to Him alone. The great deeds, the great names, the great discoveries of the world are there, telling in language not to be misunderstood, that there is an intimate connection between the progress of the human faculties, and the motives for their exertion.

To increase the quantity and to improve the quality of agricultural products, suited to the subsistence of man, and with as little labor as is compatible with the object, should be the great effort of intelligent agriculturists. The progress of tillage is one of the most interesting chapters in the whole record of human society. Its origin is lost in the obscurity of time, and fable has usurped the place of authentic history. That the food of man, in his primitive state, was the spontaneous products of the earth, we are equally assured by scripture and by reason. It would be a curious subject of inquiry to ascertain, if the means were in our power, the various steps by which the acorn yielded its place to more palatable and succulent food, and finally to the cereal grains, and to bread, their best form, the great staple of human support. But the arts of destruction have always exerted so much more power over the imagination, than those of preservation, that while battles and battle-fields and conquerors crowd every page of history, till they have become almost as undistinguished as their victims, we are compelled to grope our way amid the mist of fable and tradition, in a vain search after the progress of the human intellect, in these discoveries and improvements, which make an essential part of civilization itself, and many of which must have been as early in their introduction, as they have proved general in their use. The moralist and the philosopher have equally exposed the tendency and the folly of this love of military glory; but it is deeply enrooted in human nature, and operates as powerfully at this day as in the earlist times, when a plastic mythology gave to the victorious destroyer a niche, a statue and an altar in the Pantheon of the Gods.

It were a rash attempt to undertake to set limits to the productive power of the earth. It has been greatly increased by human knowledge and labor, and the discoveries of modern science, and especially of chemistry, have still further enlarged its sphere, and promise yet more beneficial results for the future. We know but little of the agriculture of the ancients. The scattered notices of it, which have survived the lapse of time, are meagre and unsatisfactory, and certainly inspire but little respect for the practice or practical philosophy of the cultivators of the soil. The twelve yoke of oxen that Elisha left in the furrow, when called to a higher duty, furnish a very unfavorable commentary upon the quality of the Jewish stock, as well as upon the instruments of tillage, which required such a prodigious waste of animal labor.

For ages agriculture was stationary, and probably advanced but little from the time Virgil wrote his Georgics, till the last century. We are told by the best authority, that in Scotland, within one hundred years, "there were no rotation of crops; fallows were unknown except in one or two counties; the processes and implements of husbandry were alike wretched; the occupiers were in extreme poverty, and famines were every now and then occurring, that sometimes laid waste extensive districts." That "the returns were about three times the seed," and that so late as 1727, "a field of wheat of eight acres, in the vicinity of Edingburgh, was considered so great a curiosity that it excited the attention of the whole neighborhood, and that numbers of persons came from a great distance to see it." And if such was the condition of one of the most enlightened countries of Europe, we may well believe, what indeed was the fact, that elsewhere the state of agriculture was equally deplorable and degrading; and well is it for the cause of humanity that improved systems of tillage have been adopted, and the quantity of food greatly increased to meet the augmenting demand of an enlarged population, and thus spare us from famines like those, which were the scourge of the world in ancient and mediæval times.

The causes which led to the rapid improvement of agriculture in portions of Europe, after it had been stationary for centuries, I shall not detain you by investigating. They are connected with the general progess of opinion, the melioration of political institutions, the wonderful advance in all the arts and sciences, and especially in those which relate to husbandry, and with the new spirit of intercourse which has brought the nations of the earth much closer together, and made them much more useful than at any former period in their history.

Time and reason have banished, not all indeed, but many of the idle and superstitious notions, which once prevailed, and which impeded both the theory and the practice of husbandry. Men trust more to their own observation, and to the deductions of enlightened science, and less to those legacies of a dark period, which so long held in bondage the cultivators of the soil. The moon is left to perform the proper functions assigned to her by the Creator in the firmament of Heaven, and is relieved from that supremacy over the vegetable kingdom, which made her the great calendar of the farmer, and gave to her various phases more power over the world of vegetation, than climate, soil, or cultivation.

Intimately connected with the increase of agricultural products is the important question of their relative adaptation to the purposes of human subsistence; and both of these interesting objects of inquiry have recently engaged the attention of able and scientific observers, and the progress of the investigation has been equally honorable to them, and satisfactory to all, who take an interest in this great department of human concern. Researches into the true principles of vegetable physiology have been pursued with admirable judgment and success, and have opened to us new and enlarged views of the structure and organization of plants, and of the supplies necessary to their growth and subsistence, as well as of those they furnish for the support of man.

And it is here that chemistry, especially by the improved processes of analysis, has shown itself the efficient handmaid of agriculture. It has taught us that the incombustible or inorganic portion of plants—the ashes, small indeed, not exceeding two per cent in weight, but essential in the economy of vegetation—is derived from the soil; while the organic portion, the carbon, the hydrogen, the oxygen and nitrogen are imbibed by the roots and leaves, and conveyed by the proper vessels in a manner unknown to us, to every part of the plant, and incorporated by assimilation with the new body into which they enter.

How far these inquiries may be extended, and to what hidden operations of nature they may ultimately conduct us in the progress of human knowledge, it were rash to undertake even to conjecture. That we have not reached the boundary, which divides what may be known from what must be unknown, is evident from all that is daily passing around us in the world of matter. The limit may be far distant, and no doubt important developments will hereafter reward the zeal and industry of philosophical inquiries, and perhaps give us plain views, where we have now but imperfect glimpses, of Jehovah's kingdom. All this is, however, wisely concealed from us, that presumption and despondency may be equally repressed. But we may safely assume, that many of the final causes, and of the secondary agencies also, which govern the vast creation of God, and preserve that wonderful harmony in his works by which the elements of destruction are forever pressing upon those of preservation, but can never pass the limits prescribed to them by Him, who has assigned to each its proper functions in the economy of nature, and has made these hostile principles to contribute to universal order, if ever made known, will probably only be made known in another and a higher state of existence. In the whole range of the visible creation there is no brighter spot than this for the eye of man to rest upon, nor one where the wisdom of the Creator is more wonderfully displayed. The original impulsive power and the power of gravitation are so combined, as to retain the planetary bodies in their orbits, and to conduct them along their trackless paths in the heavens with unerring precision. There they have been and are, and there they will be till their great work is ended. And yet how admirable the adaptation by which these mighty masses are guided and controlled, and the agencies in operation balanced and restrained. The predominance of one would bring these worlds together, involving by their contact the utter destruction of our system, while that of the other would release them from the bonds of mutual dependence, and send them to wander in the boundless regions of space, where even the human imagination could not follow them in their endless career. And the same wisdom is displayed in the natural operations around us, and in the midst of which our life is passing. A change in the constituent principles of the atmosphere, an undue superiority of one of its elements, would render it unfit for respiration, and incapable of performing the functions assigned to it. Fire, water, animals, plants, the whole economy of nature indeed, would feel the effects of such a modification, and of the revolution it would bring with it. The ocean and the atmosphere are kept within their respective limits; rain and sunshine follow each other in beautiful succession, and the whole order of the universe is maintained by this combination of antagonistic principles, thus harmoniously blended, and made to work in unison together. It is not the wisdom of man, but the ordination of God, which sets bounds to the fecundity of insect life, and spares us from inflictions like those which visited Egypt, when her proud monarch hardened his heart against the divine command. What would avail human might if the barriers against indefinite multiplication were broken down, and if the locusts should *cover the face of the earth, so that there should remain no green thing?* In such a warfare, powerless would be human strength against an enemy whose numbers would mock to scorn the efforts of man. And thus the hail, or the wind, or the rain, if not held in check by a superior power, would rule the ascendant, and scatter destruction over the fair face of nature. But a far mightier than they has set bounds to their destroying agencies, and each furnishes its contribution of good to the great work of all.

The impressive language of reproof addressed by the Almighty to Job may be addressed to every son of Adam: "Where wast thou when I laid the foundation of the earth? Declare, if thou hast understanding." "When the morning stars sang together, and all the sons of God shouted for joy?" The human faculties are bewildered in the contemplation of the wonderful power of the Creator, and still the more as it is made known to us by a knowledge of his works; by the knowledge that there are worlds of animated beings around us, forever invisible to the naked eye, but each enjoying his alloted share of happiness; while far beyond us, in the regions of space, are those shining orbs, whose object, to our vision, is only to deck the sky with glittering points, but which reason and analogy equally teach us are destined to far nobler ends, to the performance of important functions in the great scheme of creation, and probably peopled with intelligent beings, perhaps better fitted than we are to comprehend these miracles of creative power—a power felt in the regions of illimitable space equally inaccessible to human vision and to the efforts of the human comprehension.

> "We cannot go
> Where universal love not smiles around,
> Sustaining all yon orbs and all their suns,
> From seeming evil still educing good,
> And better thence again, and better still,
> In infinite progression."

The researches of modern chemistry have been successfully directed to two objects of inquiry, equally essential to a just and rational view of the true theory and practice of agriculture. One is the elements, which furnish the proper nutriment to the animal system, and the other the plants, in which this pabulum of life is most abundantly found. Experiments, skillfully directed and steadily pursued, have already conducted us far in this interesting field of investigation; and strange is it that it has been but recently explored in the real spirit of inductive philosophy. Scarcely a century has elapsed since much has been known of the true structure of plants; and many important points of vegetable physiology, even among those most accessable to daily observation, are yet involved in doubt and dispute. Chemistry has been far behind the sister scinces, and it is only with the commencement of the present century that commences that era of its advancement, when the labors of analysis and the principles of induction rescued it from the hands of the alchemists, and have given it the dignity of a science, contributing not less to the comfort of man, than to a true knowledge of many of the most interesting operations of nature. And it is by no other process that we can expect to obtain just conceptions of the nature and properties of the undecomposed substances, which, in the present state of our knowledge, may be called elementary principles, and by whose means all living organized bodies are constituted and perform their functions. The boldness—the blind rashness rather, of man, leads him to reject the slow process of following nature, step by step, and adding fact after fact to his store of knowledge, and then combining these together, and deducing such general laws, as it may be given to him to develop and comprehend. For centuries the theory came first and then facts were sought to support it. But though this scientific empiricism is not yet wholly banished, it is yielding to a more rational spirit of investigation, and the reward has been found in that wonderful advance of knowledge in the natural sciences, which has made the age in which we live eminently a practical one. Our first business is with the operations of nature, and after that with their causes and effects. And it is obvious, that it is far easier to penetrate the former than the latter, and that *how* the functions of organized matter go on is a problem of easier solution, than *why* they go on. The general process of vegetation meets us at every step of life, and observation and experience have revealed many of its principles. We know that seeds germinate, that the sap circulates, and that the body of the plant is gradually developed by assimilation, and we have ascertained many, both of the laws and of the organs, by which these operations are conducted, and it is in this great and useful field of inquiry that our labors can be most successfully exerted. All beyond it is a subject of speculation—of deeply interesting speculation, indeed, in its pursuit, but too doubtful in its result to encourage us in the task of exploring it. What are the means by which the great operations of nature are conducted through the vast domain of creation, and by which the principle of vitality, animal and vegetable, is infused into organic matter, by which, in fact, the Almighty Will does its work; why, for example, there is a tendency in all bodies towards one another, agreeably to a law we have discovered and denominated gravitation, but whose mode of action we cannot even conjecture—these are inquries, which, in the present state of our knowledge, we have no encouragement to investigate. Man, and the little experiments he can make in the narrow circle that encompasses him, sink into utter insignificance before the overpowering considerations that force themselves upon us in the contemplation of these manifestations and modifications of Almighty Power. Silence becomes wisdom. But whatever revelations the future has in store for us, it needs none to tell us, that the clearer becomes our vision of the wonderful scheme of creation, the greater will be our admiration of the wisdom and goodness that designed it, and of the power that made and preserves it.

But we must return from this digression to that branch of our subject more immediately before us. I have already said, that practical chemistry had been investigating the proper nutriment of animal life, and the plants, which most abundantly supply it, and that important discoveries had followed these investigations, peculiarly valuable to the farmer, who desires to conduct his operations not only profitably but intelligibly. It is ascertained that among the edible plants there is a great difference in their capacity to provide for the subsistence of man. They yield the principles of nutriment in very different proportions, and this consideration is a most important one in the choice of their culture. Of the bulbous roots naturalized in our country, the potato is by far the most valuable, analysis showing that one-fourth of it may be considered as nutritive matter. The carrot, the parsnip, the beet and the turnip are less valuable, but still important objects of cultivation.

But the family of the cereal grains is that which is best adapted to the use of man, as it is by far the richest in nutritive qualities, and has constituted the principal articles of human food among the civilized nations of the world. Analytical chemistry thus comes to add its testimony to general experience, and to make known to us one of the causes of this preference. At the head of this family is wheat, whose introduction into Greece was rewarded by the divine honors conferred on the benefactor. Like the horse and the ox, it is the universal companion of civilized man, and has accompanied him in his migrations around the globe. It possesses a greater proportion of the nutritive principle than either of its congenèrs, or than any other vegetable production known to us. Careful experiments have been made with a view to practical results, and tables prepared showing the nutritive properties of plants used as food, either for man or for the domestic animals. It is a curious, as well as interesting subject of inquiry, and one which should engage the attention of every intelligent farmer. It lies at the very foundation of agricultural improvement.

The knowledge we have gained of the food of plants exhibits one of the happiest applications of chemistry to vegetable physiology, and its investigations have opened new views in that great department of creation, leading, with other recent discoveries, to the conclusion that there is a striking analogy existing between the two living kingdoms of nature, and that simplicity is no less an attribute than wisdom in the works of Almighty power. That some soils are more favorable than others to the production of plants, and that their growth may be stimulated by adventitious substances, are facts that experience must have taught in the very earliest stages of husbandry. But it was reserved for a late period, for our own indeed, to give us a reasonable insight into some of the principles of vegetable organization, which regulate the germination, the growth and the decay of plants, and which enable us to understand how we may best aid the operations of nature. We know there are organs performing their assigned functions, common to all vegetation, and essential to its growth and development. Having ascertained that plants require food, and that they possess proper organs to prepare it, and to convey it to its destined work, where it is incorporated, by some process inscrutable to us, into a new body, we have a way opened for an inquiry into the nature and properties of this nutrimental matter, and of the practical application of the principles evolved to the purposes of the farmer. The functions of vegetable life are admirable and admirably performed, and are forever in operation, unchanged and successfully, upon a world of vegetation, which, in its vast extent and variety, and in the countless subjects it contains, beyond the faculties of man to conceive, is among the most wonderful displays of creative power. Within the range of personal vision, how vain the effort to conjecture even the number of plants that start into life and then disappear, to give place to their successors in this ceaseless round of creative organization. But who shall count, who shall even dare to imagine, how many individuals compose that mighty mass of vegetation which covers the face of the earth, and penetrates far into the recesses of the sea? It is not given to man to enter even the threshhold of such a work. It is before us and within our reach, but forever inaccessible to us. By observation of the differences appreciable in the structure and properties of plants, botanists have succeeded in arranging them into separate classes, by which they are brought, in some measure, within our grasp. These classifications are, however, artificial, though founded on natural phenomena, and are designed to introduce order into the consideration of a vast subject, and thereby to facilitate our acquaintance with it. The roots, the body, the bark, and the leaves of plants, and the flowers and fruits in the season of fructification are vegetable productions with which we are all familiar. Much of the internal structure, the wood, the pith, the sap, and the tubes are equally well known, and some of their functions have been long ascertained, or rather conjectured with reasonable certainty. But it is only recently that much progress has been made in this branch of natural history, and that we are now beginning to understand, not, indeed, the organic laws of vegetable life, but many of the functions of vegetable organization, and the processes, by which the great work of production goes on.

The ascent and descent of the sap, analogous to the circulation of the blood in the human system, the vessels through which it passes, the functions of the leaves, resembling those of the lungs, by which the vital fluid is divided between two sets of vessels, one transpiratory to exhale its watery parts, and the other secretory to conduct the residue, after having undergone the necessary change, through the proper vessels to continue its prescribed operations in the economy of nature, the presence of extraneous matter, of silicious substances, for example, in the epidermis of plants, all these and other discoveries have enlarged the sphere of our knowledge, and have given us satisfactory conceptions, instead of vain conjectures, of many operations in the progress of vegetation. They promise yet brighter rewards for future exertions. The elaboration by plants of different substances contained in the sap, and posseseed of various properties, is truly a wonderful process. We have no conception how the several vessels select from the same materials the proper alimentary matter, and reject what is unfit, nor how these elements enter into the new body, and finally offer their services to man in a new form. Investigations into the cellular structure of plants hold out the prospect of important accessions to our present knowledge of vegetable physiology, connected as they are with some of the most difficult questions concerning the organization and growth of vegetation. The cells, it is now conceded, form the basis of all plants. Their minuteness and the rapidity of their development are equally beyond our comprehension, for it is estimated that they average but 1-500th part of an inch in diameter, giving more than one hundred millions to every cubic inch, and that in some of the fungi they are generated at the rate of fifty-six millions in a minute. Such estimates are far beyond the grasp of the human faculties, producing shadowy impressions rather than adequate conceptions; but vaguely conjectural, as they must be, they are yet sufficient to satisfy us, that we move in a world of miracles, from the cradle to the grave, not one of which perhaps it is given to man fully to comprehend, and that the wonderful processes by which this mass of vegetation exists and is maintained are governed by laws that lie far beyond the present boundary of our knowledge. And these minute vesicles are membranous, each shut out from all others, and in plants of the lower order, both absorbing and assimilating, while in those of superior structure some of the cells absorb the nutriment, while others incorporate it into the body of the plant. The mind is lost in the effort to conceive the number of these microsopic agencies, even in a single herb, and yet they are the materials of which the whole vegetable creation is constructed. And each of these ministers of Almighty power, invisible to the naked eye, has a kind of independent life, with its own proper functions, the whole forming together, in some mode inscrutable to us, the life of the plant and giving to it its ligneous substance, ond thus, while it is constituted a single body, it contains within itself an infinite number of separate agents, efficiently but mysteriously operating together. Countless are the sands upon the sea shore, swept by the advancing and receding tides, but ever barren and indestructible; but what are they to this wondrous multitude of living organs, hundreds of which a grain of sand would cover, swept by a tide that ebbs not, the onward progress of creative power, coming into existence and passing from it, to be forever replaced by kindred tribes, thus calling death from life, and arresting destruction by never-ending reproduction?

And thus is the earth covered with EVERY GREEN THING; and with this uniformity of structure, all the diversified forms of vegetable organization, almost infinitely various, are brought into being and made subservient to the purposes of man. Among these gifts of nature is our sugar maple, and elsewhere is the Caoutchoue tree, yielding that most useful substance, India rubber. The poppy, whose concrete juice is opium, an important article in our materia medica; the Ajuapar, whose sap furnishes an active poison, used to impregnate the rivers in order to obtain the fish; the Bambusa Guaduas, which sometimes reaches the height of one hundred feet, and whose stem is hollow and divided by joints at short intervals, each of which contains pure limpid water, invaluable to the traveler in hot arid regions; and the cow tree, a remarkable production, which supplies a milky juice similar in its properties to the milk of animals, and extensively used by the inhabitants, where this beneficent production abounds. But the whole vegetable world offers no subject more worthy of contemplation than the Cocos Mauritia, a tree found in the tropical regions of South America. Its green shoots serve as aliment, and it furnishes bread and wine and oil and fruit, and materials for clothing, mats, hats, and sails for ships, and for dwellings. A wonderful illustration of the power of nature to evolve substances so numerous and so various by the ordinary process of vegetation, from the same alimentary matter which everywhere furnishes the nutriment of plants. The application of the principles deduced from the researches of scientific observers is not less important to the purposes of the practical farmer, than is the increased knowledge of the beautiful system of vegetable life to his intellectual advancement. That plants require proper food for their growth, and for the ultimate development of their properties, and that this food is secreted and sent on to its work by vessels variously constituted for the different functions they have to perform, and that this organization is not merely mechanical, operating by chemical affinities and laws, but physilogical, embracing a principle of vitality, low indeed in the scale of living being, but mighty in its extent; these facts, now well established, and daily opening more and more to us, are intimately connected with the whole process of cultivation. "Stones grow," says Linnæus in his gradation of inorganic and organic forms, "plants grow and live, and animals grow, live and feel." We know that where the nutrimental principle is most abundant, and judiciously applied, there it will produce a corresponding effect upon the body that imbibes it. We are thus brought to a consideration of the nature and condition of the soil and of the properties and preparation of the materials for manure, the most important part of practical agriculture. It is a subject into which I have not time to enter. I touch but the most general outline. The system of creation is one of life and death, following each other in never-ending succession; a system of production, of destruction, and of reproduction. It were idle to speculate upon the final cause of this ordination of nature, and worse than idle to endeavor to investigate the purposes of the Creator in this infinite multiplication of organic being, performing certain assigned operations, and then disappearing to give place to their successors. From the flower, that blooms but to die, to the giant of vegetation, the monument of the ages that have swept over it; from the insect which sports its brief hour in the sunshine, fulfilling the duty imposed upon it, and then vanishes from existence, to

> "that sea-beast,
> Leviathan, which God of all his works
> Created hugest, that swim the ocean stream,"

and to man, the visible head of this wonderful creation, to whom was assigned "dominion" over it, all, all proclaim the universal decree, THAT AS THEY CAME FROM DUST, THEY MUST RETURN TO DUST. When, we know not, but we know it will be soon, for brief is the interval, even when most extended, which separates us from this momentous change, ordained by Almighty wisdom. But the most superficial observation shows us, that a principle, looking to the greatest share of enjoyment by sentient beings, rational and irrational, compatible with their condition, and to its greatest diffusion, by the multiplication of individuals, pervades the system of nature; and a beautiful manifestation it is of the power and goodness of God. If death comes to all, it is but to give place to renewed life, thus fulfilling the great law of existence. We know but little, next to nothing indeed, of the true elements of matter. Those are termed such which baffle all our efforts at decomposition.

At the commencement of the present century there were twenty-nine of these elementary bodies, and now there are sixty-three, showing the great progress of practical chemistry. Their number changes with our chemical knowledge; but we must avoid the error of supposing, that their enumeration makes known to us the true primordial substances, which constitute the basis of the material world. It is a knowledge we may never reach. But however this may be, and whatever is the principle of vegetable nutriment, experience shows that it is exhausted by the supplies it furnishes, and then becomes useless, and must be replaced by substances in which it abounds. And we have thus the theory of soil and manure; and every agricultural observer should keep it in view, in the preparation of his ground, in the application of fertalizing matter, and in the various operations connected with the growth of the crop. It is by combining observations together that we give them their true value, and learn the more general laws that control the phenomena of nature.

I have already said, that the field of agricultural improvement has no prescribed limits that human vision can rest upon, and we are thus encouraged to extend our researches with rational prospects of success. And as an example of the effect of cultivation, it is stated by Bousingault, that a beet seed weighing but the fraction of a grain, has produced a beet weighing one hundred and sixty-two thousand grains, or twenty-eight pounds. But though we know not how far we can go, we know where we cannot go. We know that we can increase the size and improve the properties of plants by judicious culture, but that we cannot so change their essential nature, as to confound the established order of creation, and to destroy the boundary that divides the various species by new families, called into existence by man. Reason and analogy, and universal experience from the earliest periods, teach us that our efforts at melioration should be confined within their legitimate limits, the attempt to improve the qualities of existing plants and animals, and not to invade the province of the Creator, by rash endeavors to multiply, with the faculty of reproduction, the forms of matter, endowed by Him with life. If we have reason to be proud of the advance of modern knowledge, we have reason to be humble, when we see the presumption with which it is too often applied. Not content with the proofs of an intelligent Creative power, which accompany us from the cradle to the grave, and which revelation and experience equally announce, we are seeking with blind rashness to reject the true origin of this great scheme of Almighty wisdom, this union of mind and matter, and to find some kind of fortuitous creation, some shadowy plan of progression, by which the most imperfect organization gradually rises in the scale of being, and step by step, spontaneously and by its own inherent force, ultimately assumes the most perfect forms of animal life.

There is a tendency, not to be misunderstood, towards a cold and heartless materialism in many of the physical investigations of the present day, and there is a tendency, equally obvious in moral investigations, to be carried away by new and strange doctrines, one of which teaches us, as its Hierophant announces in a work, captivating for its boldness and novelty, that all *distinction between physical and moral is annulled*, that "grades of mind, like forms of body, are mere stages of development," and that *there is no essential difference between man and beast.*

And unfortunately the restraining power of the Christian religion upon the hearts and minds of men has been weakened by a similar spirit of presumptuous research, neither directed by wisdom nor conducting to truth. From the great storehouse of German metaphysical theology, where dialectic dexterity has more votaries than a simple and earnest spirit of enquiry, strange words have issued, *rationalism and super-rationalism, naturalism and super-naturalism, supernatural rationalism and rational super-naturalism, transcendentalism*, and many a kindred weapon of controversy, well calculated to impose upon the human judgment, and to lead captive the human imagination by the assumption of learning and by puerile legomachies: all these and similar verbal subtleties, worthy of the disciples of the Stagyrite, belong to the various schools of theology, which convert the gospel of Jesus to their own views, rejecting the true scheme of salvation, or accommodating it to the corrupt heart of man. The Christian religion becomes a myth or fable, and its Divine author a fanatic or an imposter, who healed the sick by interposing at a favorable crisis of the disorder, when nature had herself commenced the cure, and who raised the dead by calling them to life, at the very moment of recovery from the effect of a cataleptic attack, and whose other miracles were performed by similar coincidences, or not performed at all. And our faith is to be shaken in the life and death, and merits of the Savior, and the hopes of his promises to the living and their consolation to the dying, are to be sacrificed to these dreams of a morbid imagination, which belong to a world of peculiar ideas, and not to our world of action, and which have no sympathy with human nature, and no prospect beyond the narrow boundary of physical existence.

Let no man delude himself with the notion, that when he has once mastered, if master he can, these Shibboleths of a barren controversial theology, he is better fitted in head or in heart to study the history of the Redeemer's mission, or that these and other "great swelling words"—

> "For all this tedious talk is but vain boast,
> Or subtle shifts conviction to evade"—

have enabled him to gain one step forward in the path of Christian knowledge.

The human mind is strangely constituted; forever roaming in the fields of enquiry, it is too often disposed to push its investigations beyond the boundaries assigned to man, and to believe nothing it does not comprehend. It has been well said of the disciples of this school, that "to be defied to the face by any stiff-necked problem, which this poor universe can produce, is a humiliation to which they are not accustomed to submit." As the nature and attributes of the Deity are beyond their understanding, His existence is beyond their belief. They belong to the class so well described by Pope:

> "Who boldly take the high priori road,
> And reason downward till they doubt of God."

A theory of "development," as it is called, not less bold than startling, has been advanced in a work recently published, and entitled "Vestiges of the Natural History of Creation," and which, according to an eminent Scotch Review, has been received with a "sudden run of favor," while an English Review of equal authority considers this book "the best adapted of all the productions of modern literature to give a right direction to the philosophical investigations of the highest subjects of human interest." This theory authoritatively announces that "the whole train of animated beings, from the simplest and oldest forms, up to the highest and most recent, are then to be regarded as a series of advances of the principle of development, which have depended upon external physical circumstances, to which the resulting animals are appropriate." And thus the *mysteries* of the universe are unfolded by the mystifications of a false philosophy, and men who cannot believe in a Creator, believe in such a creation as this, in spontaneous generation and in the transmutation of species. Verily, scepticism and credulity are very near neighbors in these mortal frames of ours, and the old Latin saw "Credo quia impossible est," I believe because it is impossible, is as true now as it was twenty centuries ago. We have to judge between the Redeemer of man, who tells us that *the very hairs of our head are all numbered,* and that *not a sparrow falls on the ground without our Heavenly Father knowing it,* and the Prophets and Neophytes of this new faith, who assert that "the creation of a lower animal is an inconceivably paltry exercise of Almighty power." And we are called upon to believe, that our progenitors were monkeys, and that our posterity will be angels, or something higher in the order of nature, by successive transformations, the result of "a creation by law," by which "organic life *presses in,*" (or in other words creates itself,) *"wherever it has room and encouragement and accommodates its forms to suit the circumstances,* (this is no less clear than satisfactory,) and "a *speradequacy* in the measure of this *underadequacy" would enable a goose to give to its progeny the body of a rat,* and thus "one species gave birth to another till the second highest gave birth to man." And we are also told, that our horses were vast *Packy derata,* (names are things, said Mirabeau, and when they are not, they are too often their substitutes,) huge, half formed, living monsters, stretching like the fallen Archangel of Milton, "many a rood," with three toes and no claws, eating herbs as strange as themselves, and that they will march steadily onward in the scale of being, perhaps till they become able, like the winged horse of the ancient mythology, to carry us through the fields of air, instead of the fields of earth, with improved powers and enlarged faculties, suited to a higher sphere of action: and that wheat was once a fern or sea-weed, and will in time become an ambrosia-bearing plant, as far exceeding its present condition, as it is now advanced beyond its original prototype, fitting food, it may be, for the human race, when it has taken its higher position in that ascending series of progression and development, to which it is destined by this consolitory system of psycology. Till that era arrives, thus foretold by this new faith and foreseen by these new prophets, our agriculture will be guided by the true principles of observation and induction, and without the vain and impious effort to break down the barriers established by the Creator, and to usurp His province in the government of creation.

Reason and revelation equally assure us, that the Great First Cause is every where, and every where and always in operation, either primarily and by its own agency, or secondarily and by the agency of laws it has established. And it is not given to man to comprehend where the one ends or the other begins, or how the system of causation commenced its work, or still goes onward with its task:

"Earth, on whose lap a thousand nations tread,
And Ocean brooding his prolific bed,
Night's changeful orbs, bright stars and silvery zones,
Where other worlds encircle other suns,
One mind inhabits, one diffusive soul
Wields the large limbs and mingles with the whole."

The fossil bones of antedeluvian monsters in rocky strata and their footmarks upon indurated clay, strange hieroglyphics upon the monuments of nature, written in a language which man is striving to decipher, have been resorted to in proof, or in illustration of this scheme of creation, equally contradicted by the Book of God's Word, and by the Book of his Works. There are far better "vestiges of creation" in His footmarks around us, in His foot-steps upon the ocean and the land, impressed upon the whole organization of Nature, than these researches, too often merely conjectual, into the remote condition of the globe, can furnish. They are interesting, as are all the facts connected with the natural history of the planet we inhabit, but worse than worthless, when used to shake our faith in the revelations of Christianity, or in the attributes of its divine founder.

I have presented for your consideration, for your co-operation, indeed, various sugestions connected with the advancement of agriculture; but far beyond these in influence and importance, is the advancement of the agriculturist; the education, sound, practical and enlarged, of that vast body of our youth who form, and are to form, the farming interest of our country—an interest that embraces more than one half of our population, and a still greater proportion of the permanent influence to which our social and political institutions must look for support in those periods of their trial, which have heretofore come upon other nations and have come upon us. The cultivator of the soil is engaged in one of the noblest occupations that belongs to the whole circle of human employment. In replenishing the earth and subdueing it, and in multiplying every herb bearing seed, and every tree in which is the fruit of a tree yielding seed, all of which were given to man "for meat," before he left his primitive residence, where God first planted him. He deals with organic life, with its production, its improvement, its multiplication with the means of subsistence for that great family of rational and responsible beings which "has dominion" over all that the earth brings forth, as well as over every living thing that moveth upon it. His existence does not pass in crowded cities, the works of man, surrounded with the physical and moral ills, which a dense population is sure to bring with it. He walks abroad among the works of God, reading the great Book of Nature, whose every page is filled with lessons of wisdom, written in characters that no man can misunderstand, *but the fool that saith in his heart there is no God.*

The light that shines, the wind that blows, the rain that falls, the phenomena of nature, are the companions of his daily walks and works, not mere objects of curiosity or even of contemplation, indifferent or interesting, as he neglects or observes them, but ever active agents in the progress of production, co-laborers with himself in the domain of Nature, performing the functions assigned to them, "in seed time and harvest, and cold and heat, and summer and winter, and day and night," which we are told by Him who knoweth and ordaineth it, *shall not cease while the earth remaineth.*

The work-shop of the farmer is not a narrow and heated room, shut out from light and air—but broad fields and an open sky are the witnesses of his labors; and it is not mere inert matter that he deals with, calling into exertion his mechanical powers only, but one of the great kingdoms of living nature, furnishing subjects of ceaseless observation ond wonder to the highest intellect, and forever inviting the researches of man, as well by the enlarged views it presents of great natural operations, as by the effect of this increased knowledge upon the heart and the understanding, and by the rewards, which are sure to follow the exertions of the enllghtened cultivator.

From the hyssop that springeth out of the wall, to the cedar of Lebanon, from the lowliest plant that creeps into life, to the giant of the forest that rears its head above a sea of vegetation, resisting the winds of Heaven for centuries, there is a mighty mass of organized forms endowed with a principle of vitality, which proclaim the power of God, and invite the researches of man. Wondrous are its extent, its variety, the laws of its being, the purposes it fulfills, the mode of its production, its existence and its reproduction, and the admirable organization by which its functions are to be performed, and inorganic matter converted into its beautiful foliage, which covers the face of the earth, rejoicing the eye and the heart, and ministering to the wants of sentient creation. And the life of the farmer passes in the midst of this great family of nature. It is his daily care to cultivate, to increase, to improve those branches of it which are the most necessary for human comfort and subsistence; and it should be his daily pleasure, as it is his duty, to observe the processes of vegetable life, the habits of plants, and the laws regulating their organization, that he may know how to *make the earth bring forth by handfuls*, like the seven plenteous years of Egypt, and still meliorate his practice, as he extends his knowledge. Who does not see that there is scope enough for the most powerful intellect, the most enlarged understanding? The practical study of the works of Creation, admitting the application of advanced science, as well as the highest powers of personal observation; and yet since the earliest period, indeed, since the acorn gave place to wheat, as the principal article of subsistence, a delusion has been propagated, not universal indeed, for there are honorable exceptions, both in ancient and in modern days, but far too general, and so firmly maintained, that even now it exerts a powerful influence, and is but slowly yielding to the intellectual progress, which marks the age in which we live. The Book of Ecclesiasticus, though excluded as apochryphal from the canon of scripture, is of ancient origin, and no doubt depicts truly the customs and opinions of the Jewish people. It ministers to this mischievous prejudice, and presents a melancholly picture of the intellectual condition of the Hebrew husbandmen:

" How can he get wisdom, that holdeth the plow, and that glorieth in the goad; that driveth oxen, and is occupied in their labors, and whose talk is of bullocks?"

" He giveth his mind to make furrows, and is dilligent to give the kine fodder."

And so because a man is brought by his daily occupations into contact with the world of spirits around him, with organic life, animate and inanimate, the noblest works of the Creator, it is asked emphatically "*How can he get wisdom?*" as though the best school for its acquisition were not an enlarged sphere of observation and reflection. It is not difficult to diseover, in the progress of social and political institutions through the world, the true source of this deplorable error. Labor is dishonored and discredited when it brings degradation, legal or conventional, when the mark of contempt is upon the forehead of the laborer, and he feels, and all feel, that his place is considered the lowest in the scale of human employment. Thank God this state of things is unknown in our country, but it has extended beyond its immediate circle by the prejudices respecting the education of those engaged in agriculture, which it has created and fostered. Where the owner of the soil and the tiller of the soil are seperate classes, divided by legal or social barriers, where the institutions of a country operate to accumulate land into masses, still augmenting as wealth increases, and repress all tendency to distribution by laws of primogeniture, and by the other legal machinery by which the strong are made stronger, and the weak weaker, where the many sow without reaping, and its few reap without sowing, who can wonder that the intellect is without cultivation, as industry is without encouragement? The wretched institutions of the middle ages, by which power and property were wrested from the mass of the people, from almost the whole people, indeed, and concentrated in the hands of the feudal nobility, and were guarded by the iron hand, and by the sterner rule of legal exactions and penalties, made part, an essential one, too, of the political condition of Europeon nations, and yet survives in full force in some of them, while in others they have yielded to the march of events, but in all, their impress is marked upon the constitution of society, and exerts a powerful influence upon the whole social system. And thus was degradation brought upon the laborer, and disgrace upon his occupation, and the impression was left as a legacy, which time alone can remove, that he who tills the soil pursues an employment which, if not totally inconsistent with any advanced state of mental improvement, is better without it, and should oontent himself with *holding the plow and talking of bullocks*,

"Strong as his ox, and ignorant as strong."

Many deep rooted prejudices have passed away in this country, and many others are destined to follow them. There is no miracle like that of old to turn the shadow backward, which marks the progress of these events. This prejudice against the necessity of education, of an enlarged education for the agricultural class of society, though it has not wholly disappeared, has waned before the light of knowledge, and will, ere long, be remembered but among the perversions of the human intellect. In no country on the face of the globe are there such motives as here, for the early diffusion of information among the great body of our fellow citizens. Land is open to all. There are no barriers which guard it against the approach of honest labor. He who does not acquire it, is prevented only by himself. He either does not desire to live the life of a farmer, or he is wanting in that industry and good conduct which open all the avenues of property to all who seek them. The relation of landlord and tenant in the cultivation of the soil, if not unknown, is so limited that it creates no social castes divided by those conditions, and almost necessarily antagonistic to each other. Our boundless public domain invites industry by the facility of acquistion, and rewards it by such advantages as were never before offered to those who seek a moderate competence with personal independence.

But there are still graver considerations, which are connected with this subject, and which force themselves upon our attention with peculiar interest at the present time. I need not tell to an American audience that the signs of the political atmosphere are portentous and alarming. The cloud that was no bigger than a man's hand, like that seen by the prophet from Mount Carmel, has overspread the heavens, and threatens to burst in ruin upon our country. I am dealing with great national facts, and not with questions of party—with the dangers that encompass us, and not with the causes that produced them. That the attachment to the Union is weakened, almost broken indeed, in large portions of the Confederation, he who runs may read in all that passes around him. Our day of trial is come, as theirs came to our fathers. If we meet it as they met it, in a spirit of patriotism and conciliation, we may strengthen their work by our own, and transmit our great and *goodly heritage*, our hope and the world's example to those who are to follow us—the most magnificent legacy, after the religion of God, that was ever bequeathed by one generation to its successors, since human governments were instituted. And upon the intelligence, and virtue of the country, and especially of the farming interest, which constitues so large a portion of it, must we rely for that active and ardent patriotism, which is the more devoted the more perilous is the crisis; which looks to the claim of all while maintaining the rights of each, steadily seeking in the Constitution the true principles of action and forbearance, and temperiug its judgment with that "brotherly affection," to use the language of Washington, without which it were idle to expect, to hope, indeed, that the bonds that unite us can retain us together. Withdraw this power of attraction, and our system will soon be broken up, leaving its members to wander in uncertain space, or to form new combinations with their own elements of destrrction, and with a similar fate before them. God grant that we may be wise in time, and that for ages hereafter the American farmers, from the St. Croix to the Pacific, wherever situated, or whatever the products of their agriculture, may come together, as we have come to-day, assembling, not upon the battle grounds, but upon fields of husbandry,—not to contend in blood for political supremacy, aided by improvements in the art of death, but to compete together in a spirit of emulation, and not of enmity, for the advancement of the ART of life, in the production of human subsistence, that, in the words of the Patriarch of Israel, "we may live and not die."

TREASURER'S REPORT.

The Michigan State Agricultural Society, in account with J. C. Holmes, acting Treasurer.

	DR.	
1850.		
Dec. 13.	To cash paid J. Brown, expenses attending meeting of the executive committee, Dec. 11th, 12th and 13th,	$11 70
	paid Grove Spencer, expenses as above,	4 90
	paid James B. Hunt, " "	2 00
	paid F. V. Smith, " "	12 00
	paid M. Shoemaker, " "	8 50
	paid Titus Dort, " "	3 00
	paid Walter Wright, " "	11 00
1851.		
March 27.	Paid Secretary on account of salary,	250 00
June 3.	Paid balance due for printing diplomas,	20 00
Sept. 6.	Paid F. Boyt for committee books,	5 00
	paid S. S. Barrows for work on fair ground,	80 00
	paid C. C. Wright on account of medals,	50 25
13.	paid S. S. Barrows for work on fair ground,	80 00
17.	paid G. W. Pattison for printing tickets,	4 00
	paid R. W. King, balance of account, crockery,	47 63
	paid A. Richmond, for binding,	32 75
20.	paid S. S. Barrows, for work on fair ground,	100 00
25.	paid for evergreens for Floral Hall,	13 00
	paid for pumpkins for stock,	4 50
	paid Benj. Chappel, services,	33 50
	paid S. S. Barrows, for work on fair ground,	100 00

Sept. 26.	Paid L. P. Perkins, for services as clerk in Treasurer's office,	$10 00
	paid J. M. Wilcoxson, for services as clerk,	12 00
	paid O. W. Moore, for services as clerk,	12 00
	paid Mr. Church, for services in locating booths,	8 00
	paid Seers Stevens for collecting, and for services as clerk,	18 90
	paid John Palmer & Co.'s acct.,	98 38
	paid postage, express and rail road charges,	65 54
	paid sundries,	14 68
Oct. 9.	paid H. H. Brown, treasurer,	331 97
	cash on hand,	6 92
		$1,452 12

1850.	CR.	
Dec. 12.	By cash received from balance of former account,	$466 12
1851.		
Feb. 28.	By cash received of F. S. Finley,	19 00
June 3.	" " Titus Dort, for members tickets sold at Lansing,	20 50
Sept. 23.	By cash received of citizens of Detroit,	892 50
24.	" " for rent of ground adjoining the fair ground,	54 00
		$1,452 12

—

The Michigan State Agricultural Society, in account with H. H. Brown, Treasurer.

1851.	DR.	
Sept. 29.	To cash paid gate-keepers and supe'tend'ts at fair,	$211 25
	paid R. Starkey, services,	6 00
Oct. 3.	To cash paid H. O'Beirne, services,	11 00
	paid C. C. Wright, engraver,	234 33
	Trowbridge & Co., for forage,	20 36
	E. M. Church, services,	6 00
	J. K. Myrier, do	4 00

Oct. 3.	To J. C. Holmes, salary,	$250 00
	A. McFarren, books, &c.,	30 12
Oct. 11.	services at fair,	31 00
	D. Thompson, forage,	52 80
	labor and police,	49 00
	Detroit Advertiser, for printing,	76 80
Oct. 14.	A. Sheley, lumber,	285 60
	Beard & Co., provisions,	8 30
	Ash & Smith, do	29 31
	John Hamilton, expenses attending meeting of the executive committee, Dec. 11, 1850,	11 50
Oct. 21.	A. Sheley, lumber,	500 00
" 24.	Titus Dort, services,	21 63
" 29.	J. Hull, provisions,	4 00
Nov. 4.	Hovey & Co.,	4 80
	S. S. Barrows, carpenter,	300 00
" 5.	J. Williams, forage,	5 42
	Z. Chandler & Co.,	2 50
	J. Darby, engraving,	3 50
Nov. 11.	C. C. Wright, engraver,	171 05
	C. F. Davis, painting signs,	3 75
	A. Sheley, lumber,	250 00
	C. C. Wright,	1 00
Nov. 15.	B. Wright & Son, for lumber,	200 00
	George Sumner,	3 42
Nov. 25.	Campbell & Patten,	1 50
28.	paid A. Sheley, for lumber,	269 71
	H. De Graff, for hardware,	3 50
29.	D. S. Osborn, provisions,	8 32
Dec. 2.	S. S. Barrows, carpenter,	39 32
	George Heron, services,	15 00
	A. McFarren, books, &c.,	50 00
	Thomas Christian, printing,	9 00
	M. Davis, services at fair,	6 00
11.	paid W. Wright, expenses attending meeting of the executive committee at Jackson, Dec. 9th, 10th and 11th, 1851,	11 00

Dec. 11.	Paid G. C. Munro, expenses as above........	$8 00
	paid James B. Hunt, " "	8 50
	paid Andrew Y. Moore, " "	7 12
	paid Grove Spencer, " "	5 70
	paid Titus Dort, " "	4 00
	paid J. Brown, " "	11 00
	paid W. H. Montgomery, " "	I4 00
	paid J. C. Holmes, on account of executive committee, ...	10 68
	paid J. Owen & Co., account,...............	21 56
	paid premiums,*............................	987 00
		$4,279 35

1851.	CR.	
Sept. 26.	By cash received for tickets at fair,...........	$2,878 67
29.	cash received of State Treasurer,..........	1,000 00
Oct. 9.	cash received of J. C. Holmes, former acting treasurer, ...	331 97
	advances by treasurer,....................	68 71
		$4,279 35

* The whole amount of cash premiums awarded at the fair and for which drafts have been drawn on the treasurer, is $1,308 00, leaving unpaid at the time of this report the sum of $321 00, all of which has since been paid.

CATALOGUE

Of the Library of the Mich. State Agricultural Society, Dec. 31*st*, '51.

Revised Statutes of the State of Michigan, 1838, 1846, two volumes each.

Senate Journal of the State of Michigan, 1839, 1843, 1844, 1846, 1847, 1848, 1849, 1850, eight volumes.

House Journal of the State of Michigan, 1843, 1844, 1845, 1846, 1847, 1849, 1850, eight volumes.

Senate Documents, 1842, 1844, 1845, 1847, 1848, 1849, 1850, 7 volumes.

House Documents, 1847, 1848, 1849, 1850, 4 volumes.

Senate and House Documents, 1843.

Joint Documents, 1842, 1843, 1844, 1845, 1846, 1847, 1848, 1849, 1850, 9 volumes.

Laws of Michigan, 1837, '39, '40, '41, '42, '43, '44, '45, '46, '48, '49, '50, '51, 13 volumes.

Patent Office Report, 1847, 1848, 2 volumes.

" " Agriculture, 1849–50.

" " Mechanical, 1849–50.

Annual Message and accompanying documents, part 1, 1849–50.

Commerce and Navigation, report of the Secretary of the Treasury, 1850.

Journal of the American Institute, New York, 4 volumes.

Transactions of the American Institute, 1842,–43–45, 1 volume.

do do do 1846, '47, '48, '49, 4 volumes.

American Poultry book, by M. C. Cock.

The American Poulterer's Companion, C. N. Bement.
" " Farm Book, Allen.
" " Shepherd, Morrell.
Ladies' Companion to the Flower Garden.
Downing's Fruits and Fruit Trees of America.
American Fruit Culturist, Thomas.
European Life and Manners, Coleman—2 volumes.
Landscape Gardening and Rural Architecture, Downing.
Trees of America, Brown.
Youatt on Diseases of the Horse, with notes by Spooner.
Youatt on the Horse.
Youatt & Martin on Cattle.
Beck's Botany.
The American Poultry Yard, Brown.
The Farmer's Companion, Buel.
Rural Economy, Boussingault.
Allen on the Grape.
Gardner's Farmer's Dictionary.
Coleman's Tour, 2 volumes.
Memoirs of the Board of Agriculture of the State of New York, volume 1, 1821.
Encyclopedia of Domestic Economy.
American Encyclopedia.
Domestic Animals, Allen.
Johnston's Agricultural Chemistry.
Liebig's do do
Diseases of Animals, Cole.
American Husbandry, Gaylord & Tucker, 2 volumes.
Downing's Country Houses, part 1.
Report of Board of Agriculture, State of Ohio, 2 volumes.
Transactions of the New York State Agricultural Society, 2 volumes.
do do Michigan State Agricultural Society, 1849, '50, 2 volumes.
Michigan Farmer, 7 volumes.
Wool Grower, 2 volumes, T. C. Peters, Buffalo, N. Y.
Reports of Agricultural Societies in the State of Massachusetts.

J. C. HOLMES,
Sec. Mich. State Agricultural Society.

REPORTS

OF

COUNTY AGRICULTURAL SOCIETIES,

FOR THE YEAR 1851.

REPORT

Of the Secretary of the Berrien County Agricultural Society, to the Secretary of the Michigan State Agricultural Society, for the years 1850 *and* 1851.

At the instance of a great number of the citizens of Berrien Co., a meeting was called, and held at the court house, on the 22d day of February, 1850, to take into consideration the propriety of forming a Society, having for its object, to promote the improvement of agriculture, manufactures, and the mechanic arts. The society was organized, and on the thirteenth day of March, following, a constitution was framed for, and adopted by the Society. On the 4th day May, 1850, an election of officers for the year took place, to wit: a president, a recording secretary, both ex-officio members of the executive committee, five other members of the executive committee, a treasurer, a corresponding secretary; also a vice president and a corresponding secretary for each township in the county.

The first annual fair of the society was held at the village of Berrien, on the 19th day of September, 1850. It was attended by great numbers of our farming community and citizens generally, and, by the number, variety, and excellent quality of the articles exhibited, gave clear evidence that a spirit of improvement in agriculture, horticulture, and domestic economy, was already, to a considerable extent, awakened amongst our farmers, and in their households.

A very able, instructive and interesting address, was, on that occasion, delivered before the society, by Warren Isham, Esq., editor of the Michigan Farmer, which by its highly practical as well as scientific character, was well calculated to effect the most important and beneficial reforms in the business of agriculture. On the 19th day of September, 1850, the society numbered 90 members.

From the report of the treasurer, made February 22, 1851, it appears that the receipts of the society, for the year ending at that time, inclusive of $150 raised by tax authorized by the law of 1849, amounted to $266.17, and that the expenditures for the same period were $70.88, leaving at that time a balance on hand of $195.29.

The second annual fair of the society was held at the village of Berrien, on the 18th and 19th days of September, 1851. The board of supervisors of this county, had again, generously, as well as wisely, granted a tax of $150 in aid of the funds of the society. An amount considerably more than that of the former year, had been appropriated to pay premiums; a more general attendance of our citizens was had, and a very decided improvement was apparent, in the number, variety, and quality of the articles exhibited.

All which is respectfully submitted.

EBENEZER McILVAINE,
Recording Secretary.

Berrien, Dec. 26, 1851.

ADDRESS

Delivered before the Berrien County Agricultural Society, at their Annual Fair, on the 18th day of September, 1851.

BY E. MCILVAINE, ESQ.

Gentlemen of the Berrien County Agricultural Society:

Your committee, appointed to procure an address for this occasion, having invited me to prepare and deliver one, I accepted their invitation; and after considering in what manner, by what I might say, I could best subserve the interests of this Society, I have concluded to address you on the knowledge most requisite to the practice of agriculture, and the best sources from which it is to be derived, with such remarks on practical farming as may occur to me.

Agriculture derives the principles on which it is based, chiefly from the applications of those of other sciences, and especially of chemistry, to this pursuit.

Chemistry makes known to us, what distinct elements or principles exist in each of the various substances, which together constitute the natural world,—in all the different earths, metals and minerals; in all the various fluids; and in all aeriform substances—one of the most important of which latter, is, the common air we breathe. It teaches us the peculiar powers of each element or principle, and gives us the law governing and directing all its operations. It shows us that some chemical elements will unite together, and that others refuse to do so;—and further, that, when the nature of the elements brought together is such as to admit of a union between them, they will unite only under certain conditions or circumstances, and in certain fixed and invariable proportions—and that the result of these combinations, is the formation of compounds or substances entirely different from either of the elements which go to form them; that certain of these elements will unite with others in two or more differ proportions—each proportion being at the same time fixed and definite—and thus, by every new combination, compounds will be formed, not only different from their several elements, but just as unlike in their properties, to either of the other compounds made up of the same elements. It further makes known to us, that each element, in its combinations with others, invariably chooses its companion—that when one element is in a state of union with another, on presenting to it a third element for which it has by its nature a stronger preference, it will leave the second, and combine with the third—thus forming a new compound entirely different in its character from the first.

To illustrate this part of my subject, I will here give a few examples: If we bring together one pound of the gas called oxygen, and four pounds of the gas called nitrogen, they will unite and form five pounds of the air we breathe; but if we take two pounds, instead of one, of oxygen, and only four pounds of nitrogen, there will remain one pound of oxygen, which will not combine with the nitrogen, but will remain oxygen still. Now, common or atmospheric air, as we all know, sustains life; but neither of its elements, by itself, possesses this property, and death would inevitably ensue, from breathing either of them in its uncombined state. Indeed, a different, but definite combination of these same elements, would form nitric acid or aquafortis. Again, if we mix in a close vessel eight pounds of oxygyn with one pound of the gas called hydrogen or inflammable air, and apply fire to them, they will unite and form nine pounds of water —a substance entirely different from either of these gases. And again, pure lime has a strong preference, or as it is termed by chemists, a strong affinity for carbonic acid gas, one of the ingredients, although not one of the elements of the atmosphere. By the agency of this affinity, whole mountains of lime have been crumbled, during successive ages, into fertile beds of chalk. But lime has a stronger preference or affinity for sulphuric acid, than it has for carbonic acid —and hence, if chalk be subjected to the action of sulphuric acid, the lime will leave the carbonic acid, combine with the sulphuric acid, and thus form alabaster or common gypsum.

Now, we learn from chemistry, that the mass of matter which composes the whole surface of the earth, including the water and the air, consists of about 55 of these elements or simple substances; and that about 14 of these, constitute the productive power of soils, enter into the composition of plants and animals, and are therefore immediately connected with agriculture. These 14 are oxygen, hydrogen, carbon, nitrogen, chlorine, sulphur, phosphorus, iron, alumina or clay, silica or flint, potash, soda, lime and magnesia, the last 4 of which, are not, strictly speaking, elementary or simple substances; but as they are never found in their elementary condition, they may therefore be conveniently considered as simple bodies.

From the same source we learn, also, that although these simple substances or elements enter to some extent into the composition of all plants, yet, that they exist, in different kind of plants, in very different relative proportions; that some plants contain a large amount of certain kinds of these elements, whilst others contain very little. Wheat, for example, contains a large amount of potash and soda; potatoes and turnips, on the contrary, contain very little. One hundred parts of wheat straw yield fifteen and one-half parts of ashes; the same quantity of barley straw eight and one-half, and of oat straw only four; and the ashes of the three are chemically of the same composition; hence, upon the same field, which will yield only one harvest of wheat, two successive crops of barley may be raised, or three of oats.

And further, we learn that the food consumed by vegetables and animals, is similar in the chemical substances or elements which compose it; that the same substances, out of which are formed all the organized parts of plants, compose the entire organized structure of animals; that by the agency of these elements, existing in the earth, the water and the air, a perpetual supply of nourishment is furnished, first, for the support and growth of plants; and next, for the growth and perfection of animal life; and, that, from the decay of plants, as well as from the death of animal matter, are derived in rich abundance, the means for the existence of succeeding generations.

We also learn that in the vegetable, and equally in the animal economy, some of these simple substances are chiefly employed for one purpose, and others, for another; that some are used to form the solid portions, others, the liquid portions, and others again, the seeds of plants; that some are appropriated to the formation of the harder portions, others, of the softer portions, and still others, to form the liquid portions of the animal structure. The results of experiments made by chemists, prove that the different organs, whether of plants or animals, are composed of these substances in very different relative proportions. Some organs containing a large share of certain of them, and others, very little, but a large share of other and different substances. Thus, the seeds of plants and trees, contain a larger portion of nitrogen, their leaves more potash, and their trunks and branches more phosphate and carbonate of lime. Carbon enters largely into all their solid parts. Phosphate of lime composes the greater portion of the bulk of the bones of animals; and that chemical compound of oxygen and iron, called red oxide of iron, is always present in the blood, is indispensably requisite to its healthy action, and even to the very continuance of animal life. Thus, the science of chemistry gives us a knowledge of the nature and operations peculiar to each of the several substances which together compose all the various organs of the different plants and animals. It further makes known to us the several sources from which the animal and vegetable kingdoms derive these substances. That atmospheric air, composed, as it is, of oxygen and nitrogen, contains two of the most important of these; and that, besides, it has mechanically mixed up with it, others indispensably requisite to the growth and maturity of plants and animals; that, for instance, carbon, in the form of carbonic acid gas, is present in it, and that this gas constitutes about the one thousandth part of its entire weight, which same substance, carbon, united with oxygen, forms carbonic acid, which compound constitutes about fifty per cent of both animal and vegetable matter; that ammonia, a compound formed by the union of hydrogen and nitrogen, in certain definite proportions, and owing to its nitrogen, essential to the growth and maturity of the seeds of plants, is found in it; and that several other chemical substances, equally necessary in the animal and vegetable economy, are found mixed up in it, in the state of gases; that some of these substances owe their presence in the air, to the perpetual decay and decomposition of animal and vegetable matter; others, to the result of constant chemical action going on throughout the inorganic regions of nature; and again, others, to the existence of like substances in the waters of the ocean; and all of them, to the constant evaporation going on over the whole surface of the globe, by means of which they rise into the atmosphere in watery vapour, and carried by the winds over the whole earth, and again restored to it by the rains, in which they are found dissolved. In this manner, sea-water, in immense quantities, rising every year into the atmosphere, contributes the saline matters it holds dissolved in it, such as common salt, muriate of potash, muriate of magnesia, sulphates of soda and lime, together with small portions of the phosphates of lime, ammonia and iron, to give fertility to the soil.

That water is another important source from which these substances are derived, and a very useful agent in their application to the roots of plants; that besides the oxygen and hydrogen, by the union of which it is formed, other valuable substances, existing in the portions of the earth through which it passes, are often found dissolved in the water from springs; and rain water, as has already been shown, has in it ammonia and other elements entering into the constitution of plants. Indeed, water, when impregnated with the carbonic acid of the atmosphere, and aided by those changes of temperature incident to different climates and seasons, or resulting from chemical agency, will decompose rocks, and make them yield the potash, soda, and lime they contain, to fertilize the soil.

Heat, whether caused by the sun or produced by chemical action in the soil, quickens the circulation of the fluids of plants, aids them in taking up and appropriating the chemical elements they require, and thus promotes vegetable growth.

And light is equally essential to plants; when excluded from it, they never require their natural color, their leaves are white or pale, and their juices watery and insipid; when exposed to the influence of light, they decompose the carbonic acid of the atmosphere, absorb the carbon which it contains, and give off the oxygen; but when withdrawn from this influence, and surrounded by darkness, they absorb the oxygen of this compound, and give off the carbon, and hence it is, that house plants poison the air of sleeping apartments.

From Geology, we learn further, that the earth contains inexhaustable supplies of mineral substances of every variety, many of which, through chemical agencies, assisted by great and sudden changes of temperature, and the various influences of climates and seasons, are crumbling in pieces, forming thereby new soils, separating into finer parts those already formed, imparting to them new and valuable chemical elements, and better fitting them to convey and apply to the roots of plants, other elements requisite for their growth, and which it is in the power of the air and water to bestow.

But these natural causes are hardly sufficient to perpetuate the races of the uncultivated plants and wild fruits of the earth, at least not on the same spot of ground; and are manifestly insufficient to furnish the quantity of the chemical substances, or in other words, the nutriment required by plants when improved by cultivation. Wheat, for instance, in its natural state, was a very small grain, and potatoes and turnips, unfit for food. Now, the difference between a plant in a state of nature and a cultivated plant, is not merely an increase of size, but it consists in the more perfect development of certain of its parts useful to man; for which development, more of certain chemical substances are required, than it can, unaided by artificial means, obtain.

The first step, then, for the agriculturist, is to analyze or obtain the results of the analysis of each kind of cultivated plant; and thereby ascertain what chemical elements or compounds it consists of, and the amount of each contained in it, and by comparison of these different amounts with each other, he will find out the proportion in which each element exists in it.

Next, he should in like manner analyze or procure an analysis of the soil of the field he designs for any given plant, and if the result of such analysis shows that this field contains, in its different parts, all the elements or compounds found in the plant, and in a proportion of each equal to that of the same element as found to be contained in the plant, he may safely conclude that this soil is every way fitted for growing in perfection the plant in question.

But, if, on the contrary, his soil should be found to be wanting in any of these elements or simple substances, he may be equally certain that the plant, if attempted to be raised on this field in its present state, will be defective in those of its parts which the substances deficient in the soil go to form. And, in this case, he must either supply by artificial means the substances needed, or he must raise on his field some other kind of plant, not containing of each of the substances composing it, a greater share than the soil possesses. Suppose now, he decides to add to his soil, by artificial means, enough of the deficient substances or elements, to make it correspond with the relative quantity of each, as already found by analysis to be contained in the plant he chosses to raise, what means must he use? He must again consult chemistry, through the medium of works on the application of its principles to agriculture; in these works he will find a list of all artificial fertilizers or manures, with an analysis of each; and from its results there given in full, he can ascertain of what distinct substances each fertilizer is composed, and the relative quantities it contains, and thus he will be able to select and apply with certainty the fertilizer required.

And when, by the application of the proper fertilizer, he has brought his soil to the condition required, and has succeeded in raising on it in perfection the plant in question, he will naturally be led to expect, that the same plant may be raised in equal perfection in the given soil, every year, the seasons being favorable, just so long as all those portions of its chemical constituents, taken away from it in the harvest of one year, shall have been restored, as preparatory to putting in the crop of the next. But here, again, chemistry will inform him, that there exists a natural and insuperable obstacle to such a practice—that every kind of plant and tree, in the progress of its growth and maturity, throws off from its roots, into the earth around it, a kind of matter, which is poisonous to succeeding plants and trees of the same kind; and that, hence, any spot where a fruit tree has once grown, is rendered, at least for some years, incapable of bearing another tree of the same kind; and that, although, in the case of plants of yearly growth and maturity, the deleterious influence from this cause, may not be very perceptible in its effects on them, during the first few years of their successive growth in the same soil, yet that this matter gradually accumulates, until in course of time, the soil in question becomes entirely unfitted for the production of this kind of plant.

To rid himself of this nuisance, he must resort to deep and thorough plowing, thereby bringing up the fresh sub-soil charged with new supplies of chemical fertilizers, separating and exposing this matter to their action, and to the restorative influences of those chemical elements, which either compose, or are found mixed up with air and rain water, as already stated.

But, even if this obstacle did not exist, still, the agriculturist would find it more for his interest, to adopt, in his farming practice, a judicious rotation of crops, especially in this country, where land can be had at a comparatively low price, and where many of the most powerful fertilizers can be had, if at all, only at such high prices, compared with those of agricultural products, as will hardly justify their expense. Let him remember the beneficial influences which his soil is continually receiving from the chemical substances existing in atmospheric air, in rain water, and in his sub-soil; let him bear in mind, that the deeper and more thoroughly he plows and lightens his soil, the more will it be subjected to, and receive of, these beneficial influences; let him not forget that owing to the differences as already stated, in the comparative proportions of the chemical elements composing the different kinds of plants, and therefore requisite for their nourishment, one kind will live and thrive on what another leaves, a second on what is left by the former, and that in the meantime his soil is gradually regaining the chemical elements of which it has been deprived by cultivation; let him consider that each kind of plant has its peculiar tribe of predatory insects, which would accumulate in a succession of the same crop; and finally, let him take warning by the old worn out soils of some parts of Maryland and eastern Virginia, which have been robbed of their fertility and reduced to barrenness, by incessant and alternate cropping with wheat and tobacco, both heavy consumers of the same chemical substances—and he will at once introduce and rigidly adhere to a judicious rotation of crops.

But even this, of itself, unaided by other means, will not be sufficient to preserve a soil in its present state of fertility, much less to increase its productive powers. For this purpose, bearing in mind that whatever has been grown in his soil, whether grain, grass or weeds, has extracted from it, directly, by means of its roots, and indirectly, through the agents of the atmosphere and of rain water, the chemical substances of which it is composed; and that, to restore its fertility, he must return to it as far as possible, what has thus been taken away—let him carefully preserve all his straw, stalks, husks, weeds, and refuse vegetable and animal matter of every kind; let him properly prepare and feed to his domestic animals whatever portions of these they will eat, such as straw, stalks, husks, weeds, and refuse portions of vegetables, and add the rest directly to his manure heaps; let him take proper care, by the addition of muck and surface soil, and by carefully protecting his manures from the wasting influences of the sun and rain, to fix and retain in them the various invaluable chemical substances they contain, either in the form of gases, or in liquids and solids ready, by increase of temperature and under the influence of fermentation, to become decomposed, assume the form of gases, and escape into the air, or liable by the action of solar heat to evaporate, or be dissolved and washed away by the rains—let him, I repeat, take suitable care of these agricultural treasures, and use them judiciously, and whilst he sees others in disregard of this admonition, wasting and destroying these rich and lasting sources of fertility, and starving their soil into barren and desolate wastes by incessant tillage, he will, under such a course of management as I have recommended, find his soil becoming every year richer and more productive.

In addition to these sources of fertility, geology makes known to the agriculturist other resources which soils present, both for their chemical and mechanical improvement. By its aid, he can predict the general quality of the surface soil, and more than this, of the unseen sub-soil in the several parts of entire countries; and if a soil be of inferior quality, and yet susceptible of improvement, he may learn whether the means of improving it are likely, in a given locality, to be attainable at a reasonable cost. For instance, it often happens that very stiff clay soils are found in the immediate vicinity of poor, light, sandy lands. In such cases each will benefit the other, by mere mechanical mixture, thus rendering the clay soils more open for the admission of air to the roots of plants, and the sandy lands more retentive both of water and manure.

But, the agriculturist must not expect to derive from the mere general presence of the requisite fertilizers in his soil, all the beneficial effects to his plants, which these fertilizers are capable of producing. He must recollect, that weeds take up and consume these fertilizers as well as plants, and, if not exterminated, will deprive the latter, in a great measure, of their nourishment; that his plants do not possess, like his animals, the power to go from place to place, and collect the food they require; that they receive their nourishment, in the earlier stages of their growth, chiefly through their roots, and can only get it when it is brought within their reach; that the fertililizing properties of every well prepared soil, are equally diffused throughout its entire extent; and that, in the case of certain plants, for instance, corn, potatoes, rutabagas, field beets and carrots, which, in cultivation, are set at considerable distance apart in the ground, great portions of these fertilizers are too remote to afford nourishment to their roots; and that if he would avail himself of the whole productive powers of his soil, he must begin at an early period in the life of these plants to use his plows and hoes amongst them, in exterminating weeds, in removing from the roots of his plants, so far as may be practicable, those portions of his soil which these plants have exhausted of nutritious properties, and in bringing within their reach other successive supplies of the food they require, contained in fresh portions of his soil; and this he must continue to do until his plants are too far advanced in the progress of their growth, to need his further aid for their successful maturity.

Hence it appears how indispensibly necessary to the agriculturist who seeks to perfect himself in his art, is a knowledge of the scientific principles on which that art is based. Yet geology has been generally regarded as having no connexion with agricultural pursuits. And, whilst in Great Britain, France and Germany, the principles of chemistry have been extensively applied to agriculture with the most signal success, in this country, these principles are as yet very far from receiving that earnest and general attention which their importance demands. Now, as an American citizen, actuated by that pride of country which we all justly feel, I sincerely regret that truth requires of me this admission. And I trust, that all that is necessary to give us that superiority in this, which we have already attained in so many others of the useful arts, is, that the attention of our agriculturists generally should be called to this subject, and that its various most important advantages should become more fully known and more justly appreciated.

Now the means for obtaining a knowledge of these principles and applying them to agricultural practice are within the reach of every one. Plain, practical works on geology, on chemistry, and on the applications of the principles of these sciences to improvements in the various departments of agriculture, with other well written works devoted to the full and familiar explanation of the process of chemical analysis of soil and plants, and containing accurate tables of the results of such analysis obtained in a great number and variety of cases, together with the apparatus and chemical substances requisite for analyzing all kinds of soils and plants, can be procured for a few dollars; and the illustrious example of Franklin, of Burritt, the learned blacksmith, and of a host of other distinguished men, proves what vast treasures of knowledge can be acquired merely by improving in study those leisure moments which every one has, and leaves without excuse those who will not profit by this example.

But mere theoretical knowledge of the scientific principles applicable to agricultural operations and improvements, is not, of itself, sufficient to make a successful practical farmer. Very far from it, indeed. It furnishes the only sure foundation for accurate practical knowledge, and the only safe guide for its attainment; but you might just as reasonably expect a student of law or medicine, whose only professional knowledge is derived from the principles laid down in his elamentary books, to suddenly become an accomplished practical lawyer or physician, as that a student of agricultural science, possessing a mere theoretical knowledge of its principles, should, without experience in their practical applications, at once go forth a successful practical farmer.

Indeed, to success in agriculture, no less than in other pursuits, practical ability in the prompt and correct application of principles, is indispensably requisite. This is to be derived from the thorough study of the best works on the various branches of agricultural practice; from reading the most valuable agricultural newspapers and other periodicals; from carefully considering the cases of practice found reported in these books and periodicals; from a knowledge of the experiments tried by other intelligent agriculturists and their results; and from the agriculturist's own experience. But, let him ever bear in mind, that cases of practice should invariably give all the circumstances under which the experiments they relate were made; the minute particulars of the experiments themselves, and the exact results obtained; otherwise they can be of no practical value; their obvious tendency is to mislead, and he will act wisely in entirely rejecting the uncertain light they afford.

Amongst the means for improvement in practical farming, agricultural societies deservedly occupy a prominent position. These societies, originating in the desire of agriculturists for mutual improvement, have existed many years in the different parts of Europe, and have been formed in most of the more settled portions of the United States. And wherever these societies have exerted their influence for a series of years, on the farming business of a country, old habits and prejudices have gradually, but steadily yielded to it; new and more successful methods of cultivation have been generally adopted; new varieties of plants and fruits, of superior excellence, and better adapted to the climate, soils and seasons of that country, have been successfully grown; improved breeds of domestic animals have been introduced and raised with increased profit; and the whole aspect of that country bears abundant testimony to the increased prosperity of the enterprising farmer. In it, well conducted agricultural schools have, in a great measure, taken the place of other seminaries of learning; and even old establisded Colleges and Universities have found it necessary, in order to retain their influence and support, to add to their law and medical departments, another, under competent professors, to instruct their students in the principles of Scientific Agriculture.

Rightly understood and practiced, this profession yields to none in importance, honor and profit; the illustrious Liebig has well said, that there is no profession which can be compared in importance with that of agriculture, as to it belongs the production of food for men and animals, and on it depends the welfare of the whole human species, the riches of States, and all commerce.

And this beautiful region, possessing as it does great natural advantages of soil, climate, timber, running streams, navigable waters, and internal improvements, needs only the general and skillful application of the principles of scientific agriculture, to perpetuate its fertility, develop its resources, and thus furnish rich and never failing sources of wealth to its enterprising inhabitants. I say enterprising, for actuated by a settled purpose and reasonable prospect of improving their condition in life, they voluntarily gave up the numerous comforts and conveniences of older and more settled places, and came and cheerfully submitted to all the hardships and privations of a new country. And here let me add, the energy of character which they have displayed, in clearing away dense forests, and converting the soil they once occupied into those beautiful farms which meet the eye in all parts of this country, merits the highest praise.

Now all that is wanting in order to effect the most rapid advancement in agriculture, is to direct this same spirit of enterprise, and this same energy of character, to improvements in this pursuit.

And an occasion like the present is well calculated to awaken an earnest and general desire for this improvement. Here are exhibited the best products of our agricultural industry; here an opportunity is afforded for ascertaining the methods pursued by each of the competitors for premiums; here each one can see his own deficiencies, and learn the best methods of remedying them; and he will naturally be stimulated to renewed efforts for his own improvement. And thus will the exhibitors of each annual fair surpass those of the preceding; and this county will soon, by its rapid progress in agricultural improvement, give satisfactory evidence of the beneficial effects to be derived from societies of this kind.

And now, gentlemen of this Society, in conclusion, allow me to add, that in forming this association, you have commenced a good work; you have laid a sure foundation for the prosperity of your county; go on as you have begun, and success will reward your efforts. And when, in future years, you shall come together on occasions like this, you will find in the evidences of steady and general improvement, afforded by the superior number, quality and variety of the agricultural products exhibited before you, abundant cause to rejoice in having, by your united efforts, established and sustained an institution productive of such great and good results to all within the sphere of its influence.

REPORT

Of the Calhoun County Agricultural Society.

LIST OF PREMIUMS

Awarded at the Third Annual Fair of the Calhoun County Agricnltural Society.

HORSES.

Best stallion, W. Johnson.
Second best stallion, M. H. Crane.
Best brood mare, L. L. Downs.
2d do Paul Broat.
Best team of draught horses, S. G. Pattison.
2d do do Devillo Hubbard.
3d do do M. Mann.
Best 3 year old colt, A. L. Hays.
2d do Gilbert Knapp.
Best 2 year old colt, Paul Broat.
2d do Stephen J. Smith.
Best 1 year old colt, Abram Bennett.
2d do Clark Brockway.
Best sucking colt, Levi Eaton.
2d do H. D. Hall.

CATTLE.

Best grade dairy cow, S. G. Pattison.
2d do S. Mason.
3d do C. T. Gorham.
Best native dairy cow, James A. Way.
2d do Philo Dibble.
3d do Leonard Cleaveland.
4th do James Connelly.

Best bull 2 years old and over, S. G. Pattison.
2d do do James Connelly.
3d do do H. Polhemus.
Best yearling bull, Thomas Johnson.
2d do Lemuel Downs.
3d do John Hadden.
Best yoke working oxen, Azar Rowley.
2d do J. Connelly.
3d do Josiah Freed.
Best pair 3 year old steers, L. Maynard.
2d do do J. Otis.
Best 2 year old steers, A. D. Morton.
Best 2 year old heifer, Seldon H. Gorham.
2d do C. B. Turner.
Best 1 year old heifer, Philo Dibble.
2d do S. G. Pattison.
3d do Abram Bennett.
Best bull calf, Lemuel Downs.
2d do Leonard Cleaveland.
3d do J. Stiles.
Best heifer calf, James Connelly.
2d do Frank Dickey.
3d do J. C. Frink.
Best yoke fat oxen, L. Maynard.
Best 5 yoke of oxen, Town of Fredonia.

SHEEP.

Best full blood merino French buck, J. Otis.
do Pauler merino buck, Joab Polhemus
2d do do H. Polhemus.
Best grade buck, Z. Tillotson.
Best 5 ewes, Franklin B. Wright.
2d do do
3d do do
Best lambs, F. B. Wright.

SWINE.

Best boar, Azar Rowley.
2d do I. B. Woodcock.
3d do Elias Alley.

Best breeding sow, James Connelly.
2d do Wm. J. Thornton.
3d do Azar Rowley.
Best pen of 5 pigs, under 6 months old, I. B. Woodcock.
2d do do do Wm. J. Thornton.
3d do do do Azar Rowley.

DAIRY.

Best sample of butter, Lewis Wilson.
2d do Eleanor M. Chisholm.
3d do Julius Wright.
Best 20 pounds of cheese or over, Thomas Chisholm.
3d do do David Aldrich.
Butter 2 years old, Theron Hamilton.

GRAIN.

Best acre of corn, Cornelius Orsborn.
Best bushel of winter Wheat, Thomas Knight.
2d do do J. Connelly.
Best 12 ears dent corn, Geo. Hentig.
do flint corn, Jeremiah Brown.
do sweet corn, do

ROOTS.

Best one-half acre of potatoes, James H. Kerr.
2d do do Thomas Chisholm.

FRUITS.

Best variety of apples, Joseph Otis.
2d do Levi Eaton.
Best 10 pounds of grapes, John Faulkner.
2d do do O. C. Comstock, jr.
Best variety of pears, Dr. Hudson.
2d do Thomas Knight.
Best 12 quinces, C. P. Dibble.
2d do Joseph Otis.
Best sample peaches, Hr. Hudson.
2d do R. E. Hall.
Best sample of plums, O. C. Comstock, jr.

GARDEN VEGETABLES.

Best sample of cauliflower, George Woodruff.

Best 10 stalks rhubard, J. Brown.
" 3 heads of celery, Robert Gardner.
" 3 heads of cabbage, C. P. Dibble.
" 5 carrots, F. Danforth.
" 5 beets, Granville Beardsley.
" 5 rutabagas, F. Danforth.
" 5 parsnips, O. C. Comstock, jr.
" 5 salsify, Robert Gardner.
" 10 onions, Asahel Hawkins.
" 5 tomatoes, James A. Way.
" 3 squashes, Daniel Woolsey.
" 3 pumpkins, Thomas Chisholm.
" 10 peppers, J. A. Way.
" 2 quarts of Lima beans, O. C. Comstock, jr.
" ½ bushel white beans, George Hentig.

DOMESTIC MANUFACTURES.

Best 10 yards full cloth, C. Mason.
" wool carpet, R. Church.
" rag do do
" pound stocking yarn, C. Mason.
2d do do Thomas Chisholm.
Best pair wool stockings, F. C. Rathbone.
" do socks, Mary D. Chisholm.
Fringe gloves, John Harris.
" mittens, F. C. Rathbone.
Best made shirt, Mrs. McDonald.
" sample of needle work, Mrs. J. A. Way.
2d do do Miss Hahn.
Best sample of worsted work, Mrs. I. E. Crary.
2d do do Mrs. J. A. Way.
Best sample of crotchet work, Mrs. J. A. Way.
2d do do Jeremiah Brown.
Best straw hat, F. C. Rathbone.
" bed spread, C. B. Webster.
2d do P. Dibble.

Best parlor and cook stoves, Smith & Pratt.
" bee-hive and box of honey, J. Wright.
" worsted stockings, Ida Polhemus.

POULTRY.

Best pair of turkeys, George Hentig.
do geese, John Potter.
Best and greatest variety of fowls, John Potter.
2d do do Lewis W. Tillotson.
3d do do Mrs. McDonald.

FINE ARTS.

Best 10 daguerreotypes, A. Kendall.
" oil painting, Miss Churchill.
" 5 drawings, Miss Agnes McClure.
2d do George Woodruff.

FLOWERS.

Best boquet of flowers, Miss Susan Jones.
2d do do do
3d do do do
Best and greatest variety of dahlias, Jeremiah Brown.
2d do do do J. J. Bardwell.
3d do do do O. C. Comstock, jr.
Best dahlia and best seedling dahlia, Mrs. J. A. Way.
" and greatest variety of green house plants, Mrs. Moore.
do do flowers, Jeremiah Brown.
2d do do do O. C. Comstock, jr.
Best and greatest variety of roses, Mrs. J. A. Way.

FARM IMPLEMENTS.

Best farm wagon, G. Hentig.
" ox-cart, C. T. Gorham.
" plow, J. Connelly.
" cultivator, Geo. Hentig.
" ox yoke and bows, J. R. Hendryx.
" grain cradle, Wm. H. Kerr.

BOOTS, SHOES AND HARNESS

Best pair of fine and coarse boots, Charles Eberstein.
" set of farm harness, Ladd & Mills.

Best pleasure harness, W. P. Sutton.
" single do do
" saddle, Joab Polhemus.
" halter, Ladd & Mills.

COOPERAGE.

Best flour barrel, J. Conklin.

BREAD.

Second best loaf of bread, David Aldrich.

PLOWING MATCH.

Best specimen of plowing with horses, L. L. Downs.
2d do do do
Best do do oxen, Thomas Knight.

MISCELLANEOUS ARTICLES.

Best divans, book-case and bed-stead, Beach & Card.
" hearth rug, Mrs. H. N. Banks.
" Indian boxes, William Pendleton.
" painted spreads, Joab Polhemus.
" rug and mats, Mrs. H. D. Hall.
" knit shawl, Mrs. Jeremiah Brown.
" grained door, John Hodge.
" 3 hearth rugs, Silas Osborn.
" sausage cutter, T. Bement.
" sausage stuffer, H. D. Hall.
" churn, E. Woodruff.
" basket potatoes, P. Dibble.
" force pump, Benj. Estes.
" worsted flowers, A. Thompson.
" moss basket and mat, Mrs. C. F. Haskell.
" barrel of flour, T. F. Craigg.
" monochromatic drawings, James Monroe.
" buggy, Seymour & Co.
" specimen of printing, H. C. Bunce.
" calf-skin, H. A. Woodruff.
" book-binding, H. Gill.

Several articles of merit were exhibited, worthy of premiums, which were not entered by the owners on the books of the Secretary, and the names of the owners not being known, it was impossible to award premiums to them.

OFFICERS OF THE SOCIETY.

President—Charles Dickey.
Secretary—James A. Way.
Treasurer—Charles P. Dibble.

VICE PRESIDENTS.

Clarence—John R. Palmer.
Lee—Daniel P. Wood.
Convis—Jasper Haywood.
Penfield—Alexander Gordon.
Bedford—John Meachem.
Battle Creek—E. C. Manchester.
Emmett—Jeremiah Brown.
Marshall—O. C. Comstock, Jr.
Marengo—S. G. Pattison.
Sheridan—D. Billinghurst.
Albion—Marvin Hannahs.
Eckford—Elisha Gilbert, Jr.
Freedonia—David Aldrich.
Newton—Harvey Smith.
Leroy—David Miller.
Athens—James Winters.
Burlington—Theron Hamilton.
Tekonsha—Tracy H. Southworth.
Clarendon—James Humeston.
Homer—Arza Lewis.

REPORT

Of the Cass County Agricultural Society.

J. C. Holmes, Esq., *Sec'y Mich. State Ag. Society.*

I report herewith a condensed view of the proceedings of the Cass County Agricultural Society, for the year A. D. 1851:

1. At the annual meeting of the Society, Mr. H. J. Redfield, from the committee appointed to report the names of officers for the Society for the year, submitted the following list of persons, which was finally adopted:

President—Justus Gage.

Treasurer—Joseph Smith.

Secretary—Geo. B. Turner.

Corresponding Secretary—H. R. Close.

Executive Committee—John S. Gage, C. F. Harrington, Wm. Clyburn, H. Jewel, John Nixon, Ira Warren, O. N. Long, Wm. Allen, S. T. Reed, Pleasant Norton, H. Heath, A. Reading, Peter Truitt, F. Patrick, B. Hathaway.

A committee of two for each town in the county was appointed, to collect funds for the Society.

The thanks of the Society were unanimously tendered to B. W. Phillips, of Lagrange, for the energy and zeal displayed by him in procuring members to the Society, and for other valuable services.

A meeting of the executive committee of Cass county Agricultural Society was held at Cassopolis, on Saturday, the 24th of May; present a majority of the committee. Justus Gage, President.

On motion, Nathan Aldrich was appointed one of the executive committee for the township of Ontwa, in the place of Oliver Drew, who resides in Milton.

On motion, a committee of three was appointed to prepare a list of Premiums to be awarded by the Society at its first annual fair.

Messrs. Wm. Allen, P. Norton and S. T. Reed, were appointed said committee.

On motion,

Resolved, That a committee of three be appointed for the purpose of preparing and publishing the proceedings of the society, procuring diplomas, and the Transactions of the State Agricultural Society.

Wm. Allen, G. B. Turner and Geo. Sherwood, were appointed such committee.

On motion, the following Judges were appointed:

Cattle—Moses Joy, Reuben Allen, B. W. Philips.

Horses—Arch. Jewell, P. Norton, Wm. Jones.

Sheep—A. Redding, John Nixon, George Redfield.

Swine—James Bonine, O. Drew, Jonathan Gard.

Agricultural Implements—Gideon Allen, Nathan Aldrich, Jessee G. Beeson.

Grain and Vegetables—Hiram Jewell, M. Sherrill, W. G. Beckwith.

Plowing Match—David Bradly, Joseph Carpenter, T. M. N. Tinkler.

Miscellaneous Articles—Wm. Allen, B. Hathaway, S. T. Reed.

Fruits and Flowers—Heman Redfield, E. S. Smith, D. Jewell, Mrs. E. S. Smith, Mrs. G. Sherwood, Mrs. J. Gage, Mrs. G. B. Turner.

Domestic Manufactures—Lewis Edwards, A. B. Copley, Cyrus Bacon, Mrs. G. Allen, Mrs. A. Redding, Mrs. S. F. Anderson, Mrs. L. Edwards.

Committee of Arrangements.—Asa Kingsbury, G. B. Turner, Jas. Sullivan, Joseph Smith, E. B. Sherman.

Ladies Committee of Arrangements—Mrs. Jas. Sullivan, Mrs. W G. Beckwith, Mrs. Jacob Silver, Miss A. M. Redfield, Miss E. Sherman, Miss Sarah Lindsey.

Resolved, That the First Annual Fair be held at Cassopolis, on Thursday, the 18th day of September next, provided the citizens of that place will, at their own expense, prepare the ground, pens, &c.;

and in case the citizens of Cassopolis shall refuse to make the preparation aforesaid, free of cost to the Society, then the committee of arrangements are authorized to select some other township or village for the holding of said Fair, provided the citizens of the place selected shall make all the arrangements necessary, free of charge.

Resolved, That the President of the Society be requested to procure some suitable person to deliver an address at the First Annual Fair of the Society.

The committee appointed to prepare a list of premiums, made a report, which was adopted.

The executive committee then adjourned, to meet again on the second Saturday of September next, for the purpose of perfecting all the arrangements necessary for the First Annual Fair.

G. SHERWOOD,
Secretary, pro tem.

Dowagiac, Jan. 21, 1852.

LIST OF PREMIUMS.

The following are to be awarded at the First Fair of the Cass Co. Agricultural Society, to be held on Thursday the 18th day of September, 1851:

CATTLE.

Best Durham bull, over 2 years old,	$2 00
2d best,	Diploma.
Best under two years old,	1 00
Best Devonshire,	2 00
2d best,	Diploma.
Best grade bull and cow.	
Best native breed,	2 00
Best milch cow,	2 00
2d best,	Diploma.
Best yoke working oxen,	2 00
2d best,	Michigan Farmer.
3d best,	Diploma.
Best fat ox, cow or steer,	2 00

HORSES.

Best stallion,	$3 00
2d best,	1 00
Best breed mare,	2 00
2d best,	1 00
Best span of work horses,	2 00
2d best,	1 00
Best colt under one year old,	1 00
Best colt under three years old,	1 00
Best single horse in harness,	1 00

SHEEP.

Best merino buck,	2 09
2d do	Diploma.
Best grade buck,	Diploma, or 1 00
2d do	Diploma.
Best pen five merino or Saxon ewes,	2 00
2d do do	Farmer, or 1 00
3d do do	Diploma.
Best pen of five ewes, grade,	1 00
2d do do	Diploma.
Best pen of five native ewes,	1 00
2d do do	1 00

SWINE.

Largest hog,	2 00
Best boar,	2 00
2d "	Diploma.
Best breeding sow,	1 00
2d do	Diploma.
3d do	Diploma.
Best lot of pigs not less than four,	1 50
2d do do	Diploma.

FOWLS.

Best twelve of any kind,	1 50

FARMING IMPLEMENTS.

Best plow for breaking,	$2 00
" subsoil plow,	1 00
" cross plow,	Diploma, or 1 00

Best farm wagon,..........$1 00
" harness,.......... 1 00
" cultivator,.......... 1 00
" horse rake,.......... 1 00
" fanning-mill,.......... 1 00
" corn-sheller,.......... 1 00
" hay and straw cutter,.......... 1 00
" churn,..........Diploma, or 1 00
" cheese-press,.......... 1 00
" bee-hive,.......... 1 00
" farm gate,.......... 1 00
" grain drill,.......... 2 00

LEATHER.

Best side of leather, sole and upper,.......... 1 00
" pair of men's boots,.......... 1 00
" pair of women's shoes,.......... 1 00

PLOWING WITH HORSES.

Best plowing,.......... 2 00
2d do ....Diploma.

PLOWING WITH OXEN.

Best plowing,.......... 2 00
2d do ....Diploma, or 1 00

BUTTER.

Best 10 pounds,.......... 2 00
2d do ....Diploma, or 1 00

CHEESE

Best, one year old,.......... 1 00
" under one year old,.......... 1 00

BREAD.

Best six loaves, ...Diploma.

HONEY.

Best 10 lbs. taken up without killing the bees,.......... 1 00

SUGAR.

Best 10 pounds maple sugar,.......... 1 00
2d do do ....Diploma.

DOMESTIC MANUFACTURES.

Best pair of woolen blankets,.......... 1 00
2d do do ....Diploma.

Best five yards of white flannel, ... $1 00
2d do do .... Diploma.
Best 10 yards woolen cloth, ... 1 00
2d do do .... Diploma.
Best 10 yards home made carpet, ... 1 00
2d do do .... Diploma.
Best pair of woolen stockings, ... 50
2d do do .... Diploma.
Best pair cotton stockings, ... 50
" knit woolen gloves, ... 50
" knit cotton gloves, ... 50
" knit woolen mittens, ... 50
Best linnen sewing thread, one pound, ... 50
" specimen ornamental needle-work, ... 50
" table cover, ... 50
" woolen shawl, ... 50
" woolen quilt, ... 1 00
" worked quilt, ... 1 00
" white quilt, ... 1 00
" patch-work 1 00

NEEDLE WORK.

Best specimen, ... 1 00
2d do .... Diploma.

PAINTING AND DRAWING.

Best oil painting, ... 1 00
2d do .... Diploma.
Best in water colors, ... 1 00
2d do .... Diploma.

FRUITS.

Best and largest variety of apples, ... 2 00
" fall apples, ... 1 00
2d do .... Diploma.
Best winter apples, ... 1 00
2d do .... Diploma.
Best peaches 1 00
2d do .... Diploma.

Best pears,..$1 00
2d do ....Diploma.
Best quinces,..1 00
2d do ....Diploma.
Best grapes,..1 00
2d do ....Diploma.

FLOWERS.

Best exhibition,..1 00
2d do ....Diploma.

VEGETABLES.

Twelve best table turnips,.................................Diploma.
Six best blood beets,.................................... "
Six best parsnips,.................................... "
Twelve best onions,.................................... "
Three best head cabbages,.................................... "
Twelve best tomatoes,.................................... "
Twelve best sweet potatoes,.................................... "
Best half peck beans,.................................... "
Largest squash,.................................... "
" pumpkin,.................................... "
" and best variety of Irish potatoes,............... "
Best and greatest variety of vegetables,...................2 00
2d do do ....Diploma.

RULES AND REGULATIONS.

1. Premiums will be awarded on all miscellaneous articles of merit not included above.

2. All articles intended for competition must be entered before 10 o'clock, of the day of the fair.

3. When but a single article of a kind is exhibited, it will not receive a premium unless of decided merit.

4. In case the amount of money in the treasurer's hands be insufficient to pay all the premiums, then the same be paid in proportion to the amount received.

5. Plowing match at 2 o'clock.

6. Address at 3 o'ciock.

7. Report of judges, after the address.

MEETING OF EXECUTIVE COMMITTEE OF CASS COUNTY AGRICULTURAL SOCIETY.

The Executive Committee of the Cass County Agricultural Society, met at the Court House in Cassopolis, Saturday, Sept. 13th, inst., and agreed upon the following rules and regulations to govern the proceedings of the first annual fair of the society, to be held on the 18th Sept., inst.:

1. A register will be opened at the Court House, in the Sheriff's office, where members of the different committees will please register their names immediately on their arrival, and if possible, before 11 o'clock, A. M., in order that all vacancies may be filled by the executive committee.

2. The Executive Committee, the Secretary, Treasurer, and other officers of the Society, will meet at the Shesiff's office at 10 o'clock A. M., of the 18th.

3. All exhibitors who intend to compete for premiums, must become members of the society by paying 50 cents and receiving a ticket, which will entitle the bearer to all the immunities of the Society.

4. Exhibitors will be careful to have their animals and articles arranged in their appropriate places by 11 o'clock A. M., otherwise they will be overlooked by the committee.

5. All articles intended for exhibition must be entered with the Treasurer at the Sheriff's office, in the Court House. Upon such entry, each person will receive a card with the number, as entered by the Treasurer.

6. Exhibitors will be careful to have their cards placed near their animals or articles, so that the judges may have no difficulty in finding them.

7. No animal or article will take more than one premium.

8. Stock should be accompanied with a concise statement of pedigree, feeding, &c., and must be exhibited by the owner or agent.

9. Plowing match at 2 o'clock, P. M.

10. Address at 3 o'clock, P. M.

11. Report of committees after the address, as per large bill.

12. Cash premiums will be payable on demand, by the Treasurer.

13. Diplomas will be ready for delivery afrer the first Monday following the day of holding the fair.

14. The address will be delivered in the court room.

15. The hall of the court house will be used for the exhibition of fruit, vegetables, and articles of domestic manufacture, and will be under charge of the ladies' committee.

16. Farming implements and other bulky articles of domestic manufacture, will be exhibited under the shed on the fair grounds.

17. Discretionary premiums will be awarded to individuals living out of the county, when the article exhibited does not come in competition with that raised or manufactured in Cass county.

18. The show grounds for stock will be north of Mr. Lofland's premises, and east of Mr. Root's.

19. The Ladies Commmittee of Arrangement will prepare the hall of the court house and the middle room on the south side, for the exhibition of fruit, vegetables, and articles of domestic manufacture.

20. Mrs. Barak Mead, and Mrs. S. F. Anderson are added to the Ladies Committee of Arrangement.

G. B. TURNER,
Secretary.

The following is a list of premiums awarded at the Cass County Fair, held at Cassopolis, on the 18th of September, A. D., 1851:

The committee on cattle award as follows:

B. W. Phillips, Lagrange, for best Durham bull.... Cash Premium.
Joseph Smith, Jefferson, for 2d do .... Diploma.
Jas. E. Bonine, Penn, best bull under two years.... Cash premium.
Thos. Tinkler, Wayne, best grade bull.................. Diploma.
Wm. Jones, Penn, for best milch cow.............. Cash premium.
Jas. E. Bonine, Penn, 2d do .... do
David Brady, Lagrange, for best yoke work oxen... do
B. Bullard, Mason, 2d best do do .... Diploma.
Jessee Jones, " 3d best do do .... do

The committee on horses award:

B. W. Phillips, Lagrange, for best stallion..........	Cash premium.
Lewis Rinehart, Porter, 2d do	do
Archibald Jewell, Wayne, best brood mare........	do
Daniel Frasier, " 2d do	do
A. J. Luther, Ontwa, best span matched horses....	do
James Townsend, Penn, 2d do do	do
Isaac A. Huff, Lagrange, best colt under 2 years....	do
David Finch, " do 3 "	do
M. Rudd, Penn, best single horse in harness........	do

The committee on swine award:

Joseph Smith, Jefferson, largest hog..............	Cash premium.
James E. Bonine, Penn, best boar...............	do
Justus Gage, Wayne, 2d do	Diploma.
Daniel McIntosh, Penn, best breeding sow.........	Cash premium.
Edward Beech, Lagrange, 2d do	Diploma.
Nathan Aldrich, Ontwa, 3d do	do
Daniel McIntosh, Penn, best lot pigs................	do
Nathan Aldrich, Ontwa, 2d do	do

The committee on grain and vegetables award:

Benj. Gage, Wayne, best wheat......................	Diploma.
Archibald Jewell, Wayne, 2d best lot of wheat..........	do
Wm. Allen, Mason, best lot of white beans............	do
do do corn,.................	do
D. T. Nicholson, Jefferson, best lot sweet potatoes........	do

The committee on agricultural implements award:

Morris Custard, Lagrange, best two horse wagon...	Cash premium.
Nathan Aldrich, Ontwa, do plow.....	do
Heman Redfield, Mason, for best bee-hive..........	do
H. Redfield, " best straw cutter.........	do
C. Smith, " best cheese press........	do

The committee on sheep award:

John Gage, Wayne, best Spanish merino buck.....	Cash premium.
J. E. Bonine, Penn, two best do do	do
F. Brownell, " four best merino yearlings..........	Diploma.

The committee regret that the regulations of the society prevent their awarding a higher premium for these yearlings, as they are of superior quality.

The committee on miscellaneous articles award:

Daniel Carlisle, Lagrange, best ten lbs. maple sugar....... Diploma.
Amos Northrop, Calvin, best lot of honey........... Cash premium.
Philo White, Wayne, 2d do .... Diploma.
Mrs. E. Thomas, Ontwa, best worsted work.............. do
Mrs. E. Thomas, Ontwa, best paintings................. do

The committee on articles of domestic manufacture award:

H. Thompson, Ontwa, best embroidered shawl.......... Diploma.
Mrs. E. Thomas, " best linen hose................. do
" " best table spread................ do
Mrs. Beckwith, Jefferson, best quilt.................. do
Mrs. E. Thomas, Ontwa, best bureau cover............. do
" " " patch work.............. do
Mrs. Sullivan, Lagrange, best hearth rug.............. do
Mrs. A. B. Copley, Volenia, best five yards flannel....... do
Geo. Meecham, Porter, three best cheese................ do

The committee on plowing award:

Benniar Tharp, Calvin, best plowing, with oxen.......... Diploma.

The committee on fruit and flowers award:

Heman Redfield, Mason, best and largest variety of apples, thirty-four varieties.......................... Cash premium.
Miss Julia A. Redfield, Ontwa, best fall apple...... do
A. A. Goddard, Mason, fourteen varieties apples.......... Diploma.
Miss Julia A. Redfield, Ontwa, best winter apples......... do
D. T. Nicholson, Jefferson, four varieties winter apples..... do
Mrs. McKyes, Wayne, best lot peaches............ Cash premium.
A. A. Goddard, Mason, 2d best lot peaches............. do
Heman Redfield, Mason, three varieties quinces..... Cash premium.

Fine specimens of peaches were offered by C. C. Landon and others. A bunch of very large and fine looking apples by D. T. Nicholson, and a fine variety of pears by Nathan Aldrich, and the committee regret that the rules of the Society would not admit of premiums therefor. Mrs. W. G. Beckwith presented a beautiful flower-pot and wreath, but not being marked, the committee were unable to decide upon it.

G. B. TURNER,
Secretary.

The following remarks from the editor of the National Democrat, published at Cassopolis, express my own views and feelings with respect to the character of the exhibition, at the first annual fair of the Cass County Agricultural Society:

"It is admitted by all with whom we have conversed on the subject, (and they are not few,) that the fair on the 18th, vastly exceeded their expectations, not only in regard to the quantity of stock, and number of articles exhibited, but the superior quality and excellence of both.

"We venture the assertion that no one county in the State can bring forward as good stock as Cass county: this is saying much for her, but not more than she is able to back up by an actual showing.

"The number of persons in attendance on that day is variously estimated; we have heard no one, who had an opportunity of judging, put it at less than two thousand.

"The address delivered by Heman Redfield, Esq., of Mason township, is spoken of as a vigorous and masterly production; we regret that we had not an opportunity of hearing it.

"The committees in making their reports expressed many regrets that the imperfect arrangements for conducting the fair, precluded them from a more efficient discharge of their several duties; many articles of merit, if not entirely overlooked, were but partially examined. As this was the first fair ever held in the county, it could not be expected that it would be conducted with all the order and system incident to counties more experienced in such matters. We flatter ourselves that the arrangements for the next fair will be free from all the objections that might be urged against the late ones. We hope those experienced in these matters will prepare a well digested plan for our next."

All which is respectfully submitted.

JUSTUS GAGE,

Cor. Sec. of the Mich. State Ag. Society for Cass Co.

ADDRESS

Delivered at the Cass County Agricultural Fair, September 18th, 1851.

BY HEMAN REDFIELD.

Ladies and Gentlemen:

The partiality of your President having selected me as your speaker upon this occasion, it gives me much pleasure to offer you my sincere congratulations upon the interest and ambition this day manifested for the cause for which we are associated.

That the experiment has been successful, and that our society is established upon a permanent foundation, has been most amply demonstrated. May we not now indulge the aggreeable conviction, that each returning exhibition will continue to derive additional interest and value, until our county shall assume that position to which by nature it is entitled as the first among the agricultural districts of our beautiful State.

The variety and fertility of our soil, the abundance of our water privileges, and the unlimited markets almost surrounding us, in connection with the energy and enterprise of our population, as this day witnessed, would seem to indicate the possibility of such an event at no distant period.

A reference to the statistical report of the Secretary of State, to the last Legislature, discloses the fact that few counties in the State, in proportion to the territory and number of population, produce an equal amount of wheat and other grain, and very few, if any, excel us in this respect.

It has with much truth been remarked, that by the holydays and anniversaries of a nation can its character best be judged; and that it is an evidence of wisdom to encourage the pastimes and divertisements of a people. Indeed, it may well be doubted whether the ancients and all uncivilized nations have not been wiser in their day and generation than the Anglo-Saxons and their American children. Relaxation of body and mind is necessary to the health and elasticity of both. When the primitive fathers of New England banished the Merry Christmas festival of the mother country, they soon found the necessity of creating a substitute, which was discovered in the more sanctimonious feast of thanksgiving. Aside from the excercises of that day, at best but partially observed, and illy adapted to recreation, there is but one day in our calendar regarded as a set time for amusement and relaxation, and that is the "Glorious Fourth," consecrated by our birth as a nation. From the sun-set of our national rejoicing to its rise after the lapse of fifty-two long weeks, not one day is devoted to a genuine and suitable indulgence in national festivity. Thus the "harvest home" is forgotten, and the rustic enjoyments of Christmas and New-Year are almost unknown. No joyous groups dance around the May-pole, or twine the garland for the blushing brow of their youthful queen. No anniversary is observed for reinvigorating the system by wholesome, athletic exercises, or dispelling care among flowers, groves and fountains. The refined and poetical pastimes of Greece and Rome were marked moreover, by that indelible impress of moral purity, so essential to the full development of the physical and mental powers of a free and happy people.

The wild sports and robust exercises of the former denizens of our plains, are seen in the expanded chest, the symetrical form, and the agility and grace of each movement, while the unremitting labors of their white successors in the workshop or the field are such as to draw the body forward, allowing no exercise for the antagonistic muscles, and consequently the body is so often deformed and bowed down to the posture of the brute creation. Contrast the tall, erect posture and elastic tread of the Indian, with the plodding pace and inclining gait of the white man. But the spirit of our population is too utilitarian to be suited with any association which does not blend with relaxation and amusement the elements of utility and gain; and the purpose for which we have this day met seems peculiarly appropriate to add to our enjoyments, while it contributes largely to our prosperity.

This laudable movement is begun, too, at a period which ensures us the most substantial benefits. At no time in the history of the world have so many great and important discoveries been made as in the last quarter of a century. Results more wonderful than the grandest conceptions of human genius ever anticipated, have been witnessed and perfected. Human intellect, in all its researches in every department of science, has been amply and wonderfully rewarded. And though for a long period of time considered subordinate in its character, the science of agriculture has now attained its proper dignity, and shines out the highest and noblest of all arts. And thus it must ever remain.

The wisdom of man having seemingly expanded itself in the contemplation of the heavens and the immaterial world, and turned at last to the study of the earth and its productions, has found a field wide enough for the exercise of all its faculties, and more potent for good than all the reflections of philosophers, the researches of alchymists, or the deductions of theologists.

Agricultural chemistry has already attained to that state of perfection which obviates the necessity of those blind experiments and rude conjectures to which our predecessors were forced to resort in the study of nature and the beautiful and wonderful manner in which her laws are developed and displayed. A great saving of time and money has thus been created, and the farmer of to-day need be at no loss to elucidate and understand the cause or results of those operations in his profession, but a little while ago considered anomalies or mysteries. He can find at once, by turning to the different works upon agricultural subjects, information in regard to any particular subject he may wish to investigate, whether in respect to his crops, his soil, or his stock.

Heretofore having no guide, he has oftentimes been the dupe of strolling vagabonds, palming upon him their worthless trumpery, under high sounding names, as new and useful inventions in domestic husbandry. And I presume there are many among you who can point to some piece of *pine crackers* lumbering up your garrets or sheds in the shape of churns, cheese-presses, seed-sowers, corn-shellers, &c., &c., wholly unfit for the purpose represented, but very good *tooth cutters,* and standing monuments of gullability and our ignorance of mechanical principles.

In the purchase and selection of stock likewise has the farmer often been the victim of the same species of fraud and deception. Lantern-jawed Yankees and cute Varmounters have infested our highways and by-ways, disposing of animals with extravagant names, and at still more extravagant prices, generally proving of spurious pedigree, and often inferior to our native breeds. It is devoutly to be hoped that their day has passed, and that hereafter those suspicious and seedy looking individuals poking along our dusty roads, with their still more suspicious cargoes of pine and varnish, with attachments of cogs and cranks, or big horned sheep, with fleeces of two year's growth, and black with the application of sand and tar to make it weigh, be regarded as we look upon polecats and sheep-dogs. For now the business of supplying the farmer with the best and most valuable improvements in stock and farming utensils has been assumed by men of long known and well tried integri y, and who are willingly responsible for the worth of every article they vend. They are found in every State, and their articles in nearly every county. Their enterprise has already given a strong impulse to the cause of agriculture, and has largely contributed to stamp the present as the age of progress in the culture of the arts of peace.

In this connexion comes up the question, are agricultural implements useful? Are sufficient benefits and advantages derived from them to compensate for all the trouble and expense of their operation? The best answer I could give would be perhaps to refer you to the commendable spirit of emula ion and rivalry manifested here to-day. The very fact of so large and intelligent number of individuals meeting together upon an occasion where all political and sectarian feelings are carefully avoided, will necessarily produce upon ingenious and well regulated minds a spirit of ambition and comparison, a desire to investigate, and a determination to excel.

When a man finds himself a member of a numerous, prosperous, respectable and peaceful association, heartily engaged in a rational competition and exhibition of the productions of their own industry and skill, his pride is stimulated and his relative importance increased; his opinions are elevated and his respectability augmented. When this point is gained, nothing is wanting but to collect and bind them together by fair and impartial regulations, in common and united effort, for the attainment of whatever the object might be. By such means have the amount and value of the productions of the old country been doubled in the past few years, and the exhausted lands and inferior stock of our eastern States so rapidly renovated and improved; and though our own beautiful Peninsula has but cast its garments of maiden flowers, fairer than diamonds and gold, and donned its full woman's dress—the harvest of labor by man. No State in the Union offers a wider field for the exercise of energy and zeal, or is more susceptible to improvement in all the various branches of husbandry. In the growing of wheat, in proportion to population, no State has begun to equal, and of other grains, none has surpassed her. The Commissioner of Patents has stated our crop of 1848 at ten millions of bushels, an average of 23½ bushels to each person. By the census of 1840, our population was in round numbers 212,000, and the amount of wheat raised 2,157,000 bushels. The population of 1849 was a little less than 350,000. Thus, while our population had not doubled in that time by more than fifty thousand persons, we have, according to the above estimate, more than quadrupled our production of wheat and increased it at the rate of one million of bushels per year for eight consecutive years, making the amount we raise to each individual more than double that of any other of the wheat growing States.

Considering that this great amount of the "staff of life" has been very generally produced in the cheapest and most slovenly manner, upon land half cleared and half tilled, may we not safely calculate that the spirit of advancement now animating many of the counties, and the general attention being paid to the selection and use of improved tools, teams and seed, together with the advantage of one of the very best agricultural papers in this or any other country, will enable us in 1860 to exhibit a still more flattering evidence of wealth and prosperity.

The writer from whom I gather the above statistics, expresses a doubt if the returns for 1850 would exhibit an equally favorable picture. Well, let us see: Our population last year is stated in round numbers at 398,000, and the amount of wheat raised 9,000,000 bushels, which will, as before, give about 23 bushels to every man, woman and child in the State.

We need but the right spirit of enterprise, and the hands of industrious numbers to make our lovely prairies, our swelling plains, and rich heavy timbered bottoms, one of the fairest agricultural districts upon which the sun ever shone. We need but to retrieve our high position and show to the world by facts and figures that cannot lie, that our productions are steadily increasing.

> And away to the west where the primeval flood
> Yet throws its dark fringe on the Michigan flood,
> Where pale in their beauty the forest flowers bloom,
> And the earth is yet mantled in forest land gloom,
> With the bounds of an empire, the dark virgin soil
> Full of treasure awaiteth the husbandman's toil.

The tide of emigration will again send its living current hitherward; our borders will be extended, our waste places made fruitful, our boundless resources developed, and the despondency and discredit which has partially retarded our onward course for the few past years, (from causes upon which it is not my province here to dwell,) will be forgotten, and our growing State become what it should be, the home of the famishing legions, the exiles of hunger and seekers of bread.

The cornucopia is ours; it is our just emblem, and there is every incitement to retain it; and although it has been said that many of our fair fields are exhausted and now lay clad in sorrel—nature's grave clothes—fit emblem of a tired and somewhat deterioated soil. Yet they are not lost. The plow, which has served partially to exhaust, will also suffice, with generous husbandry, to redeem them, and the experience gained will hereafter serve to retain them smiling and fruitful. How can we render the greatest amount of good in behalf of the cause, to ourselves, our State, and the world at large, is a question which naturally arises. I answer, by united exertion, by observation, and by a fair and honorable competition; and we should commence by acquiring all the knowledge within our reach, from the experience of others who have effected that for their soils which we propose to accomplish for ours.

Twenty or thirty dollars expended in the purchase of the standard agricultural works in the language, and a careful perusal of the Michigan Farmer, will save any good practical farmer thousands in the end, and double his profits in two years.

Now we have in our county about 60,000 acres of improved land, something less than a quarter of our territory, and the total value of our property of all kinds is, as assessed, about $800,000, with a population of 11,000. In 1849, we raised from 18,000 acres about 160,000 bushels of wheat, something over thirteen bushels to each individual. And yet an average of about ten bushels per acre for a soil of the most productive character; not over half a crop at the best calculation. I believe it is generally admitted that our soil must be deepened before it can be permanently improved, and that one acre of soil twelve inches deep is worth more to make money from by cultivating it than four acres six inches deep. Admitting that under the best circumstances an acre of soil six inches deep will produce fourteen bushels of wheat, and that twelve bushels will pay the expenses, and we have two bushels as profit. Now double the depth of the soil and the amount of the crop, making the former twenty-eight bushels instead of fourteen, and the latter twelve inches instead of six; fifteen bushels instead of twelve will now pay all expenses, and leave a net profit not of two but of thirteen bushels per acre. Manure well, plow deed, sow in good season, then trust to Providence, and instead of selling $60,000 worth of wheat we can market three times that amount.

There was raised in our county two years ago about 600,000 bushels of other grain, of which at least one-half was a surplus, worth as much as the wheat crop, and susceptible by good husbandry of equal augmentation in amount and value.

We own three thousand horses, worth on an average say $40, or $120,000. Now it costs no more to raise a oolt worth at 4 years old $80, than one hard to jockey off at $40. And a little reflection will convince any one that the above value can be doubled in five years.

The enterprise of a fellow citizen offers you a stock of as good blood and reputation as can be found, and which he has, I think, safely ehallenged the State to equal. And there are several other excellent breeders of that noble animal among us. We certainly should exert ourselves to patronize and sustain them.

We possess 8,000 head of cattle, generally of an inferior size and quality, and are selling the average of our young cows and steers at from eight to ten dellars, when in good condition, and I am fully satisfied that the value of this stock can easy be doubled by an importation of thorough breeds, the judicious patronage of those we have, and a more general attention to care and keeping.

We have likewise 17,000 sheep, shearing in 1849, 44,000 lbs. of wool, about 2½ lbs. per head, and worth that year an average of 30 cents per pound; a gross value of about $14,000. We have in our limits as good stock sheep as can be found in the country, and a general attention to this department of our industry will enable us to increase the weight of the fleece to four pounds, worth 40 cents per pound, and the value of the carcass proportionally.

In addition to the above list, we have among our grubs and in our puddles about 10,000 things which Wolverine audacity have denominated swine—variously known as narragansetts, alligators, land sharks and flea breeders. In one sense indeed this class of our domestic animals have received much attention; but that attention has resulted from wonder and disgust, and has been expressed in unmeasured ridicule, sarcasm and invective. It is well known that a well bred and well kept hog can be easily made to weigh in 18 months, 400 lbs., worth $3 per cwt., while it is a hard matter to make the critters I speak of ever weigh 200 lbs., and a harder matter to dispose of the compound of acorns, ground-nuts and carrion for $2 per hundred.

We have thus glanced at the field which is open to us, and we are called upon by every principle of interest and policy to enter it. Let us devote our best energies to the study and cultivation of our soil. The earth is always a bountiful mother, and none more so than that of Michigan. For many years she has rewarded your very indifferent attentions an hundred fold. But as there is an Alpha so is there an Omega to all things. The word *finis* is written upon everything in the material world, and this unrequited liberality must have an end. At any rate, let us not test it further. Let it be sufficient for us to know that we have our past experience, the example of the eastern States, and other countries, as well as the indications of theory, all concurring to convince us that with good management the limit of its productive powers can neither be predicted nor foreseen.

As far as the principles of chemistry have been applied to agriculture, the result of each improvement has been more and more astonishing.

The invention of improved plows, and other agricultural implements, and the more economical modes of using them, the use of lime, the application of gypsum, bone manure, clover, &c., have all thus far tended to the raising of crops, not only at far less cost, but in greater abundance. Nothing repays the labor of the husbandman more freely than the willing soil, nothing is more grateful to his attentions or offers surer reward than patient industry. Yet how few practical men are acquainted with what is already known of the important art by which they live—trained up in ancient methods, attached generally to conservative principles in every shape—farmers, as a body, have always been more opposed to change than any other class of community. They have been slow to believe in the superiority of any method of culture which differs from their own, from those of their fathers, or of the district in which they live; and when the superiority could be no longer denied, they have been almost as slow to adopt them.

But these old prejudices are fast dying out, and a healthy impulse has been given to the cause of agriculture in every State, and the cultivators of the most ancient, most honorable, and noblest of all arts, are generally anxious for information and eager for improvement. True, indeed, there are a few specimens of the hold-back and stand-still class occasionally seen, full of prejudices, and stiff-necked as some old coarse-wooled patriarch, who looks upon the intrusion of the large, sleek and fine-fleeced members of his kind with an air of contempt, and stalks majestically round the pasture, priding himself upon the bigness of his horns, and challenging comparison with the thickness of his skull—good patterns of the old "sled-length" regime; with one knotty log at the rickety bars, an old blunt axe and slivered up helve by its side, and not *ary* stick cut for *mornint.* The staked and ridered domicil, lopped over like some old lame hen, the place left for glass well stuffed with various unmentionables of various colors; while a band of tow-headed, juicy-nosed and ragged tatterdemalians are capering and rolling in the dirt, with Bull and Tige basking by the side of their lord and master in the sun, and some unfortunate, sallow, wo-begone victim in dilapidated attire, pipes her shrill notes and gives fierce battles to numberless chickens and pigs at the threshold, with the old stub of a splint broom, while the old sow digs the potatoes, the old cattle toll the corn, and grubs choke the wheat.

Thanks to the spirit of improvement which makes such a spectacle rare; but time was when such a sight was neither rare nor the picture overdrawn.

The public mind is becoming awakened to the establishment af agricultural societies in all parts of our country, affording undeniable indications of the interest universally indulged. We may therefore safely predict, that neither the practice or theory of this art will be suffered in future to experience that want of encouragement under which it has heretofore been permitted to languish. It requires only the general exhibition of such an interest as we have seen to-day, and the adoption of some general means of encouragement, to stimulate both practical ingenuity and scientific zeal to expend themselves upon this most valuable branch of national industry. Knowledge and science are never unwilling to shed all their light upon the practical arts; on the contrary, they are ever willing to proffer their assistance. Need I advert in proof of this to the unwearied labors of the vegetable physiologist or the many valuable observations recorded by scientific chemists. Need I speak of botany, which is, as it were, the very foundatian upon which the first element of agriculture rests; or to zoology, which alone can throw light upon the nature of the numerous insects, which so often prey npon our crops and ruin our hopes, and the study of which alone can be reasonably expected to arm us against their ravages and instruct us how to extirpate them. The recent contributions of geology are the best proofs of the readiness of the sciences of observation to give their aid to the promotion of agricultural knowledge. The geologist can best explain the immediate origin of your several soils, the cause of the diversities which even on the same farm they not unfrequently exhibit, the nature and difference of your sub-soils, and the advantages you may expect from breaking them up and bringing them to the surface.

It is thus that all branches of human knowledge are bound together, and all the cultivators of them mutually dependant, and it is by lending each to the other a helping hand, that the success of all will be accelerated and secured, while with the progress of the whole the advance of each individual is made sure.

As I have before said, there never was a time more peculiarly favorable to the increase of agricultural knowledge than the present. The growth of our population requires it. Practical men are anxious to receive instruction, and scientific men are eager to impart what they know, and to make new researches for the purpose of clearing up what is unknown. Human science is ever progressive, and to refuse to follow the indications of existing knowledge, because it may be to some extent uncertain, would be as foolish as to refuse to avail ourselves of the morning light because it is to some extent less brilliant than that of the noon-day sun. Looking back upon the vast strides which organic chemistry has made in the past few years, and is still continuing to make, some visionaries have ventured to anticipate the time when the art of man shall acquire a dominion over that principle of life by which plants now grow, and alone produce food for man and beast, but have predicted that the time will come when man shall manufacture by art those necessaries and luxuries for which he is now dependant upon the vegetable kingdom. Can this be so? Having conquered the winds and the waves, is man really destined to gain a victory over the uncertain seasons also? Shall he come at last to tread the soil beneath his feet an idle and a useless being—to disregard the genial shower—to be indifferent alike to rain or drouth, to clouds or sunshine—to laugh at the thousand cares of the husbandman, and pity the ceaseless toil and sleepless anxieties of the ancient tillers of the soil? In fine, is the order of nature through all past time to be reversed, and the habits and pursuits of the whole human race to be altered by the progress of knowledge? No! there is not—there never can be anything excellent without labor. Labor is holy. By placing before man so many incitements to the pusuit of knowledge, the will of the Deity is, that out of the increase of wisdom he should extract the means of increased happiness also. Set man free from the necessity of tilling the earth by the sweat of his brow, and you rob him at once of all the calm and tranquil pleasures of life. Obviate the necessity of labor and you deform the body and vitiate the mind. Lazy poets may sing, and idle sentimentalists prattle of the beauties of a rural life. They may tell of the sighing of the breeze, the singing of birds, the waving of grain, the perfume of new made hay, looing of kine, the milkmaid at the stile;

they may find "sermons in stones, books in the running brooks, and good in everything." To you who are practical men and women, such stuff, from such a source, and so expressed, sounds, doubtless, all trash and moonshine. To you these are but incidental to years of hard drudgery with the axe and scythe, the spinning-wheel and churn, the dungfork and plowtail, the cheese-press, soap-barrel and wash tub. They are purchased only at the price of industry, economy and virtue, and by the exercise of those virtues only, can they be honestly appreciated and enjoyed.

There are few sights indeed more pleasing than a neat and well tilled farm; but all the admiration or romancing in the world will not make one head of wheat or length of fence, but

> "There the hand of hard labor but raiseth its wand,
> And the harvest all golden springs up from the land."

Then join our Society, lend us a helping hand, put your plow deep in the soil, save, gather and apply all the manure you can make, and you will soon find then a

> "Tithe of the labor that now dampens your brow will place you in plenty;
> A tithe of the toil make you chief of the manor and lord of the soil."

In marketing your crops, sell only clean grain, keeping your best for seed. In the words of the Constantine miller, "Keep all the chaff, chess, cockle, smut, sticks, nails, gravel stones, rat and hen dung, out of market, and out of sight." Breed only the best of stock, dispose only of the poorest. Feed and shelter every thing well. Prune, graft, and nurse well your orchards. Mind the "place for every thing, and keep it there when not in use." Keep an account with your farm as you would with a rascally merchant. Take the Michigan Farmer, it will pay forty fold; and your county paper, too, if you think it good for anything, (and, in my humble opinion, there are few better of its size,) and you will excel. The better feelings of your nature will govern and control. Your minds will expand, and intelligence, prosperity, contentment and wealth, will necessarily follow. And you will look forward to each returning anniversary of this day with eager anticipations, and come up here with an honest and manly pride. You will bring with the fruit of your orchards and fields, your well kept and well bred stock, all testifying to your industry, your energy and skill. You will see what your neighbors have done, and in the association the asperities of your natures will be softened, your social feelings improved, your acquaintance extended, your happiness increased, and you will gain many a useful lesson for after life. Your wives will share your pride, and gladly bring forth the rich products of the dairy and the loom, telling in language too plain to be misunderstood, of the attractions of those homes illuminated by their cheerful industry and maternal care. It is only when thus enabled to be useful and to do good, that the mothers and sisters and wives of our affections are enabled to display the depth of their kindness and the boundlessness of their love. So surely as man in the cultivation of the soil, or any honorable calling, strives to fulfil the high destiny of his race, so surely will the patient wife of his heart partake of his anxieties or joys and add dignity to his efforts. So surely as man fails to adorn the sphere in which he moves, so surely does he entail mortification, sorrow and sickness, upon her he has sworn to protect. Your daughters and your sons will be eager to participate with you in the pleasures here presented. All fresh from the hand of nature, fragrant in the beauty and excellence arising from pure simplicity of life, strong in the proud consciousness of rectitude, of morals and honesty of heart, with none of the hypocritical cant of miscalled polished life upon their lips, or the ruinous shackles of artificial society upon their bodies or minds, they will shine out your fairest offerings, an honor to themselves and a blessing to you. The one in the gentle modesty of her nature will blend ornament and utility, the flowers of their rearing and the fanciful productions of their hands, with more solid offerings of their mother; their department will be the centre of attraction, and they the only rivals to dispute or divide our admiration. Nor are the enjoyments of these occasions restricted to the farmer alone. The kindred arts are cordially invited to participate, and specimens of the industry and skill of all who labor are heartily welcome. There can never be too much happiness, too much cheapness for men. His wants will ever continue to grow with what supplies them. As long as men live on earth they will continue to call for more, still more, from the bosom of their common mother.

Transporting ourselves to some far distant period in coming time, then with a world sufficiently peopled to develop its varied resources, yet not too densely crowded, when all organisms shall be improved far beyond their condition, when every family may enjoy the comforts and luxuries confined to the wealthy few of the present day, when each individual may exercise a refined taste in the daily enjoyment of the bath, of books, and of works of art, and educated mind extends further and wider its dominion over sense and matter. Who shall say the end is yet? Labor, the industrial arts its right hand, and freedom, the vital element it breathes, in the plentitude of its power striking out new fields of action, calling to its aid now undiscovered and yet subtle elements, will then develop of earth's known and yet unknown treasures richer stores, then mould them all to newer forms, more elegant, until the refined and more refined material combines in all its intimate connexion with the immaterial of man's nature to elevate him to a loftier state, when looking back on this age he will regard it as we view the barbarism of old. And this more immaterial, will it not with all the advance of science, letters, morals christianity, have learned to joy in high appreciation of the moral, beautiful, and have kept its even pace in the progressive order of the world. It will when labor walks the earth a God in dignity and power, and every man shall be a sovereign king in his worth and greatness, and one of God's own chosen in knowledge and all moral excellency. Christ furnished us the type of poverty and labor, rising to the God in heavenward ascension, and we must cherish faith in man, created image of Divinity, advancing ever with the stride of progress upon this teeming fruitful footstool.

"Therefore pass on and reach the goal,
  And win the prize and wear the crown:
Faint not, for to the steadfast soul
  Comes wealth and power and renown.

To thine own self be true, and keep
  Thy hands from sloth, thy heart from sin,
Push on and thou shalt surely reap
  A heavenly harvest for thy toil."

REPORT

Of the Genesee County Agricultural Society.

J. C. Holmes, Esq., *Sec. Mich. State Ag. Society:*

Dear Sir—In compliance with the 11th article of the constitution of the Michigan State Agricultural Society, the following is respectfully submitted as the report of the proceedings of the Genesee county agricultural society, for the year 1851.

The annual meeting of the Genesee county agricultural society was held on the 8th of January, when some amendments to the constitution were adopted, and the following officers of the society were elected for the year 1851:

President—Benjamin Pearson.

VICE PRESIDENTS.

Argentine—Isaac Wixom.
Atlas—Reuben Goodrich.
Clayton—James E. Brown.
Davison—Goodenough Townsend.
Fenton—Samuel N. Warren.
Flint—George Crocker.
Flushing—Nelson W. Butts.
Forest—John Crawford.
Gaines—James Van Vleet.
Genesee—Josiah W. Begole.
Grand Blanc—Gurdon Watrous.
Montrose—John Farquharson.
Mundy—William Smith.
Richfield—Josiah King.
Thetford—William N. Van Tyler.

Vienna—Grovernor Vinton.

Recording Secretary—Francis H. Rankin.

Corresponding Secretary—Levi Walker.

Treasurer—Augustus St. Amand.

EXECUTIVE COMMITTEE.

Caleb S. Thompson, George M. Dewey, Charles N. Beecher, Jonathan Dayton, Charles D. W. Gibson.

AUDITORS.

Daniel N. Montague, Julian Bishop.

The executive committee met on the 15th March, and prepared a list of premiums to be offered at the annual fair in October. In accordance with the provisions of the 14th section of the constitution, this list was published and circulated in the several towns throughout the county.

On the 30th of August, they again assembled and appointed a marshal of the fair, viewing committees to adjudicate upon the merits of the products offered in competition for premiums, also a committee to select a suitable place to hold the fair and a piece of ground for the plowing match.

It being deemed best for the interest of the society to secure for a term of years a piece of land upon which to hold its annual fairs, a judicious location was selected, and a lease made for a term of five years. This ground has been permanently fenced in, and suitable buildings and pens erected, for the exhibition of stock and other articles. The cost of enclosing and fitting up the fair grounds has been a considerable draft, this year, upon the limited funds of the society; but they have the satisfaction of reflecting that the greater part of this expenditure will be saved them for several years to come by the durable nature of the fixtures provided.

The second annual fair was held on the 1st and 2d days of October. The weather was propitious, and a large concourse of citizens assembled. The arrangements for the exhibition, although far from perfect, were a very decided improvement upon those of the previous year; and it is but reasonable to expect for the future, the continuance of a gradual amendment in the system of management, as the experience of each successive year shall enable the committee to avoid the difficulties and guard against the defects to which every such society is liable in the first stages of its existence. The number of entries for exhibition was 312, upon which 155 premiums were awarded or recommended.

The expenses of preparing the ground for the fair already alluded to, having left in the hands of the treasurer barely funds sufficient to pay the regular premiums offered by the society, the executive committee were obliged to pass over those articles upon which *discretionary* premiums had been recommended, without conferring any pecuniary reward upon the meritorious producers. However, as far as possible to obviate this disadvantage to that class of successful competitors, the committee caused the following report to be published in the county newspapers:

SECOND ANNUAL FAIR OF THE GENESEE CO. AGRICULTURAL SOCIETY, OCTOBER 1st AND 2d, 1851.

REPORT OF THE EXECUTIVE COMMITTEE IN RELATION TO PREMIUMS AWARDED.

Upon an examination of the several committee books of the judges at the recent fair, the executive committee find that some of the viewing committees have recommended premiums on a considerable number of articles not specified in the printed list published by the society; also upon some articles specified in the published list, but where the conditions upon which the premiums were offered have not been complied with by the exhibitors; and premiums have also been recommended upon a few articles not manufactured in this county. Neither of the two latter classes would be eligible to premiums under the rules of the society, in any circumstances; but the committee have pleasure in publishing the recommendations of the viewing committees, as the best means of enabling the exhibitors to derive the benefit intended by the awards of the judges.

In relation to the first class mentioned, the committee regret that the expense attendant upon the permanent fencing and fitting up of the fair grounds, have left no funds at their disposal this year, for *discretionary* premiums, and here also they extend to the exhibitors the only alternative in their power—that of giving publicity to the favorable notices of the judges.

The treasurer is now prepared to pay the premiums offered by the society, on all articles and stock specified in the published list, upon which awards have been made, and where the published conditions of competition have been complied with by the exhibitors. The committee subjoin two separate lists of the successful competitors; one showing those to whom premiums are payable, and the other, those who exhibited articles which the viewing committees have favorably recommended, but upon which the executive committee have no power, for the reasons above stated, to make payments.

B. PEARSON,
President.

Flint, Oct. 4, 1851.

PREMIUMS AWARDED IN CONFORMITY WITH THE SOCIETY'S PUBLISHED LIST—ON CATTLE.

For bulls, 2 years old or over, 1st premium to Emer Woodin, 2nd Thomas Wolfret, 3d Grovenor Vinton.

Yearling bulls, 1st Alanson Munger, 2d Lyman G. Buckingham, 3d Nelson H. Chittenden.

Bull calves, 1st Stillman F. Grow, 2d Grovernor Vinton.

Milch cows, 1st James B. Walker, 2d Peabody Pratt, 3d Porter Hazelton.

Working oxen, 1st William Schram, 2d Jeremiah Kelsey, 3d Porter Hazelton.

Three year old steers, 1st B. F. Olmsted, 2d R. H. Wallace, 3d Asahel Robinson.

Three year old heifers, 1st Samuel R. Atherton, 2d Rowland B. Perry.

Two year old heifers, 1st Oscar F. Forsyth.

Heifer calves, 1st Jonathan Dayton, 2d George S. Hopkins.

HORSES.

Stallions 4 years old and over, 1st John Hamilton, 2d John Hill, Jr., 3d Benjamin Goyer.

Brood mares 4 years old or over, 1st C. C. Hascall, 2d C. C. Hascall, 3d William Eams.

Matched horses 4 years old or over, 1st J. H. Alger, 2d Warner Lake, 3d Nelson Main.

Three year old geldings, 1st D. H. Seeley, 2d Grovernor Vinton.

Three year old mares, 1st John Glass, 2d Charles N. Beecher.

Two year old colts, 1st Isaac Schram, 2d Lewis Cumings.

Yearling colts, 1st James Lacy, 2d Nelson H. Chittenden.

Sucking colts, 1st William Barnhart, 2d Eber Adams.

SHEEP.

Fine wool bucks, 1st Rowland B. Perry, 2d Elbridge G. Gale.

Pen of ewes, 1st Charles Bates, 2d Rowland B. Perry.

Pen of buck lambs, 1st E. B. Dewey, 2d Rowland B. Perry.

Pen of ewe lambs, 1st Rowland B. Perry.

SWINE.

Boars 6 months old or over, 1st Simeon Simons.

Sow and 5 pigs, 1st Jonathan Dayton.

POULTRY.

Cock and two Hens, Daniel S. Freeman.

Lot of poultry, Mrs. C. S. Payne.

FARMING IMPLEMENTS.

Breaking-up plow for general purposes, King & Forsyth.

Plows for single team, for general purposes, 1st, E. Rockafellow, 2d, E. Rockafellow.

Farm wagon, David Watson.

Fanning mill, C. H. Rockwood.

MISCELLANEOUS ARTICLES.

Wagon harness, R. L. Sheldon.

Carriage harness, W. W. & J. Booth.

Bureau, Solomon Stone.

Box of saleratus, Warner Lake.

Window sash, Solon C. Bliss.

Panel door, William Eddy.

Fine boots and shoes, John Quigley.

CARPETING, &C.

Pair woolen blankets, Grovenor Vinton.

Woolen carpet, Mrs. D. Curtis.

Rag carpet, E. Schram.

Tow cloth, Paul Davison.

Woolen shawl, Mrs. E. Woodin.

BUTTER AND CHEESE.

Butter—1st, J. Eldredge; 2d, Mrs. J. Barret; 3d, Mrs. Stafford.

Cheese—1st, Alanson Payson; 2d, Grovenor Vinton; 3d, Rowland B. Perry.

HORSE SHOEING.

Span of horses best shod, Wicks and Failing.

HOSIERY.

Pair woolen knit stockings, Miss Mary Gage, (12 years old.)

Pair linen knit stockings, Mrs. E. Woodin.

Pair woolen socks, Miss Jane Gage, (10 years old.)

Buckskin mittens, David Mather.

Buckskin gloves, David Mather.

FLOUR.

Best barrel flour from least wheat, E. & R. Goodrich.

ORNAMENTAL NEEDLE WORK.

Ornamental needle work, Mrs. J. B. Walker.

Worsted worked portfolio, Mrs. P. Pratt.

Greatest variety of worsted work, Mrs. A. T. Crosman.

Fancy chair work with needle, Holmes & Moeller.

Worked collar, Mercy Webster.

Worked quilt, Mrs. Cudney.

White quilt, Mrs. C. Roosevelt.

FANCY WORK.

Worsted fruit basket, Miss Hannah Foote.

FLOWERS.

Greatest variety of dahlias, William M'Clinches.

Bunch of dahlias, Miss Sarah Moon.

Collection of green house plants, Mrs. C. S. Payne.

FRUIT.

Ten varieties of table apples, A. C. Bliss.

Six varieties of winter apples, C. N. Beecher.

Seedling varieties of peaches, J. W. King.

Twelve quinces, Henry I. Higgins.

VEGETABLES.

Turnips, Lyman G. Buckingham.

Beets, Steward H. Webster.

Onions, Benjamin Pearson.

Italian pointed cabbage, Benjamin Pearson.

Lima beans, Steward H. Webster.

Largest Pumpkin, Wm. M'Clinches.

Pink-eye potatoes, Daniel Curtis.

FIELD CROPS.

Sample winter wheat, Rowland B. Perry.

Crops of Indian corn—1st, Kowland B. Perry; 2d, Rowland B. Perry.

PLOWING.

With horses—plow and team owned by George W. Thayer, held by Clark Roads.

F. H. RANKIN, *Secretary*.

Flint, Oct. 4, 1851.

LIST OF ARTICLES

Recommended by the Viewing Committees for Discretionary Premiums, or otherwise favorably noticed by them.

CATTLE.

A two year old bull owned by Asahel Robinson.

Two yoke of 4 year old steers, 1st, Asahel Curtis; 2d Francis Brotherton.

Yoke of two year old steers, Horace Bristol.

HORSES.

Stallions two years old, 1st, M. J. Barrett; 2d, Judah Butler; 3d, Samuel Atherton.

A pair of matched three year old mares, P. Hicks.

A two year old mare colt, D. W. Stoel.

A two year old gelding, Isaac Schram.

SWINE.

A sow, James Delbridge.

Three barrow hogs, Alanson Payson.

A sow, (sister to the premium boar,) well worthy of notice, Simeon Simons.

MISCELLANEOUS ARTICLES.

Forest queen cooking stove, (premium for workmanship on fixtures and style of stove,) E. H. Hazleton & Co.

Folding door parlor stove, (premium for style of article,) E. H. Hazleton & Co.

A smoked ham, (premium for its flavor and manner of keeping during the summer. This ham was covered and every crevice filled with black pepper when taken out of the brine,) C. N. Beecher.

One chair, (premium,) Holmes & Moeller.

Case of jewelry, (premium,) William Stevenson.

Agricultural books, (an excellent collection, and worthy of recommendation,) A. B. Pratt.

A cooking stove, (worthy of notice as home manufacture,) King & Forsyth.

Lot of castings, (worthy of notice,) King & Forsyth.

Specimen of Irish cabinet ware, (best foreign manufacture,) Mrs. F. H. Rankin.

A rifle gun, (good workmanship, and worthy of notice,) C. W. Murray.

A three panel door, Solon C. Bliss.

Dressed deer skins, (first rate workmanship, worthy of notice,) David Mather.

CARPETING, &C.

Pair coverlets, (best blankets, premium,) Grovenor Vinton.

ORNAMENTAL NEEDLE WORK.

A bell rope, (premium,) Miss Brent.

FANCY WORK EXCEPT NEEDLE WORK.

Five oil paintings, (best oil paintings, premium for the "View near Florence.") Henry L. Brent.

Architectural drawings, (this piece of house drafting is well deserving of a premium,) Henry Stanard.

Drawings, (best drawings in water color and pencilling,) Mrs. Wm. Crocker.

Many pieces of rare merit have been handed in, but not numbered, and consequently cannot be granted premiums.

The committee have found it extremely difficult to decide upon pieces, there being so great a variety, and all so deserving of praise.

FLOWERS.

Oleander, (a fine Oleander of two years' growth, though not entitled to a premium,) Peabody Pratt.

FRUIT.

Lot of apples, (committee recommend a gratuity to owner, for best specimen of seedling apples,) Caleb S. Thompson.

The committee on fruits have found it extremely difficult, from the number of specimens of superior apples, most of them of surpassing excellence, and well worthy a premium, to make an award; the difficulty being not a little increased by a confusion of names, which evil can be best removed by a comparison at such exhibitions as the present. This consideration alone should induce all lovers of good fruit to bring along their choice specimens, that a comparison may establish their names and comparative merits, and thus induce the cultivation of the best varieties, and prevent the disappointment that after arises from the cultivation of inferior varieties under erroneons names.

The same difficulty of selection from many worthy specimens was experienced in regard to Quinces; and those who did not take a premium may console themselves that it was not from the deficiency of their specimens, but the remarkable superiority of another.

The great deficiency of good pears is a subject of regret, which it is hoped will be remedied in future.

VEGETABLES.

One doz. large pumpkins, Benjamin Pearson.
Three large turnip beet, George S. Hopkins.
Six sugar beets, Steward H. Webster.
Sample of Dutch top onion seed, Steward H. Webster.
Twelve pie-plant stalks, William McClinches.
Four Belgian carrots, R. W. Dullam.
Two blood beets, George Crocker.
Peck of flesh-colored potatoes, L. W. Thatcher.
Lot of mixed onions, William H. Crocker.
Lot of red onions, Asahel Curtis.
Two large drum-head cabbage, B. P. Foster.
Basket of red peppers, Caroline Pearson.

FIELD CROPS.

Specimen of corn, James B. Walker.
Best field wheat, Levi Preston.

PLOWING.

A premium recommended to plow and team owned and held by Rowland B. Perry.

The show of horses upon the ground was remarkably fine, and speaks well for the interest which is taken by the farmers of Genesee County, in the breeding and improvement of that noble animal. We regret that so much cannot be said for the display of neat cattle, which was far behind what our county ought to have produced. The animals of this class exhibited, were much fewer in number than might have been expected, and, with some creditable exceptions, not superior in quality. In this department, our farmers are certainly behind the advancement they evince in all other class of stock; and we hope the evidence of backwardness afforded at this exhibition, will serve to stimulate them into exertion to have their cattle keep pace in improvement with their horses, sheep, and swine.

Sheep husbandry begins to attract considerable notice throughout the county; and already much attention is being paid to the introduction and propagation of the best varieties. The sheep pens at our Fair afforded a display which was highly creditable and gratifying. The same may be said of swine, of which some very superior specimens were exhibited.

In the fruit department, the display of apples and quinces is deserving of notice, the specimens in competition being of surpassing excellence, such as could hardly be exceeded in any part of the world, and serving to demonstrate not only how well adapted are our soil and climate to fruit culture, but also, that great progress has already been made by our citizens in availing themselves of these advantages.

Upon the whole, our Second Annual Fair has been highly successful. The Society is now fairly established, and liberally supported by the farmers and other citizens of the county. The advantages to be gained from the friendly conference, generous rivalry, and interchange of experience, which can only be elicited by such annual gatherings, begin to be realized and appreciated; and we trust the Genesee County Agricultural Society has now entered upon a career of useful prosperity.

With the exception of the potato, which has suffered extensively from the rot, the crops have yielded well and been secured in good condition. The yield of corn, per acre, for which premiums were awarded under the head of field crops, was respectively as follows:

1st premium to Rowland B. Perry, 130 bushels.
2d " " " 120 "
3d " James B. Walker, 104 "

As respects the general system of agriculture throughout the county, it may be regarded as steadily improving, and keeping pace with other sections of the country, similarly circumstanced, where the land is new and but partially cleared.

An admirable address was delivered on the second day of the fair, by Levi Walker, Esq., of which a copy is herewith transmitted.

For the executive committee.

F. H. RANKIN,
Recording Secretary.

Flint, Dec. 27, 1851.

ADDRESS

Delivered at the Second Annual Fair of the Genesee County Agricultural Society.

BY LEVI WALKER, ESQ.

Mr. President and Gentlemen Farmers:

The primary occupation of the human race, so far as we can judge from the only record we have of man's primeval state, was agriculture. The first man, when he first opened his eyes upon the blessed light of the sun, found himself set down in the midst of a beautiful garden. And although that garden was planted by the hand of the Almighty, and though he "made to grow out of the ground" "every tree that was pleasant to the sight and good for food," yet I apprehend it would be a great mistake to suppose that while Adam occupied the garden of Eden, it brought forth the choicest fruits of the earth, in richest profusion, without any effort on his part. We are told that "the man" was "put into the garden," "to dress it, and to keep it," by which I understand that he was to cultivate it. That its productiveness and beauty, and the convenience and elegance of its arrangements, depended, in some degree, on the industry, diligence and skill of him who was permitted freely to eat of every tree of the garden. But however this may have been, we know that when he was ejected from that delightful residence, he was "sent forth to till the ground," and was told that in the sweat of his face he should eat bread until he returned to the ground whence he was taken; which could mean nothing less than that he should cultivate the earth and raise his bread. We are also distinctly told that his two eldest sons were agriculturists; one was a tiller of the ground, the other a keeper of sheep.

If we trace the history of the human family from that time to the present, we shall be led to the conclusion that a large majority have always been engaged in agricultural pursuits. True, this majority has not at all times and in all nations borne the same ratio to the whole; neither have their agricultural labors always been conducted with equal skill, nor crowned with like success.

Some nations have occupied themselves chiefly in hunting and fishing and in war, to the almost entire neglect of agriculture. These have always been found in a savage state, roaming over large tracts of wilderness country, few in numbers in proportion to their extent of territory, and of small and uncertain increase. Other nations, to the employments just named, have added a little attention to agriculture, and have been found one step in advance of savages. Others, still farther advanced, have been principally engaged in the rearing of flocks and herds, giving some little attention to the cultivation of the soil and the raising of grain and mechanical pursuits. While those nations, which occupy the proudest pages of history and command the most admiration for their genius, intelligence, refinement and prowess, have been alike renowned for their achievements in agriculture and its kindred pursuits, it has, I believe, proved true, in all ages and in all parts of the earth, that civilization and enlightenment and elevation in the scale of being, have been in proportion to the advancement and perfection of agriculture and the mechanic arts. And it must of necessity always be so.

Civilization—even a low degree of civilization—requires a certain density of population which cannot be sustained without agriculture and mechanics. Mankind are social and imitative and reasoning beings, capable of improvement by culture. They must reside in near neighborhood, so that they can easily and frequently come together and communicate to each other their thoughts and feelings and wishes, compare views, consider consequences, and draw conclusions; so that they can support schools and the printing press, and religious, political and social institutions. Without these, there can be no refinement, no civilization. And there can be nothing of these without a greater density of population that can be supported by any means without the aid of agriculture. The population of this county though yet too sparse to secure a high degree of enlightenment, is still so dense that it could not be supported a single month by hunting and fishing, and the spontaneous productions of the earth, within the bounds of the county. These resources would be exhausted before the expiration of that short period. And so it is everywhere. A population sufficiently dense to maintain the necessary institutions of society, must depend almost entirely upon the products of agriculture for sustenance. Hence the necessity of agriculture to the intellectual, the moral, religious, political and social elevation of man. And it is perhaps equally necessary to the full development of his physical powers. Hence also the importance of carrying it to as high a degree of perfection as possible. The highest possible degree of perfection! It may not be easy to define the limit of possibility in this direction. Probably that limit has not yet been reached any where.

The greatly advanced state of agriculture at this time, beyond that of any former period of which we have knowledge, and the fact that new and astonishing discoveries and improvements are constantly being made, admonish us to beware of setting bounds which we may not hereafter be able to overleap. There is reason to apprehend that many facts and principles which, when brought to light, will have an important bearing upon this subject, have thus far eluded the vigilance of scientific explorers. But we may safely conclude that what has been done, and is now being done in one place, and in many places, may, under equally favorable circumstances, be done in all places.

The United States are in possession of every physical advantage requisite to render it capable of being made the first agricultural nation in the world. We have a latitude extending from the 25th to the 49th degree—a climate ranging from that of constant summer to perpetual winter, with all the corresponding varieties of natural productions, vegetable and animal—a virgin soil of at least as great fertility as that of any portion of the earth, and of equal extent, abounding in all the necessary and useful metals and minerals. Our territory is traversed by numerous navigable water courses, whose capacity is more than sufficient to furnish an outlet for all the surplus products of our industry. Our extensive sea-coast is indented with harbors, furnishing unsurpassed facilities for commercial transactions.

The political institutions of our country are more favorable to the full and perfect development of all the powers of rational beings and of all the physical resources within their reach, than those of any other nation. We are the only really free people that the sun shines upon. France is a Republic, it is true, in name—but Frenchmen say it is only in name. Editors are fined and imprisoned in France for publishing their opinions. Protestant preachers and teachers are annoyed and embarrassed, and hindered in the pursuit of their respective callings. Booksellers are imprisoned for selling religious tracts. Rome, once the cradle of liberty and civilization, then the nursery of science and literature, afterwards the mistress of the world, is now under ecclesiastical dominion, governed by a Pope, the successor of St. Peter, and head of the Christian church (so claimed.) In Rome, a man is sent twenty years to the gallies for persuading another not to smoke a cigar. Thousands are imprisoned on suspicion of entertaining liberals opinions, indefinitely being denied a trial. In Florence, a noble Count is exiled for reading the Bible secretly. In Milan, a man is shot for having republican pamphlets in his pocket. In Turin, a young husband could be released from unjust and cruel punishment, by the sacrifice of his wife's virtue, not otherwise. Great Britain boasts of freedom, though not a Republic even in name. One of her distinguished poets, contrasting her condition with that of other European States, exultingly exclaims:

> "'Tis Liberty that crowns Brittannia's isle,
> And makes her barren rocks and her bleak mountains smile."

And the contrast was a fair one, and the claim to *comparative* liberty just. Yet, when one thousand of the peasantry of England all dresssed up in white frocks and a little bit of red ribbon upon their hats, are marched in a body into the crystal palace, and one of them is asked by Mr. Isham, "if he is a farm laborer?" He replies, "yes sir, Squire Pusey is our Master, and a fust rate man he is, too, and a gentleman, to pay our expenses to the great exhibition." And when asked if the laborers are prospering, he replies, "No, they're haying." Another voice at this moment bawls out "Pusey has brought these fellows out to get paraded in the newspapers, for the great show he can make of them." In answer to another inquiry, one of them said he had "hearn of America, and once petitioned his master to let him go there, but his petition was refused." Another, being asked if he belonged to Squire Pusey, said "No, I belong to the Earl of Radnor. He has brought five hundred of us here to see the great exhibition, all at his own expense." And Mr. Isham, on enquiring of an intelligent looking Englishman that stood by, how it was that these fellows called Squire Pusey, Master, and talked about belonging to the Earl of Radnor, was answered, "why they belong to the soil as much as the Serfs of Russia, and though not named in the Bond, are really transferred from master to master." That is a specimen of British liberty. A specimen of the liberty enjoyed by the freest nation on earth except our own." One thousand laborers, in the livery of Squire Pusey, owning him for their master and going up in a body to the great exhibition, at his expense; his object being to make a great show, and get paraded in the newspapers. Five hundred belonging to the Earl of Radnor, &c. And instead of "prospering," they are haying, or harvesting, or seeding, or what not, anything but prospering. Poor fellows! they don't know the meaning of the term, and probably they realize still less of the fact indicated by that charming word "prosper."

We are the only people, every one of whom is a man, and his own man; every one of whom is secure in the enjoyment of that which is his own, and in the exercise of all his natural powers, and his inalienable rights. We have no laws, either in the statute book, or in the structure and condition of society, regulating and cramping and limiting a man, in respect to the business in which he may engage, or the place in which he may reside. We can all choose our occupation and place of residence, and change as often as we think proper, without petitioning our master. We have no laws regulating or requiring religious worship. We can read the Bible in secret or openly, and worship according to our own convictious of right and wrong, or worship not at all, if we think that is the best way. We are subject to none but a voluntary tax for the support of religious institutions. We have no censorship of the Press, or of speech, no titled nobility, no landed aristocracy, no gentlemen, no castes or classes, or other divisions of society, recognized by law, no laws establishing precedence or regulating social intercourse. We all constitute one great class, standing on the same devoted platform, in perfect equality, except the distinctions which result from the differences in the gifts of nature's God, and from individual voluntary effort; distinctions which are inevitable and inseperable from a fallen state of humanity. We can think, speak, write and publish what we please. We can generally choose our associates.

We are at liberty to amuse ourselves in our own way, and generally, not only to do ourselves, but to persuade others to do or not to do any particular thing which we may imagine to be for their advantage or disadvantage. We are watched by no police, and feel no apprehensions of legal consequences while pursuing an honest course, even though it be a supremely foolish one. We are in no danger of arbitrary imprisonment, and have no occasion to submit to any dishonorable sacrifices to procure our enlargement. There is no need of any such person among us as "Luke Lawless." We have plenty of excellent land to be had for the taking, and millions upon millions, at a mere nominal price, compared with the value of land in other civilized countries. We have no laws of primogeniture. All the children inherit alike, unless disinherited by the parent. We have no enormous national debt, taxing us with its annual interest of hundreds of millions, no extravagant military or naval force, burdening us with its support. Our national government imposes no tax, except an indirect one on the products of commerce, which is scarcely felt as a burden by any one, and is held by one half of our people, to be a very profitable investment. We have, and we ought to thank God for it, and our State Legislatures and Congress, in nearly, if not quite every part of our country, laws providing for the support of schools, primary and superior, and where the population is sufficiently dense, schools are tolerably well supported, and are diffusing the blessings of light and science, to every part of our country and every portion of its inhabitants, and securing coming generations against the ignorance and poverty, which would make them content to call Squire Pusey, master, and to acknowledge that they belong to the Earl of Radnor. We have a spirit of liberality, which generally brings within our reach, the means of excellent religious instruction, and secures to us, in the eyes of the world, an honorable position for the multitude, and beneficent character of our benevolent and philanthropic institutions. We have, perhaps, more wealth, certainly a more general distribution of it, than has fallen to the lot of any othpeople. We have energy of character, perseverance and enterprise equal to any undertaking within the compass of human achievement, and we have, almost uniformly, the blessing of Heaven on all our industrial efforts.

Is there any reason why we may not make our agriculture at least equal to that of any other people? I apprehend there is none. But we have much to do to bring about so desirable a result. England is far ahead of us in this respect; also some other old countries which have not half our advantages. Agriculture is there reduced to a science, and it must be here. Modern chemistry has done much for the improvement of agriculture, and it promises much more. It has developed the structure and composition of vegetables, and in some measure the process of their growth, and ascertained their relative value as food, for men and animals. By chemical analysis it is ascertained that all soils are composed of a few elementary substances. Those substances are in different soils combined in different proportions; sometimes some of them are wanting, and others exist in superabundance. The consequence is, that some soils are excellent for some purposes, and almost useless for others. Chemistry also makes us acquainted with the properties of manures, and their mode of action in fertilizing the soil and nourishing the plant. Now the farmer ought to know in the first place, what proportion of each of the elementary substances is necessary to constitute a good soil for any particular purpose. He ought in the second place, to be able to ascertain what elements, and what proportion of each are actually present in the soils which he cultivates. And in the third place, he ought to know how to supply any deficiency, and correct any disproportion that may exist. Then he ought to make a judicious, practical use of this knowledge, in carrying on all his farming operations. This would be scientific farming. I know there are many excellent farmers who know nothing of all this, and care nothing about it. There are many men who can judge pretty correctly from the appearance of a soil, and their knowledge of its previous culture, whether it will bear a good crop of wheat, corn, oats, &c., without an analysis. But there are many others whose judgment on such subjects is poor; and all who judge from appearances are liable to frequent mistakes. Hence, it happens that we often see men labor hard to make a good crop, and fail from some deficiency or disproportion in the soil which an analysis would have detected, and enabled them to correct. But a knowledge of soils and manures; of the component parts of plants—their habits, manner of growth and relative value, is not all that is necessary to success in farming. Attention should be paid to times and seasons—to the selection of seed, and the preparation of the earth to receive it. The proper time for seeding will be best ascertained from experience, as it must necessarily be different in different latitudes and climates, and may be materially modified by other circumstances. But the young farmer need not depend entirely on his own experience for this. Much may be learned from the experience of others, of which he would do well to avail himself.

Reason and experience alike indicate the importance of using none but the best kinds of seed, and the very best of the kind that can be obtained. There is a great difference of opinion as to the best mode of plowing and preparing the ground for seeding. I shall not go into a discussion of this subject, but I think that whoever will examine it scientifically will become convinced that those who plow deepest and give their ground the most thorough preparation, will generally, other things being equal, be best paid for their labor.

The farmer's object is or should be to make the largest crops of the best quality at the least expense. To accomplish this object it is manifestly necessary to use the utmost diligence and skill. Yet it cannot be denied that, generally, in this country, farming operations are carried on rather unskillfully, often miserably so. This fact is perhaps the foundation of a great part of the complaints which we so often hear proceeding from farmers, of the hardness of their lot. How often do we see farmers brooding over the idea that they work harder, fare poorer, are worse paid, and pass a less pleasant and happy life than their fellow citizens engaged in other pursuits. In some instances this may be true; but that it is not universally, nor necessarily so, appears from the fact that many farmers, though living on the fat of the land and realizing all the comfort and happiness of which mortal man is capable, do yet accumulate wealth with a rapidity and facility which make them objects of the admiration and envy of the mechanic, the merchant and the professional man as well as their brother farmers. And why do not all prosper alike. It is not owing to a natural and irremediable difference in soil and climate, for no such difference exists in the cases under comparison. For the same reason, it is not owing to a difference in health, strength, and physical power. Nor is it because the sun shines more genially and the rain falls more seasonably on the farm of the one than of the other. But it is because one brings into exercise more judgment, more intelligence, more science, more careful attention to his business, more prudence and economy and perhaps more industry than the other. They perhaps trust in Providence alike, but one lets his powder get wet while the other keeps his dry. This makes all the difference. And yet it is very likely that the one who succeeds least, does as well as he knows how. And perhaps he is consoling himself with the very comforting reflection that badly as he has succeeded, he is not to be blamed because he has done as well as he could. But think a minute. Should your child die of a curable disease, it would afford you little consolation to be told by your physician: "True, the disease was curable—your child might have been cured—some physicians do cure all such diseases, but I did as well as I knew how." The idea would very likely suggest itself that a man ought not to hold himself out as a physician until he has learned how to cure curable diseases, the knowledge how to cure which is within the reach of every medical student. In like manner, if you do not know how to carry on your farming operations in the best way, you ought to learn how, unless you are too old to learn. If you are, you ought at least to put your children in a way to learn how to avoid the difficulties which are pressing you down. Do you doubt your obligation to do this? Let me tell you that every blessing given you to enjoy carries with it a corresponding responsibility. If our lot is cast in a land more favored by heaven than other lands, the fact throws upon us responsibilities beyond those of the people of other lands. Religious and patriotic considerations demand of us so to improve the high privileges bestowed upon us, as to do honor to the land which gave us birth, the government which protects us, and the God in whose hands we are, and to whose goodness we are indebted for all. There is another consideration which imposes the same obligation. You ought to have "respect to the recompense of reward." The idea of the great advantange which your children would derive from the knowledge which would enable them to take a stand by the side of those agriculturists who have been most successful, and who are justly ranked among the benefactors of their race, should impel you onward in this work. But how is this knowledge to be obtained? A State agricultural school has been proposed. And an agricultural bureau in one of the departments at Washington has been talked of. Either or both of these institutions would doubtless be very useful. But neither of them has been established yet, and the prospect that they will be soon is not very fair. And if they should be, they would come far short of supplying the wants of the farming community. But every good neighborhood has a school of its own and under its own control, and if the farmers please they can have chemistry, geology and botany, and every other science necessary to fit their sons and their daughters to become accomplished farmers and farmer's wives, taught in those schools, in addition to those branches which are now taught or pretended to be taught there. They have only to determine that no teachers shall be employed but such as are competent to teach and will teach them, and the thing is done. And it ought to be done as soon as possible. I apprehend there is no good reason why it should not be. Such teachers would be scarce for a time, but the supply would soon adapt itself to the demand. To set an unvarying standard is the only way to secure a supply of teachers of any defined degree of qualification. Should further legislation be found necessary the farmers have the legislative power in their own hands and can exercise it as they please.

Agricultural societies are furnishing means of much valuable information, as well as stimulants to activity and enterprise, and it is gratifying to know that people are beginning to appreciate the advantages of such associations. Can the effect be anything but salutary, for the people to assemble once a year from all parts of the county, and look upon such a splendid display of fruits, flowers, and vegetables, of animals and the products of the dairy; of manufacturing skill and mechanical genius, with which we are here surrounded? Will you not, after seeing what your neighbors have done, and learning something of the manner of doing it, return to your homes with a determination to emulate their enterprise and success? What more pleasant sight could gladden the eye, than that rich show of apples, pears, peaches and quinces, superior, I am imformed by those who have seen both, to anything of the kind offered at the recent State fair in Detroit? And those vegetables, too—those beets, carrots and onions, and turnips, &c. And did you notice those beautiful samples of wheat and corn, and beans? And may it not justly excite our pride that they were all grown in our own county? And perhaps there is not an 80 acre lot in the whole county, but what would with proper cultivation and management, produce specimens of everything that grows out of the earth, that would compare favorably with the best of those exhibited here. Such is the fertility of the soil and richness of the productions of the county, for the right use and proper improvement of which, you, gentlemen farmers, are responsible. And those exquisite productions of female industry and ingenuity, those offerings of the gentle and the fair, without whose approving smiles, concurrent action, and cheering presence, every attempt at an anniversary like this which we are now celebrating, would prove a miserable failure. Perhaps, all things considered, there is not in the State a county more highly favored than ours; nor one more capable of a high degree of agricultural excellence. And I think the right spirit is now animating our population, and that henceforth our progress will be onward.

In some places, single towns or neighborhoods have formed associations and meet together at frequent stated periods for consultation and mutual improvement in the business of their profession. Such little associations have sometimes proved highly beneficial, and it is submitted that they might be useful here. But should any of you think proper to act on this suggestion, when you get together, I advise you to do do your own talking. A man *can* talk best on subjects with which he is acquainted.

But agricultural newspapers, are probably the most available present means within our reach of obtaining agricultural knowledge; and of the multitude of papers of this kind, perhaps the Michigan Farmer is best adapted to the wants of the farmers of this county. That paper has, I believe, already a considerable circulation in the county, but I do not see how any man, rich or poor, who cultivates the earth to the extent of half an acre, can be content, or can afford to be without this or some other periodical publication of like character. I would say, therefore, in conclusion, to all professional farmers, and to all who have a garden of ordinary size to cultivate, and particularly to those farmers who have hitherto been unlucky, whose lot has been a hard one, who had found it difficult to "make the two ends meet," take an agricultural paper, read it regularly and attentively, and understand it, carry its valuable suggestions into practice, be industrious, be thorough, be careful, be prudent and economical, Do everything at the right time and in the best manner. Avoid rum shops, go to meeting every Sunday, pay your debts promptly, do your duty in all things faithfully, and then you may trust in Providence, and Providence will not fail you.

REPORT

Of the Kent County Agricultural and Horticultural Society.

J. C. Holmes, Esq., *Sec'y Mich. State Ag. Society.*

This Society is the same that was originally called the Grand River Valley Agricultural Society, the name being changed in consequence of there being other Societies formed in the Grand River Valley, and a new constitution has been adopted. The first fair was held under the name of the Kent County Agricultural and Horticultural Society, on the first day of October, 1851. There was a committee of arrangements for obtaining a suitable place for the exhibition, and putting in order a lot for the reception of animals, with pens, and other conveniences for the keeping in proper places every thing presented for exhibition, with a suitable building in the centre, about thirty by sixty feet in size, as well fitted up and arranged as the limited means of the Society would permit.

The day being a pleasant one, the attendance was more general by far than was anticipated. The course adopted of having admission tickets, was decidedly better than relying wholly upon the voluntary subscriptions, which has been the only one pursued by this Society heretofore. But few have objected to this mode, of raising funds, to meet the expenses of this Society, and I trust (after reflecting) that every one that takes any interest in the affairs of this Society, will find this the easiest method of raising the necessary means of carrying on said society.

The amount of premiums was about three times as much as last year; and I think from present appearances, we can safely calculate on a very large increase of interest another year.

LIST OF PREMIUMS.

HORSES.

1st premium, best stud horse, H. A. Dennison,	$2	00
2d do do C. S. Johnson,	1	00
1st do brood mare, D. Meach,	2	00
2d do do A. Hill.	1	00
Best pair matched horses, H. Potts,	2	00
" 2 years old colt, P. Reed,	2	00
" yearling colt, I. B. Hollinder,	1	00
" sucking colt, J. Bentham,	2	00

There were four yearling colts entered, so nearly alike, that it was difficult to decide, but finally the judges gave the Hollinder colt the preference.

There were four mares presented for examination, well worthy of a premium. Could they have given one under their regulations they would have done so with pleasure.

There was a three year old colt from Ionia county, belonging to Mr. English. which could not be entered, coming from another county as it did, but which was a beautiful specimen of the horse kind.

CATTLE.

1st premium, 2 years old bull, D. Meach,	$2	00
2d do do do J. F. Chubb,	1	00
Best yearling bull, J. J. Buck,	1	00
1st premium yoke working cattle, D. Meach,	2	00
2d do do do O. H. Foot,	1	00
Best bull calf under 9 months old, W. A. Brown,	1	00
" cow, P. Reed,	2	00
" 2 year old heifer, R. Green,	1	00
" 1 year old heifer, J. F. Chubb,	1	00
" heifer calf, Mr. Gerue,	1	00

SHEEP.

Best buck, A. Brewer,	$2	00
" 5 ewes, C. Maracle,	2	00
" lamb, do	1	00

SWINE.

Best breeding sow, O. Van Beuren,	$2	00

POULTRY.

Best Poland fowls, J. F. Chubb,	$1 00
Best lot Chinese fowls, very large and superior, and the only ones in this section of country of the kind, exhibited by Henry Hall,	1 00
Best mixed lot poultry, Dr. Platt,	2 00

HONEY.

Best box honey, H. Rhodes.	$1 00

HOUSEHOLD PRODUCTS.

Best cheese, M. Royce,	$1 00
" pail fall butter, G. M. Barker,	1 00
" pail June butter, J. F. Chubb,	1 00
" loaf of bread, do	1 00
" specimen sap sugar, do	1 00
" rag carpet, J. R. Hoag,	1 00
" wool carpet, J. Rogers,	1 00
" hearth rug, M. Royce,	1 00
" full cloth, D. Schermerhorn,	1 00
" cotton and wool flannel, J. F. Chubb,	1 00
" pair wool stockings, W. G. Henry,	75
" pair socks, J. F. Chubb,	50
" pair wool mittens, Mrs. B. Hinsdall,	50
" pair flannel sheets, J. F. Chubb,	1 00
Best patch work quilt, Mrs. B. Hinsdall,	2 00
2d do do D. Schermerhorn,	1 00

FANCY ARTICLES.

Best worsted work, top ottoman, beautifully manufactured, Mrs. D. Hollister,	$0 75
Best lamp mat, (name lost,)	50
Best specimen of ornamented oil painting, few that can excel if equal this painting, Mrs. E. T. Nelson,	2 00
Best monochromatic painting, C. F. Moore,	1 00
" daguerreotype, Mr. Proctor,	1 00

FLOWERS AND FRUIT.

Best and greatest dahlias, Miss R. A. Hatch,	$0 75
" lot fall apples, J. Ewing,	1 00
" lot winter apples, S. M. Pearsall,	1 00

Best specimen of pears, H. Rhodes,	$ 50
" do peaches, S. Wright, Jr.,	1 00
" do grapes, D. Hatch,	50
" 2 watermelons, J. Ballard,	25

VEGETABLES.

Best dozen white turnips, W. Berdsall,	$0 50
" do beets, H. H. Allen,	50
" do ears sweet corn,	50
" lot vegetables, F. H. Cuming,	1 00
" lot onions, E. F. Strong,	50

FARMING UTENSILS.

Best fanning mill, Renwick & Brother,	$1 00
" horse rake, A. R. Hoag,	50
" grain cradle, B. Hoag,	25
" set wagon wheels, G. C. Fitch,	50
" set carpenter chissels, Blain & Cook,	1 00
" hand hammer, do do	50
" adz, do do	50
" chopping axe, do do	50
" side saddle, Kruger & Tusch,	50
" double harness, do	50
" mattrass, do	1 00
" double or subsoil plow, G. S. Dean,	1 00

The articles manufactured by Blain & Cook, of this city would be hard to be beat by any of the edge tool manufactures in the eastern world.

There was an address delivered by the Rev. Mr. Hammond.

Very respectfully yours,

D. HATCH, *Secretary.*

REPORT

Of the Kalamazoo County Agricultural Society.

J. C. Holmes, *Sec. Mich. State Ag. Society:*

In obedience with the 11th article of the constitution of the State Society, the following is a brief report of the transactions of the auxiliary society of Kalamazoo, for the year 1851. With no other emotions than those of pleasure, we would communicate the complete triumph of the great cause in which the yeomanry of this county have for the last six years been engaged. It is true, there have been periods within that time, when the omens have been unfavorable, but we have outlived all these discouragements, and can now announce to the parent institution that all doubts with regard to the future, are completely silenced. It may not be traveling out of the proper province of a report to say, that to no gentlemen are we indebted so much for our success as to the untiring efforts of Andrew Y. Moore, John Milham and William H. Edgar, Esqrs. From its origin in 1845, to its present auspicious condition, they have stood by it like faithful sentinels, and have spared neither toil nor means in pushing the enterprise forward to success. It is with pride that we seize upon this opportunity of thus publicly doing them justice.

The fair for the year that has just closed, was held at the village of Kalamazoo, on the 1st and 2d days of October. Every arrangement commensurate with the prospect of an unusual exhibition of stock, manufactured articles, &c., was made, and every anticipation was more than doubly realized. Instead of an expected gathering of some 2 or 3,000 persons, 10,000 was considered a judicious estimate. "Kalamazoo county never does any thing by halves," has long been vaunted abroad; never was the saying so fully verified.

From the entrance fee, which was in single cases the small sum of one shilling and family tickets fifty cents, a sum was received sufficient to liquidate all the liabilities (present and past) connected with the promotion of the project and leave a balance in the treasury. The exhibition of horses, cattle and sheep exceeded any thing ever before seen in our midst; and so far as horses were concerned, judges, who had but just returned from the State exhibition, pronounced ours decidedly in the advance. The Devons and Durhams and indeed the grades were extensively and satisfactorily represented.

In nothing has there been such a signal improvement since the first fair of the county as in that of horses and cattle. The show of sheep, though nothing as large at Detroit the previous week, was yet highly creditable.

Manufactured articles still indicate a rapid progress and seemed to engross a great deal of attention. Indeed, every thing connected with the exhibition was beyond anticipation, and consequently satisfactory. No jar, no dissatisfaction, no complaint was heard, calculated to interrupt the great agricultural holiday.

In the afternoon of the second day the usual address upon such an occasion was pronounced by F. W. Curtenius, a farmer residing in the county. Immediately following the address the election of officers for the ensuing year took place, which resulted as follows:

President—Frederick W. Curtenius.

Recording Secretary—Amos D. Allen.

Corresponding Secretary—Luther H. Trask.

Treasurer—John Sleeper.

Executive Committee—Andrew Y. Moore, George W. Lovell, Joseph Frakes, John Milham, Samuel Clark, Alison Kinne, Edwin H. Lothrop, James Henry, Jr., Fletcher Ransom.

VICE PRESIDENTS.

Alamo—Fletcher Ransom.

Brady—Lewis C. Kimball.

Climax—Stephen B. Eldred.

Cooper—Barney Earl.

Comstock—Ezra Stetson.

Charleston—George Davis.

Oshtemo—Aaron Eams.
Kalamazoo—Isaac Vickery.
Portage—John Parker.
Pavillion—Italy Foster.
Prairie Ronde—Samuel Hackett.
Richland—Horace M. Peck.
Ross—Harry King.
Schoolcraft—Jonas Allen.
Texas—Albert G. Towar.
Wakeshma—Jacob Gardener.

On motion of Alexander J. Sheldon,

Resolved, That the executive committee be instructed to take immediate measures to procure at least not less than four acres of land, and erect a building suitable for the Society.

The following are the reports of the several committees, and a list of premiums awarded on articles exhibited:

ON FARMS.

Best prairie farm,	F. W. Curtenius	$6 00
2d best do	Joseph Frakes	4 00

ON HORSES.

The committee for the examination of horses respectfully report that they have discharged the duty assigned to them; and although the limited time allotted us for examining so large a number, made the duty very arduous, yet the animals presented were so superior that we performed it with infinite pleasure.

Your committee cannot but congratulate the people of Kalamazoo county, upon the rapid and extended improvement made in the most useful of all domestic animals, since the organization of our Society. The exhibition this day surpasses our most sanguine expectations. It is superior to that of either of our State Fairs, and has elicited the admiration of all who have witnessed it. But one cause of regret has existed, and the removal of that is so entirely within the power of our people, that we look with confidence to them for its accomplishment before our next annual fair. It is this—the small amount of funds belonging to the Society.

For this reason, the executive committee could only offer one class of premiums, whereas there should be three, or justice cannot be done to our most enterprising breeders. In a single class, horses for all work have great advantage over blood and draught horses, and your committee have consequently been compelled, in each case, to award the premiums under this restriction. The splendid blood stock of A. Y. Moore, Esq., and Gov. Throop, as well as several specimens of the draught horse presented by other gentlemen, were thus deprived of that high rank to which they would be entitled under a range of premiums embracing the three classes. For this same reason, no premiums were offered for matched and single horses, and yet your committee have rarely seen a better exhibition.

The matched pairs of Messrs. Webster, Kellogg, Bliss and Bloss, as well as the single horses of Messrs. Sager, Goss and Hogoboom, are all splendid animals, doing great credit to the taste and judgment of their respective owners, and cannot fail to please any purchaser whenever the present owners inclien to part with them. They added much to the interest of our fair, and were all entitled to premiums.

Your committee have awarded the premiums as follows:

Best Stallion over 3 years old,	Green Mountain Morgan	$3 00
2d " "	A. Y. Moore's Bucephalus	2 00
Best brood mare and colt,	D. J. Smith	3 00
2d " "	W. Beckwith	2 00
Best 3 year old gelding,	A. McCall	3 00
2d "	E. S. Knapp	2 00
Best 3 year old mare,	J. Cooper	2 00
2d "	B. D. Balch	1 00
Best 2 year old stallion,	H. Nesbit	2 00
2d "	T. H. Murray	2 00
Best 2 year old mare,	Wm. Parsons	2 00
2d "	Wm. B. Lawrence	1 00
Best 1 year old colt,	Isaac Cox	2 00
2d "	Wm. Parsons	1 00
Best yearling mare,	E. G. Robinson	2 00
2d "	A. Aldrich	1 00

All which is respectfully reported.

CHARLES E. STUART,
PELECK STEVENS,
JOHN McALLASTER,

CATTLE.

The committee on cattle, in the discharge of their duties, have been much pleased both by the number and quality of the animals presented to them for examination.

Within the last few years there has been introduced here many very fine animals of the Devon and short horn, and red Durham cattle, and our annual fairs have given pleasing evidences of the care and skill with which some of our farmers have bred them. These cattle should be very generally distributed throughout the country. The committee have no hesitation in declaring that fifty per cent more would be realized from rearing either of these breeds, than from the ordinary native stock. It may with propriety be affirmed, that the improvement so far is mainly attributable to our agricultural society. It is that which has produced that friendly spirit of emulation so indispensible to the success of our farmers, as well as all permanent improvement in stock.

We are most happy to see the society coming out of its embarrassments, and about to commence an era of prosperity. It needs only the united and fostering care of our brother farmers to secure to it the position which our county enjoys, namely, *the best in the State.*

Your committee, did time permit, would be happy to recur specially to many of the animals on exhibition. Their high merit, however, seemed well appreciated by the great number of our people in attendance, and doubtless produced a happy influence.

A few Devon grade calves, presented by the Messrs. Glynn, were beautiful animals, and showed judicious treatment. All who saw, admired them.

The various premiums have been awarded as follows:

NATIVE AND GRADE.

Best bull calf, A. Y. Moore	$2 00
2d " J. R. & H. W. Glynn	1 00
Best milch cow not less than 3 years old, A. Y. Moore	2 00
2d best milch cow, not less than three years old, N. A. Balch,	1 00
Best heifer calf, J. R. & H. W. Glynn,	2 00
" yoke of oxen, not less than three years old, D. J. Smith	2 00
2d best yoke of oxen, not less than three years old, S. Loveland,	1 00

DURHAM.

Best bull, not less than three years old, Russel Bishop,	$3 00
2d best bull, do J. F. Gilkey,	2 00
Best cow, not less than 3 years old, J. Frakes,	2 00
2d do do J. R. & H. W. Glynn,	1 00
Best 2 year old bull, R. A. Boyce,	2 00
2d do do J. Frakes,	1 00
Best 1 year old bull, John Brown,	2 00
2d do do J. F. Gilkey,	1 00
Best bull calf, John Milham,	2 00
" heifer calf, Joseph Frakes,	2 00
2d best do J. R. & H. W. Glynn,	1 00

DEVON.

Best bull, not less than 3 years old, J. R. & H. W. Glynn,	$3 00
2d best do Paulus den Blaker,	2 00
Best cow, not less than 3 years old, do	2 00
2d best do J. R. & H. W. Glynn,	1 00
Best heifer calf, do	2 00
" 1 year old heifer, M. Heydenburk,	2 00
" 2 do do	2 00
" yearling bull, F. Ransom,	2 00

DISCRETIONARY PREMIUMS.

Best milch cow, (native,) A. Eames,	$2 00
2d best do do D. J. Smith,	1 00
Best yearling bull, (grade,) S. M. Nichols,	2 00
" eight calves exhibited at the fair, J. R. & H. W. Glynn,	2 00

Respectfully submitted.

E. B. DYCKMAN,
AARON EAMES,
T. B. ELDRED,
W. B. LAWRENCE,
Committee.

SHEEP.

The committee on sheep would respectfully submit the following report of premiums awarded:

Best buck, J. F. Gilkey, ... $2 00
2d do E. T. Lovell, ... 1 00
Best yearling buck, do 1 00
" five yearling bucks, John Milham, ... 2 00
" " ewes, J. F. Gilkey, ... 2 00
2d best 5 do John Milham, ... 1 00
Best five yearling ewes, J. F. Gilkey, ... 2 00
" " buck lambs, John Milham, ... 2 00
" " ewes, do 2 00

In submitting the above, your committee feel deeply to regret that there should have been so little competition in this valuable and (at present) most profitable branch of agriculture. Although the animals presented for examination were of the very best kind—scarcely one present which would not have been worthy a premium—yet the number present was much less than at any previous exhibition—a fact much to be regretted, as your committee think no county in the State can compete with Kalamazoo in the number and quality of thorough bred sheep, and hope that another year our farmers will be more awake to their true interest, and have every premium strongly contested by competition.

J. PARKER,
WM. PRICE,
H. ROCKLAND,
Committee.

SWINE.

The committee on swine report as follows:

Best boar, not less than one year old, E. L. Knapp, ... $2 00
do under 12 months, Aug. Buel, ... 2 00
2d best boar, do do 1 00
Best sow, not less than one year old, Martin Heydenburk, ... 2 00
2d best sow, do do H. E. Fletcher, ... 1 00
Best sow pig, under 12 months old, Fletcher Ransom, ... 2 00
2d do do do P. Goodrich, ... 1 00

Best sow and pigs, Fletcher Ransom,	....	$2 00
2d do Philip Goodrich,	....	1 00

GEORGE LELAND,
JOSEPH FRAKES,
S. H. RANSOM,
Committee.

POULTRY.

The committee on poultry report as follows, to wit:

Best coop large fowls, Dr. Freeman,	....	$1 00
do Polands, Martin Wilson,	....	1 00
do Dorkins, do	....	1 00
Best lot poultry owned by exhibitor, Martin Wilson,	....	1 00

A. T. PROUTY,
JAY R. MONROE,
Committee.

FRUIT.

The committee on fruit report as follows:

Best and greatest variety of apples, Wm. Milham,	....	$2 00
do do pears, D. W. Dunham,	....	2 00
do do peaches, S. T. Brown,	....	2 00
do do quinces, S. Woodruff,	....	2 00
Discretionary, apples, W. Beckwith,	....	1 00
do do John Lyon,	....	1 00

EZRA STILSON,
SAMUEL CLARK,
CHESTER GIBBS,
Committee.

VEGETABLES.

The committee on vegetables respectfully report: That they have carefully examined the specimens entered and numbered; that but nine entries in all have been made, and in but one instance were there more than one entry of the same kind. Under such circumstances, there was but little discretion left to our committee. They have awarded as follows:

Best squashes, J. Veness,	$0 50
" onions from black seed, W. Milham,	50
" half bushel potatoes, O. Bush,	50
" pumpkins, do	50
" beets, H. Maxwell,	50
" red peppers, do	50
" cabbages,	50

SAMUEL CLARK, *Ch'n.*

HOUSEHOLD ARTICLES.

The undersigned, appointed a committee to award premiums upon household articles, respectfully report:

That they have examined with as much care as time and circumstances would permit, the several articles presented to them, and have awarded premiums as follows:

Best 20 lbs. butter, M. Heydenburk,	$2 00
" woolen carpet, Mrs. E. Warren,	2 00
2d best do Mrs. T. G. Abbet,	1 00
Best rag carpet, Mrs. Sally Dix,	1 00
" 10 yards of flannel, A. Taylor,	2 00
2d best, do do	1 00
Best 10 yards of plaid, Edward Chase,	2 00
" pair woolen stockings, Miss Sleeper,	75
" do mittens, Miss Crawford,	75
" coverlet, Mrs. Barney,	1 00
" quilt, Mrs. E. A. Howard,	2 00

This list is believed to comprise all the articles presented, on which the Society had offered premiums, but a great variety of other articles were offered upon which no premiums had been offered, of which your committee cannot withhold an expression of their admiration. Among these, we hope it may not be considered invidious to mention a few of superior excellence.

Some samples of needle work, by Mrs. S. S. Cobb, Mrs. H. G. Wells, Miss Cook, Mrs. Traver, Mrs. Krause, Miss Burrell, and others, were considered particularly worthy of commendation. Some samples of pencil sketching, by Miss Booher, were much admired. Knit comforters, by Mrs. Burrel and Miss Burrell, were fine specimens of their kind.

Several boquets of flowers, presented by Mrs. V. Hascall and Mrs. C. C. Cobb, were much admired, as was also some tissue flowers, by Miss L. Ford and Miss Tompkins.

Among the more substantial articles presented, was a lot of cassimeres and cloths from the factory of our enterprising townsman, A. Taylor, which, for fineness of texture, and beauty of finish, would do credit to any establishment in the country.

Messrs. Follet & Brown were on hand with samples of bar and fancy soap and candles, far superior to any thing we had supposed was manufactured in this county.

Messrs. Perkins & Folger exhibited specimens of house, sign, and ornamental painting and graining, which would, in our opinion, do credit to the best painters in the State.

There also was James Taylor, with a small slice of pork, being the first cut from the side, back of the shoulder, and weighing one hundred and fifty pounds; a very fair specimen of Kalamazoo county pork.

There were also many articles of the class entitled to premiums, between which and those to which the prizes were awarded, much difficulty was found in deciding, so equal were the merits. Some quilts by Mrs. Walters, Mrs. Pace, Mrs. Bush, Mrs. Avery, and others, were very fine.

We might go on enumerating articles of superior excellence, but will close with a simple expression of our admiration of the exhibition generally, as it came under our observation.

D. S. WALBRIDGE,
M. C. CHURCHILL,
I. MOFFATT, JR.,
L. CLARKE,
Committee.

MANUFACTURED ARTICLES.

The committee on manufactured articles make the following report:

Best farm wagon, E. H. Davis,........ ....$3 00
" prairie plow, Turner, White & Co.,.................. 1 00
" opening plow, A. Arms & Co.,........................ 1 00
" wheat cultivator, " " 1 00

DISCRETIONARY PREMIUMS.

Best buggy wagon, A. Healey,.......................... 1 00
" Shovel plow, E. Warren,.............................. 1 00

J. ALLEN,
L. EAMES,
J. F. GILKEY,
Committee.

All which is respectfully submitted.

AMOS D. ALLEN,
Recording Secretary.

ADDRESS

Delivered before the Kalamazoo County Agricultural Society, at the annual Fair held at Kalamazoo, on 1st and 2d days of Oct., 1851.

BY HON. F. W. CURTENIUS.

Fellow Citizens:

Had the efforts of the executive committee been successful, one better adapted to discharge the duties of a speaker upon this occasion would have occupied the platform.

While it is unfortunate for me, it is still more unfortunate for you that their exertions to procure the services of one richer in experience, and more familiar with the agricultural wants of this community, have been fruitless, and abandoning the forum, they have been obliged to betake themselves to the furrow in order to supply the vacancy.

This explanation is undoubtedly sufficient, to account for the absence of more suitable arrangements.

To day, fellow citizens, finds us once more together, for the purpose of speaking for a very few moments, upon one of the most absorbing topics,—temporal topics, which can possibly arrest the attention of man. Its magnitude is not confined merely to the few who are here assembled, nor to our adopted Michigan, nor the Union itself—it acknowledges no limits but the Globe. We come to talk of—*bread*—its best mode of cultivation—its importance—the many blessings it entails.

He who spake as never man spake, assures us that it is the fruit of toil, and anxiety and *sweat.* Its acquisition is replete with interest, inasmuch as He attached importance to it, who was familiar with our wants. If it is a matter involving so much interest, perhaps it would not be amiss to stop one moment and enquire who and how many are engaged in the production of this great element of human existence. Is it here and there only that one is found, sufficiently menial to delve in the soil, and to extract from it this staff of life?

How does this correspond with facts? In investigating this matter, let us avail ourselves of whatever statistical information may come within our reach. According to the census of 1840, (and I doubt not the census of 1850 would convey the same relative result) the adult male population of the United States, was 4,785,649, and as accurately as could be ascertained, were employed as follows:

Engaged in internal navigation,	33,076
" upon the ocean,	56,021
" in the learned professions, (including law, medicine and divinity,	65,255
" in commerce, (that is mercantiie business,)	119,607
" in manufactures,	791,749
" in agriculture,	3,719,951

To one who has never given the subject any attention, what an important and interesting fact here unveils itself. Eighty per cent—80 out of every 100 of the adult male population of our county engaged in cultivating the soil. The agriculturists have, you will see, the numerical strength of the Union, and might, were they disposed, control the government of the country. Notwithstanding they have the influence, yet they have never for one moment seemed disposed to avail themselves of it. Why is it? Evidently because there is a want of proper education among the great mass, and they are too apt to think that their appropriate sphere of action is to *"hew the wood and draw the water."* The 65,255 professional men are stronger intellectually than the 3,719,951 engaged in agriculture, and therefore rule them. It is not my intention to array one class of our citizens against another, or to institute any invidious comparisons. I am perfectly satisfied with my pursuit, and I can lay my hand on my heart and say that I am proud that I am an American farmer.

Ours is the noblest employment of life; no class in society has half the materials of usefulness, or half the claims to independence as the frugal and industrious tiller of the soil. And yet notwithstanding our avocation is one which should dignify those who embrace it, and indeed involves the great interests and welfare of the country, how stinted has ever been the pittance of legislation of which it has been the recipient.

With a few national exceptions, for the last 3,000 years, the same miserable policy seems to have been pre-eminent everywhere; and one would suppose that the Solon's and Lycurgus's of the land regarded the laborer as abundantly capable of taking care of himself without any legislative interference.

It is the glaring disparity of legislation between the physical and intellectuol laborer to which I advert, and which is a subject of regret. The great "fixed fact," that ten per cent of our population receive more legislative aid than the remaining 90 per cent, is the error to which I allude, and which calls for redress. They seem to have lost sight of that portion of the history of the past where we are made acquainted with the fact that Rome existed five hundred years without a lawyer and the city of Albany two hundred years without a physician.

It is not my design to cast any stigma upon the professions, for I regard now and have ever regarded the high-minded and honorable lawyer and physician as among the ornaments of society.

Had I a sovereign's power, agriculture should become, what it ought to be, a pursuit inviting enlistments, and not be regarded as one to be avoided. Had I the power, I say, there should be a model farm in every county of note in the Union, where agricultural experiments, at the public expense, should be constantly in progress

48

and whenever successful, placed at once within the reach of the husbandman, to subserve as far as possible his interest; nor would I permit a teacher to enter a school house, be its pretensions never so humble, who had not the ability, thoroughly to analize soil and the necessary chemical apparatus (furnished also at the public expense) necessary to impart this species of knowledge to his pupils. While agriculture is acknowledged to be the great interest of the country, no interest at the same time has received so little attention at the hands of government.

The soil—I love to contemplate it—one of the great gifts of heaven to man. Upon it hangs our happiness—aye! our very existence. While living, we repose *upon* it—*beneath* it when dead. If I wish one thing more than another, it is that I might awaken more interest in whatever pertains to its improvement, to its use and abuse; and wipe away as far as possible the stain which sometimes seems to attach to those who draw sustenance and wealth from its surface.

I would that I had the power to annihilate that "unsound public opinion which writes disgrace upon the perspiring brow of labor"—forgetting while they do so that labor is a combination of mental and physical exertion with a view to some useful result; forgetting that labor is the price of life—that it was a mother's labor that rocked our cradles, and it was that same mother's labor which contributed to rear us from infancy to manhood.

To labor we are indebted for everything that ministers to human want, save the air, which a kink Providence has seen fit to bestow upon us, and even this is breathed at the expense of exertion. Toil on thou fellow-laborer, for the time is at hand when he who walks the furrow will be more a king than he who, robed in royalty, treads with artificial dignity his palace, dependent upon you for his sustenance.

"The profit of the earth is for all; the king himself is served by the field"—is the language of Solomon. Heed not any effort to cover you with disgrace, but think of this: "that if there is one who can eat his bread in peace with God and his fellow man, it is he who has brought that bread out of the earth. It comes to him cankered by no fraud, wet with no tears, stained with no blood." The farmers gains are honest gains; what he gets is not at the expense of suffering or loss to others, but as the lawful fruit of his own exertions.

For centuries the struggle has been, not who should thrive and accumulate wealth by the greatest amount of labor, but who by the least—seeking rather to evade than boldly to confront it; thereby subjecting it to odium, rather than imparting to it merited dignity. The history of the recent past, as well as the history of the present, is beginning to exhibit something of a contrast in public sentiment, though there is much to be accomplished yet before innate worth and proportionate productiveness of human happiness will be assumed as among the standards of merit.

We can congratulate each at this time, upon the dawn of a more auspicious future. The great improvements which are daily being made in agricultural implements; the increased attention which is exhibited in the rearing of herds and flocks; the multiplied facilities for developing the capabilities of the soil, and the light which the science of geology and chemistry are throwing across our pathway, are all—all calculated to cheer and encourage us onward.

We have emerged forever, I hope, from the age, when in agricultural operations, the son reflected the sire as truthfully as ever mirror sent back resemblance. Prejudice to innovations upon the time-worn customs of our forefathers is yielding to the better way which experieuce and experiment has disclosed, and we are alive at length to the fact that agriculture, instead of being in a state approximating perfection, is yet in its infancy, and the great results which will characterixe its manhood, will be truly gratifying and astonishing.

To day, fellow citizens, finds us coming together, freighted with the experience of another year, and as an appropriate offering to our country, laying it on the altar of public good. We shall have lived to very little purpose—indeed, we shall have been faithless observers of passing events if we are unable to contribute something (though it be but little) to the general stock of agricultural intelligence. If he who can tell how to preserve the fertility of the soil, notwithstanding the repeated drafts which are made upon it, deserves well of his country, how exceeding great the claims of him who can disclose a process by which that fertility may not only be preserved, but increased under like circumstances.

It may perhaps be within the power of some humble individual occupying a place in this assembly, to make some disclosure (the fruit of watchful toil and experiment) which might earn for him the title of benefector. To come loaded down with the experience of the past is the prominent feature of this and similar organizations. It is at these annual gatherings of the people, that an unrestrained interchange of opinions in connection with our avocation is expected, and should take place.

While one recounts with earnestness the story of his crop, dwelling upon his anxiety attendant upon the precarious nature of a departure from some ordinary mode of culture, his harvest, his final success and his profits, detailing with minuteness the various stages of its production, and with a cheerful countenance making plain to you the manner in which some obstacle had been removed or surmounted—what a praise-worthy triumph is his. Another, with equal engagedness, will call your attention to his clip, and reveal to you with satisfaction the secret of his success in this particular branch of husbandry.

Here, if any where, is the proper arena where bloodless victories are to be achieved, triumphs to be gained, and laurels to be won. It is here that the advocates of the devon and the durham are to measure weapons, and battle for the mastery. It is here that the admirers of Bucephalus, and Ivanhoe and Morgan, marshaled under their respective champions, do honorable battle in behalf of their favorites. It is here that the merits of the Saxon, the Cotswold, the Escurial and the Merino's, French, Spanish and Pauler, are ably discussed, and some perhaps convinced. It is here that *that* animal known in common parlance as the "hog," but dignified with the title of "pork," when deceased, is without the least compunctions of conscience, ever and anon aroused from his dreamless slumbers, (or if he dreams perchance it is of swill,) and is made a subject of inspection and if he deserves it, of laudation.

It is here that much is to be seen, much to be learned, and much to be garnered up in the store house of memory, to be rehearsed at the winters fireside. Rightly conducted, it proves itself the farmer's jubilee, and cements together the brotherhood by ties of interest, and by community of occupation. Alone, man's efforts are feeble in the accomplishment of any great undertaking—associated, few obstacles are insurmountable.

"Divide the thunder, and its single tones serve as a lullaby for the cradled infant, but let it come in one quick peal, and it is the voice of Heaven speaking to man, and the earth trembles in reply." Originally, man was contented to rely upon his own limited experience and resources; he had no other aid in his agricultural operations. For centuries he groped his way under such disadvantages. At a later period he was satisfied with the formation of clubs, composed of his immediate neighbors, and with this improvement he toiled on cheerfully for years.

Still later, and bearing date with our own recollection, societies were organized with a view to proficiency in husbandry, limited by county lines. More recently still, and under the mania of progress, associations were formed, embracing States. Panting for still farther advancement, to-day, in the city of London, the world may be found engaged in one gigantic and unparalleled exhibition.

The conceptions of a few humble husbandmen, convened perhaps for the first time, under the shade of an oak, for the purpose of vieing with one another in the productions of the soil and of recounting their agricultural experience has been expanding and extending itself, until now, their place of convocation is a Crystal Palace—nations are rivals, and royalty itself, leaving the throne, enters the lists as competitors for the laurels.

England may proudly point at this achievement as one of the boldest strokes of policy which ever characterized a nation. This one act has done more to sow the seeds of peace and to cultivate a spirit of amity than any other conceivable one. Aristocracy for the first time condescends to smile upon labor and extends towards it an ungloved hand.

Europe for the last fifty years has done much for the promotion of agriculture, but much, much more will be accomplished before the close of the 19th century.

Fifty years ago, in England, twenty-four bushels of wheat per acre was a full average crop, and it was dwindling still. The fears of thinking, intelligent men, at this startling disclosure, were aroused;

an investigation into the causes of this appalling decline was at once instituted, and public attention in various ways directed to the evil. A great calamity was dawning—it was to be met manfully, and, if possibly averted. Mind clashed with mind, like the collision of the steel and the flint, for the purpose of emitting light and thereby presenting a remedy. *Wealth* freely threw open its coffers, and science came forward and tendered her aid.

Soils were submitted to chemical tests and duly analyzed; plants were dissected and their elements canvassed. Science and experiment working together side by side, threw light upon their pathway. The unerring crucible detected the absence of phosphorous (one of the most important of the bread bearing elements,(in this field, and demanded an application of pulverized bones, with a view to its remedy. Man submitted to the direction and fertility was restored.

That field was found wanting in carbon, and science suggested charcoal, and it was well.

This field, by reason of repeated cropping, without regard to rotation, had become sour and sorrel was the usurper; science was again consulted and she recommended ashes to neutralize the acid, and clover to sweeten; man followed the direction and science once more proved herself matchless in her remedies.

Here was a field that produced an abundance of straw but a shrunken worthless berry; science hinted at deeper plowing and told of a new soil sleeping beneath, abounding in *silicia,* and proposed that it be brought up to the sun and atmosphere, mixed moderately with lime, and then assured them the present difficulty would be obviated. Man reposing implicit confidence in her veracity, yielded himself to her guidance and smiled at the result.

And thus it will always hold true, that just so far as we yield to her directions, just so far we shall be benefitted. England, pursuing this course, has arisen from the average crop of twenty-four bushels in 1800 to the average crop of thirty-two in 1850.

Now, in some parts of England, where strict attention is paid to the laws of nature, 60 bushels of wheat is not an uncommon yield; and yet comparatively successful as she is she has not reached perfection—indeed, she is far from it.

She may plume herself upon her power, her progress and her proficiency, (and she can justly boast of much,) yet in many things she is still a scholar.

Proud though she is, still she is not too proud to learn. Watch her, and you will see her sending her agents year after year to Holland, to take lessons in the school of agriculture there. And why to Holland? Because, the world acknowledges Holland to be the focus of correct farming, or more nearly so than that of any other nation. There they apply all knowledge, chemical and geological, and all to their great agricultural work. Where there is anything to learn, John Bull is always there—set that down as a "fixed fact." She is jealous of any power that possesses any useful knowledge that excels her's. *See* her with the eyes of an argus watching the common school of Prussia—the Universities of Germany, the system of engineering of France, the infantry tactics of Austria—and the commercial thrift and the inventive powers of the United States.

Wherein is the impropriety of our pursuing a similar course? Wherever John Bull goes, let Brother Johnathan go also. In fidelity to this purpose, let not the shadow be more true to the substance.

In an agricultural point of view, nature has been more lavish of her expenditures upon our soil and climate than she has with theirs, and yet, though her land has been under the plow hundreds of years longer than ours, she produces nearly double to the acre more than we do.

The question irresistably presents itself, why is this so? Simply for the reason that when our land shows any symptoms of failure, the occupant, either ignorant of the ability to restore its pristine fertility or too covetous to expend means with a view to its invigoration, suffers it to pass into other hands, and bewildered with dreams of El Dorados and Edens, leaves all behind and penetrates the west, where a virgin soil, just as it came from nature's laboratory, invites his ploughshare. It is here that his journey is arrested, the same cruel system of husbandry adopted until another failure is but another signal for a new Canaan.

Wretched calculation—miserable policy—the idea never seems to occur to him, or if it does, makes no impression, that in emigrating he leaves behind him a more reliable market, much coveted luxuries, valuable privileges and associations and fairly expatriates himself from intellectual advantages, which may ever after be to him a source of regret.

He forsakes the laws of churches and school-houses for the verge of civilization and barbarity, doubtful for a time which shall maintain the supremacy.

I would I could say something that would curb this irresistable *penchant* for roaming and adventure; that I could instil a species of contentment within the bosom of my fellow man, and in my own. I feel that—

"I know the right and yet the wrong pursue."

The farmer, instead of being discontented, should retain his original acres, abandon the system of farming so tenaciously adhered to, ever since Standish and Brewster planted their feet upon Plymouth, cease their warfare against agricultural periodicals and book farming, listen more to nature, and my word for it he will become a better farmer, a better citizen, and a wealthier and a better man.

It is this roving, trafficking propensity, developed by too many plowmen, that detracts from their dignity, and defeats their ability to accomplish any valuable result. If the farmer must roam, let it be in quest of such knowledge as shall promote the pursuit to which he is wedded.

To be truly great a man must be really useful. It is the useful man after all, who is the true patriot; not the soldier, who exposes his life upon the battle field for twenty-three cents a day, a bed in the open air, and perchance, a grave that a dog might spurn. It is he who displays the most moral, not physical courage, who best promotes the interests and welfare of his fellow man and his country.

But we have wandered; we were speaking of the fact of England's learning in any school which was capable of imparting instruction to her. It should so be with us. If we are indisposed to learn of Holland, let us avail ourselves of the agricultural knowledge of our father land. This we may do with propriety and at the expense of less treasure and trouble. A community of language and paternity, present all the facilities for promoting the object. If England, by dint of perseverance and patience has acquired from other nations information which is valuable to her, why may it not be valuable to us, and add to our greatness as it does to hers. She has long ago learned the fact that she cannot afford to have poor land within her boundaries, neither can we. For the last fifty years she has been ransacking the United States in pursuit of bones—the bones of our dead cattle and horses, and having gathered them together for a mere song, she has shipped them to England, ground them up to powder, and by application, added double to her wheat crop.

They have demonstrated beyond the possibility of a doubt, that a half bushel of these bones is worth to them at least two loads of ordinary barnyard manures; and indeed, most readily supplies the absence of a cereal element in no other way so economically applied. If we consult our own interest in future, we will dispute possession with them of the very bones which are now suffered to expend their ability to fertilize upon the atmosphere, instead of the soil. We'll let them know that the "school master is abroad" here, as well as there, and that armed with his primmer and his atlas, he is cultivating the art of peace, and doing much to give to the thought of man a wise and salutary direction. Let us watch her with an eagle eye, just as she watches the useful operations of other nations, and whenever her movements are calculated to confer benefits, avail ourselves of them just as she would any movement of ours. If she introduces any well arranged, labor-saving agricultural implement, let us adopt it at once, and improve upon it if possible, and not universally spurn it because it is the production of a rival nation.

Occupying a nook in the crystal palace a few months since, was McCormick's American Reaper, and even as John Bull passed by it, he was sure to make some unkind expression, or at least treating it as a proper subject for a joke. More recently it became necessary to test the merits and pretensions of the various reapers by actual experiment, when to the astonishment of all who witnessed the operations of the different machines, McCormick's was head and shoulders above everything of the kind, and as such received the medal. Ever after, instead of modestly occupying an undistinguished position, it has become the observed of all observers; and England has the generosity and manliness to come forward and acknowledge the premature display of her wit, and signifies her change of opinion by ordering as many as can possibly be supplied.

It will be found as one of the great results of the *World's Exhibition*, after curiosity shall have been gratified, prejudice silenced and judgment suffered to pronounce her verdict, that while in very many respects other nations have far excelled us in the display of such things as contribute largely to promote a famine, *our country surpasses them all* in whatever contributes to the staying of one.

To be wise in one's own conceit is a barrier to success, of which every day furnishes abundant examples. It is well enough to be gratified with our proficiency, indeed, it is right; but to be perfectly satisfied, and consequently cease any further efforts, is not evidence of sound policy. Onward! onward! onward! is our proper direction. Let the mind ever go out in pursuit of higher attainments, and progress be but the incentive to still further progress, until plans shall have been consummated worthy of a giant's powers.

He, who from an elevated position, shall look back upon time as it now presents itself, will wonder at the contrast. We who now live, and pride ourselves upon our greatness, will then be regarded as being but in the alphabet of the sciences and mere apprentices in the arts. This idea may seem to be visionary and far-fetched, but after all, does not the last quarter of a century reveal many things as unlooked for and as startling? The great thought designed to be impressed is, to let success pant after still further success, until we attain a stature somewhat commensurate with the intention of our creation.

In this vast scheme of progress, no field presents stronger claims than that of agriculture—scientific and practical agriculture.

The Great Creator has been kind enough to bestow upon us a soil of exceeding fertility and a climate suitable to a full development of its productiveness. With this we have no reason to complain, and if we fail to extract from it that which will furnish an ample supply for all our temporal wants, we alone are wrong. We have no reason to be dissatisfied with whatever comes to us from the hand of Supremacy. When we compare our condition with others, less blessed in natural advantages, we are admonished at once to hush any such feeling. If you are determined to be discontented, just go with me (in imagination) to sterile Switzerland, and as we climb with difficulty, her bleak and soilless mountains, just mark the toil and hardship with which her dense population contend, and notwithstanding all these adverse circumstances, see how nobly they buffet them; trace the gleam of sunshine and contentment which is shadowed forth on every feature, and tell me honestly what reasons you can give for repining.

Are you yet inclined to murmur? then let us go abroad once more. Let us push our way up the Rhine and penetrate the interior of Germany. What is it we behold? As far as the eye can reach, do you not see women making their way up the mountain sides, with toil bearing up baskets of earth and depositing their burthen where-ever the rocks are sufficiently shelving to reeeive it? and their constructing rude beds in which to plant and partially rear the vine, receiving in return a scanty yield upon which to subsist until another harvest?

Does all this meet your vision, and will you not settle down thankful for your allotment? If not, let us direct our steps to the moors of Lincolnshire, and there realize with what an immense expenditure of treasure and labor they are pumping the water from one level to another still higher, and then again to one yet above until it flows by artificial channels into rivers, and thus disposed of, or made to irrigate other lands suffering from the absence of this important element.

After viewing all this, if your disposition to repine is unallayed, then to no purpose would the gates of Eden be thrown open for your entrance. To be successful, fellow citizens, it is necessary to be contented with our condition, and by a system of progress and improvement rendering that condition still more desirable.

Few people have more reason to rejoice than we have. Engaged in an honorable avocation, one as well calculated for intellectual as physical developement, we have claims to usefulness and happiness transcending those of any other class of our fellow creatures. Toil, judiciously expended upon the lands by which we are surrounded, may cause them to vie with the soil of Italy or with that even of the Nile itself. And in addition to all this, we have a kind Providence, who smiles upon all our efforts to acquire an honest livelihood and ever condescends to work with and instruct us.

Who does not consider it a source of pleasure, indeed, an unmerited honor, to have *him* for a co-laborer, who did but speak and out of confusion, a world leaped into existence. Who did but conceive the idea, and healthful showers, and the genial rays of the sun fell softly upon it.

In the cultivation of the soil for bread—the great staple product of the world—our Creator furnishes 95 per cent of all the elements which enter into its growth, exacting of man the inconsiderable, though necessary 5 per cent balance. In the production of peas and other plants of a kindred nature, he is still more profuse in his contributions to alleviate the labors of man, inasmuch as he causes the atmosphere to furnish all the ingredients which their growth demands.

Clover, also, (unless we permit it to go to seed) draws all its aliment from the air, embodying large quantities of nitrogen in its formation, which, when turned under by the plow, transfers that gas to the soil, to be devoted to the perfection of some crop to which (like wheat) it is so essential.

To farm it successfully, therefore, it is obligatory upon farmers to become careful observers of the operations of nature, and by so doing we shall derive immense advantage from her co operations. Be assured of one thing, that you may not violate her laws with impunity. We are taught by common sense, that if we grow the same crop year after year, upon the same soil, that particular quality of fertility which is so important to its maturity, will be entirely absorbed by the plant, and the land left valueless, unless artificially fertilized, perhaps at a cost consuming the entire profit of the yield. Better submit to the tuition of that *invisible* agent and by so doing become wiser and wealthier. Under her instructions, you will sow or plant this year, that which will draw from the atmosphere and transfix into the soil, whatever fertilizing qualities are absolutely necessary to the production of the succeeding crop, which you may have in view, differing in its nature.

Take for instance, by way of illustration, a field which has been sown year after year with wheat, (as has been too much the system of farming in Michigan,) until the crop is insufficient to discharge the expense attending its culture. What does that soil require to restore its wonted fertility? We answer that it requires one of three things—either an application of certain composts, that is (a variety of enriching manures incorporated into one,) with a view of restoring those wheat-producing qualities of which the land has been despoiled by repeated cropping; or else it calls for rotation, that is, the production of some other crop which demands a particular quality of the soil which has not been used in the growth of the wheat, such as a root crop, or corn, if the planting is preceded by a body of manure, or else in the third place, it demands rest. By rest I do not mean perfect idleness. I do not mean to be understood as saying that you must put up the fences around the field, close the gates, and give it over to the tender mercy of noxious weeds, to become a garden for the production of all manner of foul seeds—I do not mean any such thing—I mean just what I say—rest. Land may become fatigued, by continued exactions upon its power to produce—by having its surface vexed year after year, with the plow, the harrow, the cultivator, the drill and the hoe. Its energies, like yours, become paralized by continual calls upon its powers—its heart is broken and it ceases to remunerate the husbandman for his toil. *I want to impress this thought*, and therefore I repeat it, your land when it fails to repay you for your labor, requires *rest*. Did it never occur to you that soil should rest once in seven years at least, and that a strict adherence to this great natural law, should ever be inviolable. It does not become us to demand the why or the wherefore; it is sufficient to know that it is the legislation of Heaven, and therefore is unerring.

When land rests, it should rest upon a bed of clover, because then it is continually receiving, and imparting virtually nothing. Through its leaves which are its lungs, it is all the while inhaling from the atmosphere various enriching elements, in the form of gases, and incorporating those qualities into the soil, quickening and rejuvenating it. The leaves and stem of the clover are conductors of whatever is valuable in the air to the root, and thence by decay of the root to the soil, contributing in no other way, so rapid a process for the formation of a vegetable loam.

Experiment tells us that every acre of good clover will produce four tons of roots, of a most valuable character. Here let me state that to grow clover as a fertilizer, never grow it for the seed, because, in the formation of seed great quantities of phosphorus are required, and this is robbing the soil of an ingredient, which is a *sine qua non* in the production of bread. Clover may be mowed without any great damage to the land, and disposed of in that condition, but the far better way is, to depasture it with sheep, if you would derive the greatest amount of good from it.

I find it difficult to dismiss this branch of agriculture without once alluding to the interesting and indeed benevolent idea of permitting the land to sleep occasionally. You will find your interest best promoted by this humane system of improvement. Instead of being a source of expense, it will be a source of revenue; instead of losing any of its valuable properties, the soil will come back to you in due time, as fresh and as vigorous as when nature first dedicated it to the plow. The earth gathering fertility from repose, awakes to repay the husbandman for his considerate kindness, by an increased ability to produce.

By dint of physical exertion, man may accomplish wonderful results in agricultural operations; indeed, by the application of artificial stimulants, compassed it is true at a vast expense of toil and means, he may preserve measurably, the productiveness of his soil; but after all, exertions should not be unnecessarily wasted–nerve and muscle should not be permitted to do every thing—brain should be called upon to furnish its proportion of labor, and by so doing you will have enlisted an associate who will lighten your efforts, prove an agreeable companion and withal a profitable one.

Mind has something to do in the field as well as in the forum. Mind controls matter, moulding and governing it at its will. The head guides the hand, and that the hand should be guided aright, policy demands that the head be enlightened.

After all, industry avails little unless it is enlightened industry. What dependence may be placed upon judgment unless it is enlightened judgment. Toil, that spurns intelligence, is too much like the attempt to row a boat with one oar; instead of making headway, the course of the navigator is circuitous, pursuing one beaten round—ever toiling, but never accomplishing.

The great effort should then be with us, to associate body and mind in our pursuits for subsistence and usefulness—for if there is one appellation which I would covet above another, it would not be so much that I am an *American citizen* but that I might be an intelligent *American Farmer.*

The day has gone by, in this country at least, when merely to discharge the duties of a beast of burden is all that is essential to the cultivator of the soil. It is a matter of surprise that such an erroneous sentiment should have ever had an advocate.

Educated toil has decided advantages (depend upon it) over mere animal force, other things being equal. Capital invested in muscle never declares such dividends as capital invested in mind. Ignorance not only paralizes energy, but often exacts two days to accomplish what education would consummate in one.

These few desultory remarks are designed to impress upon my brother farmers the necessity, nay! the duty of educating our children for the plow and the dairy, and not turning them off upon the world with the mere ability to read a little and write less, because forsooth our parents did not appreciate the value of a greater proficiency. By education, I would not insist, nor indeed do I expect that farmers as a general thing will pass their children through the University, (as desirable as that would be,) but I do insist upon our obligation to confer upon them, if possible, an academic course of instruction.

In order to accomplish this, it is true, in many instances sacrifices must be met, and to them we must cheerfully submit. Some luxuries must be curtailed, perchance a rigid system of economy instituted, having in view this desideratum. A subscription to one or more well conducted agricultral periodicals, will contribute largely toward carrying out this suggestion, and as appropriate, it were well here to remark that *that* farmer who deprives himself of a paper devoted to his calling, and of one or two devoted to general intelligence, robs himself of a means which as much as any other is calculated to make him intelligent, prosperous and happy.

I am reminded, fellow citizens, by the time which I have already consumed, that my remarks should be brought to a close. I have endeavored, under great disadvantages, to publish some ideas calculated to give dignity to our profession, to impart to it something like its proper importance, and to divest it at the same time of that dryness and want of interest which attaches too often to addresses of this kind. If I have *but partially* succeeded, I have more than effected what I expected.

I am as well aware as any of my audience, of my irrelevancy on the one hand, and of my want of analogy on the other; and I confess that during the brief space which was allowed me to prepare for this occasion, death has entered my domestic circle and taken from it my idol, and too often as I have attempted to bestow upon my subject thoughts, my thoughts have wandered, and I have only overtaken them at the grave.

In closing this address, fellow citizens, permit me once more to congratulate you on your goodly heritage and your prosperous condition. To congratulate you upon your citizenship of a nation, than which none other under heaven has stronger claims to greatness. At its birth, our Union sprang into existence a model nation. Its foundation laid mid sacrifice and cemented with blood, is faithful to remind us of its cost.

The prosperity which has ever followed in its wake and its progress from one point of elevation to another, and then to another still, furnishes irresistible evidence of its having received the uninterrupted smiles of Supremacy. Centuries will roll away, and we who stand here may long since have gone back to the dust, but a nation having such a foundation as ours will stand—it will stand and live on in the midst of increasing glory, notwithstanding that there are those who will tell you now, and it is but a repetition of the same stale story which has been told ever since our Independence was nobly achieved—that we are on the eve of dissolution, and consequently of national ruin.

"Dissolve the Union never!
'Twere e'en a madman's part,
The golden chain to sever,
Which girdles Freedom's heart.

"What! Faction rear her alter,
And discord wave her brand,
And hearts from duty falter,
At party's base demand?

"Look up!—'tis Freedom's temple,
You long to overthrow;
And if your arms uplifted,
A demon prompts the blow.

"Think! every radiant column,
Has cost a Patriot's blood,
And would you have them shattered,
Where long in pride they stood.

"Dissolve the Union! Never!—
You may not if you would,
Go! traitor, go forever,
And hide you where you should.

"For he who breathes dissension,
To shake a people's trust,
Should cower back to nothingness,
Or crumble into dust."

REPORT

Of the Lenawee County Agricultural Society.

J. C. Holmes, Esq., *Sec'y Mich. State Ag. Society.*

Sir—The officers of the Lenawee County Agricultural and Horticultural Society, for the year 1851, are,

President—J. W. Scott.

Vice Presidents—Samuel Rapplege, J. D. Thompson, Gilbert Gage, Richard Kent, T. J. Faxon, H. M. Dayton, and Jonathan Berry.

Recording Secretary—F. C. Beaman.

Corresponding Secretary—M. A. Patterson.

Treasurer—L. G. Berry.

Executive Committee—George E. Pomeroy, William Ten Brook, Walter Wright, Amos A. Kinney, Thomas Chandler, Henry B. Pomeroy, Samuel Rapplege, E. A. Baker, F. A. Kennedy, John C. Gile, Nathan Ramsdell, William Beal, Charles Blair, Fitch Reed, Isaac Miller, R. W. Knight, Abram Knapp, D. R. Daniels, Ephraim Hicks, Joseph Rhodes, and Amos Hoag.

The annual fair of this society for the present year was held on the first Wednesday and Thursday in October.

The list of articles presented for premiums was very large. There were 77 entries of grade cattle; 26 of blood cattle; 67 of horses; 21 of sheep and wool; 14 of swine; 4 of farms; 30 of farm implements; 53 of butter, cheese, sugar, honey and vegetables; 48 of fruits; 26 of miscellaneous articles; 32 of crops and seeds; 5 for the plowing match; 65 of domestic manufactures; 6 of ornamental trees; 1 of nurseries. Of many of the above mentioned entries, each contained a numerous list of articles, and there was beside a very fine display of articles of needle work, drawings and paintings, flowers, &c.

The list of premiums awarded amounted to $629 25, of which the sum of $111 was awarded upon grade cattle; $120 upon blood cattle; $58 upon horses; $48 upon sheep and wool; $18 upon swine; $23 upon farms; $33 upon farm implements; $25 upon butter, cheese, sugar and vegtables; $46 upon fruits; $9 upon miscellneous articles; $23 50 upon crops and seeds; $10 upon plowing match; $6 upon nurseries; $30 50 upon flowers; $11 50 upon needle work, $5 50 upon paintings and drawings; $37 25 upon domestic manufacture; $4 upon poultry; and $9 upon ornamental trees.

The amount of funds in the treasury at the close of the last year, was,	$234 84
The amount received from the county treasurer the present year, was	200 00
The amount received as membership fees, was	347 00
The amount received from visitors on the days of the fair, was	194 90
The amount received from the sale of lumber, the property of the society, was	9 40
The amount received from subscriptions of citizens for defraying expenses of the fair,	158 97
Making the sum of	$1,145 11

for the use of the society during the present year.

The expenses of the society during the current year, cannot now be precisely stated, as the bills have not all been passed upon; but they will probably amount to about the sum of $246, which, added to the amount of premiums, makes the sum of $875 25. This last amount deducted from the aforesaid sum of $1,145 11, leaves still in the treasury the sum of $269 86.

The Lenawee county agricultural and horticultural society is in high favor with our citizens generally, exciting a laudable spirit of emulation amongst all classes of producers; and the large concourse of persons present at our late fair, variously estimated at from six to eight thousand, sufficiently evinces that the importance of the society is losing nothing in the estimation of the people.

An address at the fair had been expected from the Hon. Joseph R. Williams, of Constantine, but indisposition of that gentleman produced a disappointment—a disappointment that, from the known ability of the orator, was felt deeply; but which, happily, was much relieved by the courtesy of our fellow citizen, Dr. M. A. Pattison, who, though called upon unexpectedly and without preparation, addressed the audience in a manner to give both pleasure and instruction.

I regret that I am unable to make you such a report, as is evidently intended by articles 8 and 11 of the constitution of the Michigan State Agricultural Society. Owing, doubtless, to the shortness of time allotted to the judges for preparing their reports, they have furnished us with little, if anything, that can be of interest or utility to any one beyond the limits of our own county. The reports consist mainly of an award of premiums, and are generally entirely destitute of the facts and circumstances, which could in any way subserve the general interests of agriculture or horticulture. For example, P. J. Van Vleet is awarded a premium on 5 acres of wheat, yielding 55 bushels per acre, but as to the *nature* of the *soil* upon which the wheat was raised, the *kind* of wheat sowed, the *time* of *sowing*, and the manner of cultivation adopted, we are not informed. The bare announcement of the fact that so large a crop has been produced in any given place is productive of little benefit, but the *manner* of producing it, if given, might greatly benefit the public. I trust that means will be adopted to prevent the recurrence of similar defects in future reports.

I send you herewith a list of the premiums awarded at the Fair.

Respectfully,

W. C. BEEMAN,
Recording Secretary.

LIST OF PREMIUMS.

The following is a list of the premiums awarded at the Annual Fair of the Lenawee County Agricultural and Horticultural Society for the year 1851:

GRADE CATTLE.

Best yoke working oxen, 4 years old and over, George C. Coats, ... $8 00

2d best yoke working oxen, 4 years old and over, John Wheeling,	$5 00
Best 5 yoke working oxen from one town, J. E. Webber, of Madison,	10 00
2d best 5 yoke of working oxen from one town, Norton Baker, of Adrian,	8 00
Best pair of fat oxen, William Older,	5 00
2d " " John Baker,	3 00
3d " " Garret Ten Brook,	2 00
Best fat ox, Gilbert Gage,	3 00
Best pair 3 year old steers, E. P. Graham,	5 00
2d " " W. D. Page,	3 00
3d " " A. Saxon,	3 00
4th " " John M. Merritt,	2 00
Best pair 2 " Myron Brown,	3 00
2d " " Atwater Bogert,	2 00
3d " " Wilmarth Graham,	1 00
Best pair yearling steers, J. D. Thompson,	2 00
2d " " "	1 00
Best milch cow 3 years old and over, J. W. Scott,	8 00
2d " " " O. Ferguson,	7 00
3d " " " M. N. Halsey,	5 00
4th " " " Luke Wood,	3 00
Best 2 year old cow, Jesse Cram,	5 00
2d " Joseph Moreau,	3 00
3d " John Wheeling,	2 00
4th " H. B. Pomeroy,	1 00
Best bull calf, Luke Wood,	3 00
2d " D. Hutchinson,	1 00
3d " R. R. Beecher,	1 00
Best heifer calf, John Potter,	3 00
2d " H. B. Pomeroy,	2 00
3d " W. D. Page,	1 00

BLOOD CATTLE.

Best Devon bull 3 years old and over, Geo. E. Pomeroy,	8 00
" Durham cow 3 " " "	8 00
" Durham calf,	3 00

2d best Durham cow 3 years old and over, J. D. Thompson,	$7 00
3d best Devon 2 " " "	5 00
3d best Durham 3 " " Jacob Jackson,...	5 00
Discretionary premium on blood calf, " " ...	1 00
2d best Durham bull 3 years old and over, William D. Page,	7 00
Best " " " Isaac Adams,...	8 00
2d best Devon 2 " Q. W. Pomeroy,	7 00
2d best Durham heifer 2 " Luther Bradish,..	5 00
Devon cow and calf, (discretionary) Samuel Myers,.........	8 00
Best Devon heifer 1 year old, W. B. Rolland,.............	5 00
2d best Durham heifer 2 years old, S. A. Holland,........	7 00
" " bull 2 " A. D. Smith,...........	7 00
3d " " 3 " Hiram Gilbert,........	5 00
Blood calf, discretionary premium, Levi Wilson,..........	1 00
Best Durham bull 2 years old, Jas. B. Wells,..............	8 00
" " heifer 2 " "	8 00
2d " cow 3 " Rufus Baker,............	7 00

HORSES.

2d best stallion 4 years old, David Smith,................	3 00
3d " " N. McLouth,................	2 00
Best blood stallion 4 " Wm. Taxer,................	5 00
2d " " B. Herrington,..............	3 00
3d " " H. Randolph,...............	2 00
Best brood mare for all work, O. F. Church,.............	5 00
2d " " Robt. Morris,..............	3 00
3d " " S. Knapp,.................	2 00
Best blood mare, G. M. Hubbard,.......................	5 00
2d " E. Holloway,.......................	3 00
3d " D. Mead,...........................	2 00
Best stallion 3 years old, S. Rapplege,..................	3 00
2d best stallion, 3 years old, William Hunter,............	2 00
Best mare, 3 years old, William Moore,.................	2 00
2d do do A. Lowe,........................	2 00
Best span matched horses for all work, P. R. Adams,......	5 00
2d do do do Charles Pratt,......	3 00
Best single horse, Dr. Owen,..........................	3 00
2d do R. C. Newton,..........................	2 00

SHEEP AND WOOL.

Best merino buck over 2 years old, S. A. Hubbard, ... $5 00
2d do do Russel Skeels, ... 3 00
3d do do William Palmer, ... 2 00
Best Saxon buck over 2 years old, S. A. Hubbard, ... 5 00
2d do do do T. H. Bailey, ... 3 00
3d do do do S. A. Hubbard, ... 2 00
Best 5 ewes, T. H. Bailey, ... 5 00
Best long wool buck, Gilbert Gage, ... 5 00
2d do do N. N. Woodford, ... 3 00
Best 5 long wool ewes, do 5 00
Best 5 long wool ewe lambs, do 5 00
2d best 5 fat wethers, G. Ten Brook, ... 2 00
Best fleece wool, reference had to quantity, quality and condition for market, Walter Wright, ... 2 00
Best pen 5 merino lambs, H. B. Pomeroy, ... 5 00

SWINE.

Best boar, D. D. Meech, ... $5 00
2d do A. J. Carpenter, ... 3 00
Best sow and 5 pigs, Newton Mitchell, ... 5 00
2d do do Jacob Jackson, ... 3 00
3d do do William R. Porter, ... 2 00

FARMS.

Best farm, Stephen Perkins, ... $10 00
2d do Samuel Rapplege, ... 7 00
3d do Newell Mitchell, ... 6 00

FARM IMPLEMENTS.

Best harrow, Abraham West, ... $2 00
" horse rake, T. Ludlum, ... 2 00
" farm wagon, Marshall & Sayer, ... 3 00
" and largest collection agricultural implements owned by exhibitor, A. B. Palmer, ... 10 00
" ox yoke, J. Dodd, ... 2 00
" half dozen axe helves, J. Dodd, ... 1 00
" wheat drill, P. T. Voorhees, ... 3 00
" wheat cultivator, do 3 00

Best straw cutter, James Steel, ... $2 00
" roller for use, Samuel Rapplege, ... 3 00
" hay rigging, H. Randolph, ... 2 00

BUTTER, CHEESE, SUGAR, HONEY AND VEGETABLES.

Best pie plant, Samuel Lothrop, ... 50
" squashes, J. C. Wainright, ... 50
" onions, William Buel, ... 50
" pumpkins, Jacob Jackson, ... 50
" cabbage, Milton Eddy, ... 50
" carrots, Stephen Aldrich, ... 50
" parsnips, William Ten Brook, ... 50
" turnips, Rufus Baker, ... 50
" beets, Wm. Burt, ... 50
" sweet potatoes, James R. Webb 50
" table potatoes, Stephen Aldrich, ... 50
" lima beans, Wm. Ten Brook, ... 50
" and largest variety vegetables, Stephen Aldrich, ... 3 00
2d best do do Wm. Buel, ... 2 00
Best specimen butter, Ezra Abbott, ... 2 00
2d do do B. J. Harvey, ... 1 00
Best cheese, Thos. C. Swift, ... 3 00
2d do Chas. Pratt, ... 2 00
Best speciman maple sugar, N. A. Ramsdell, ... 2 00
2d do do G. D. Bascom, ... 1 00
Best specimen of honey, Benj. Williams, ... 2 00
2d " " Jona. Hare, ... 1 00

FRUITS.

Best and largest variety fall and winter apples, T. H. Bailey, .. 3 00
2d " " " " B. J. Harvey, .. 2 00
3d " " " " C. Crane, ... 1 00
Best 6 varieties winter apples, B. J. Harvey, ... 3 00
2d " " J. S. Clark, ... 2 00
3d " " C. Crane, ... 1 00
2d best and largest variety peaches, Jno. M. Merritt, ... 2 00
" " " plums, B. F. Strong, ... 2 00
Best single variety plums, Sam'l Lothrop, ... 2 00

Best and largest variety cherries, A. S. Cornell, ... $3 00
2d " " " B. J. Harvey, ... 2 00
Best single variety cherries, A. S. Cornell, ... 1 00
" and greatest variety strawberries, Benj. Steer, ... 1 00
2d " " " A. S. Cornell, ... 2 00
Best single variety " Benj. Steer, ... 1 00
" and greatest variety raspberry, Sam'l Lothrop, ... 2 00
" " " currants, A. S. Cornell, ... 1 00
" single variety " B. F. Strong, ... 1 00
" and greatest variety grapes, " 3 00
2d " " A. S. Cornell, ... 2 00
Best " " quinces, Walter Graham, ... 2 00
2d " " " J. S. Clark, ... 1 00
Best and largest variety pears, J. S. Harvey, ... 2 00
" single " " F. H. Bailey, ... 2 00
Discretionary premium for fine lot strawberries, C. B. Backus, 50
" " " " Geo. E. Pomeroy, ... 50

MISCELLANEOUS ARTICLES.

Best wagon harness for farm, Robt. Bidelman, ... 2 00
" carriage harness, Abram Crittenden, ... 2 00
" barrel flour, P. Stone & Son, ... 2 00
2d " William Beal, ... 1 00
Best bee hive, W. McLouth, ... 2 00

CROPS AND SEEDS.

Best 5 acres wheat 55 bush. per acre, P. J. Van Vleet, ... 5 00
2d " " 34½ " L. H. Cook, ... 3 00
Best 5 acres oats 54 " Jacob Jackson, ... 5 00
" ½ bush. seed wheat, Edward Hoag, ... 1 00
" 5 acres corn, 77 bush. per acre, A. Woolsey, ... 5 00
" ½ acre broom corn, G. Ten Brook, ... 3 00
" doz. ears seed corn, H. Randolph, ... 50
" acre beans, Jessee Dodd, ... 1 00

PLOWING MATCH.

1st premium, Andrew Carpenter, ... 5 00
2d " E. P. Graham, ... 3 00
3d " William Crane, ... 2 00

NURSERIES.

Best nursery, Joseph S. Kies,........................... $6 00

FLOWERS.

Best and largest collection hardy roses, B. W. Steer,........ 3 00
2d " " " J. S. Kies,.......... 2 00
Best collection climbing roses, B. W. Steer and J. S. Kies,.... 3 00
" " perpetual roses at fair, B. W. Steer,.......... 3 00
Discretionary premium, perpetual roses, J. S. Kies,.......... 2 00
" " for vase dried grapes, M. Satterthwaite, 1 00
Best collection of Dahlias at fair, C. B. Backus,............ 3 00
2d " " S. Lothrop,.............. 2 00
3d " " J. S. Kies, and A. Truax,... 1 00
Best boquet, Mrs. V. Spalding,......................... 3 00
" and greatest collection green house plants, Mrs. S. Lothrop, 3 00
" variety flowers presented during season, Mrs. V. Spalding, 2 00
Boquet, discretionary premium, Miss E. Truax,............ 1 50
Best and greatest collection hardy perenial roses, W. Ten-Brook,...................................... 2 00

NEEDLE WORK, &C.

Best ottoman cover, Miss C. Lennell,.................... $1 00
" group flowers, do 1 00
" worked collar, Mrs. Watson,........................ 50
" silk bonnet, Mrs. Fish,........................... 1 00
" lamp mats, Mrs. Lothrop,........................ 1 00
" shell work, Miss H. M. Combs,.................... 1 00
" wax flowers, Mrs. Isaac Trench,.................. 1 00
2d best do Miss F. Webster,..................... 1 00
Discretionary premium on box worked with beads, Mrs. S. Lothrop,.. 50
Discretionary premium for chair cover, Miss O. Pomeroy,.... 50
Best specimen needle work, Mrs. R. Wheeler,............. 1 00
Port folio, worked, Mrs. F. M. Cooley,.................. 1 00
do do Miss E. Ingersoll,.................. 1 00

PAINTINGS AND DRAWINGS.

Excellent daguerreotype, Geo. W. Menich,............... $0 75
do do J. L. Harned,................. 75

Best water color paintings, Mrs. Watson,	$1 00
Two paintings in oil, C. Smith,	1 00
Best Chromatic paintings, W. B. Wilson,	1 00
Drawing, discretionary premium, Miss H. S. Woodbury,	25
do do do " A. S. Van Pelt,	25
do do do " C. Horton,	25
do do do " Eloise Britain,	25

DOMESTIC MANUFACTURES.

Best pair woolen blankets, G. Ten Brook,	$2 00
2d do do John Miller,	1 00
Best woolen cloth,	2 00
Best hearth rug, M. E. Palmer,	2 00
2d do R. Wheeler,	1 00
Best 6 calf skins, B. J. Bidwell,	1 00
Best side sole leather, L. Martin,	1 00
Best side upper leather, do	1 00
Best rag carpet, H. Randolph,	2 00
2d do R. Wheeler,	1 00
Best tow cloth, A. Beckley,	2 00
Best flannel, A. West,	2 00
2d do Charles Angel,	1 00
2d best woolen carpet, Mrs. S. Perkins,	2 00
Best woolen stockings, Mrs. E. W. Kelley,	1 00
" do socks, Mrs. A. Bradish,	75
" do mittens, Mrs. H. Randolph,	50
" do shawl, S. Allen,	2 00
2d do do S. A. Hubbard,	1 00
Best bed quilt, Mrs. Anna Marlett,	2 00
2d do A. F. Brown,	1 00
Best coverlid, E. Satterthwaite,	1 00
2d do Mrs. S. Perkins,	1 00
Best dozen corn brooms, Joseph Hoag,	3 00
2d do do E. W. Kelley,	2 00
3d do do S. W. Fenton,	1 00

POULTRY.

Best lot turkeys, Samuel Pond,	$1 00
" Polands, Allen B. Chaffee,	1 00

Best Cochin China fowls, S. Pond, ... $1 00
" large fowls, do 1 00

ORNAMENTAL TREES.

Best ornamental trees planted this spring in highway by one person, not less than 10, B. G. Ellis, ... $3 00

Best and greatest number ornamental trees in or near the road so as to shade it, saved on any farm and trimmed this spring, Jona. Harned, ... 3 00

Discretionary premium on same, Smith Laing, ... 3 00

F. C. BEAMAN, *Rec. Sec.*

Adrian, Dec. 1, 1851.

REPORT

Of the Macomb County Agricultural Society.

J. C. HOLMES, Esq., *Sec. Mich. State Ag. Society:*

SIR—In obedience to the constitution of the Macomb County Agricultural Society, I transmit herewith an abstract of doings of said society for 1851.

The executive committee held a meeting at Romeo, January 22, 1851, to make arrangements for the year, and among other things resolved to hold a fair at such place in the county as should offer the largest sum for defraying the expenses thereof, (the time and place to be fixed at a future meeting,) and offered the sum of five hundred dollars to be paid in premiums.

At a subsequent meeting, holden at Mt. Clemens on the first Monday of June, it was resolved to hold the fair at Mt. Clemens on the first and second days of October. A list of premiums was prepared and published, together with rules and regulations for the fair.

The fair was largely attended and the number of articles of stock of all kinds, manufactures, the produce of the farm, the dairy, and of all other departments, showed a large increase over the previous year.

An interesting address was delivered before the society on the second day, by A. S. Robertson, Esq., of Mt. Clemens.

The following officers were elected for the ensuing year:

President—James B. St. John.

Secretary—R. P. Eldredge.

Treasurer—Norman Perry.

EXECUTIVE COMMITTEE.

Reuben R. Smith, Ira H. Butterfield, Linus S. Gilbert, Dexter Muzzey, W. Canfield.

W. CANFIELD, *Secretary.*

Mt. Clemens, Dec. 13, 1851.

LIST OF PREMIUMS.

CATTLE.

Best Durham bull 3 years old, Ira Phillips, Armada, ... $4 00
2d " " Robert Warner, Ray, ... 2 00
Best " 2 W. Canfield, Clinton, ... 3 00
" " 1 G. W. Phillips, Armada, ... 2 00
Best bull calf, Lewis D. Owen, Bruce, ... 2 00
2d " J. C. Snover, Richmond, ... 1 00
Best Devon bull 3 years old, P. K. Leech, Utica, ... 4 00
2d " " Henry S. Courtes, Armada, ... 2 00
Best grade and native 2 year old bull, Luther Shaw, Chesterfield, ... 3 00
Best bull calf, Lemuel Sackett, Clinton, ... 2 00
" Durham cow, Ira Phillips, Armada, ... 4 00
2d " S. Knight, Clinton, ... 2 00
Best 2 year old heifer, Ira Phillips, ... 3 00
" 1 " " " 2 00
2d 1 " " Wiley Bancroft, ... 1 00
Best Durham calf, Ira Phillips, ... 2 00
2d " Lewis D. Owen, ... 1 00
Best Durham cow, Ira Phillips, ... 4 00
2d " L. D. Owen, ... 2 00
3d " J. C. Snover, Richmond..Treatise on milch cows.
Best grade and native cow, Jas. B. St. John, Utica, ... 4 00
2d " " Samuel Shears, Harrison, ... 2 00
3d " " John Stephens, Clinton, Treatise on milch cows.
Best native and grade 2 year old heifer, L. M. Phillips, ... 3 00
2d " " " " 2 00
Best " 1 " William Canfield, ... 2 00

2d best native and grade 2 ys. old heifer, E. C. Gallup,....... $1 00
3d " " " J. S. Campbell, Treatise on milch cows.
Best calf, J. B. St. John, Utica,........................ 2 00
2d " L. D. Owen, Bruce,.............................. 1 00
Best yoke working oxen, Calvin Pierce, Utica,............. 4 00
2d " " George Macomber, Clinton,........ 2 00
3d " " Lewis Drake,.......American Farm Book.
Best yoke four year old steers, H. S. Courtis, Armada,...... 4 00
2d " " A. Pierce, Utica,............ 2 00
3d " " John Broctor, Armada,...American Farm book.
Best yoke three " Joseph Cole, Clinton,........ 3 00
2d " " Wiley Bancroft,............. 2 00
3d " " Lemuel Sacket, Buel's farmer's companion.
Best two year old steers, P. K. Leech, Shelby,............. 2 00
2d " " J. C. Snover, Richmond,........... 1 00
3d " " E. Dunham,...Buel's farmer's companion.
Best yearling steers, William Canfield, Clinton,............ 2 00
2d " Joseph Cole,........................ 1 00
3d " E. Dunham,........Buel's farmer's companion.
Best broke yoke oxen, Lewis Drake,..........Farmer's Dictionary.
" 5 yoke oxen from one town, Clinton,.................10 00

HORSES.

Best stallion, R. R. Smith, Ray,......................... 4 00
2d " Jeremiah Sabin, Richmond,................. 2 00
3d " M. Chamberlin,.................Mason's Farmer.
Best 3 year old horse colt, Thadeus Hazleton, Ray,.......... 3 00
2d " " Eneas Snay, Harrison,........... 2 00
3d " " W. C. Ruby, Horses Foot, how to keep it sound.
Best 2 year old horse colt, W. W. Andrus, Shelby,.......... 2 00
2d " " Enoch Crawford, Ray,........... 1 00
Best 1 year old horse colt, William B. Wright, Ray,......... 2 00
" colt, T. J. Summers, Shelby,........................ 1 00
2d " James Crawford, Armada,...................... 75

3d best colt, Abel Warren, Shelby, ... Horse Doctor.
Best matched span of horses, Philip Price, Washington, ... 4 00
2d " " Wallace M'Chesney, Chesterfield, 2 00
3d " " Mr. Washer, Wash., Stable Economy.
Best single horse, T. J. Rutter, Clinton, ... 3 00
2d " Horace Bogart, Washington, ... 1 00
3d " Lenard Lee, Armada, ... American farm book.

SHEEP.

Best merino buck, 2 years old and over, P. K. Leech, ... 4 00
2d " " Jacob Scrambling, Shelby, 2 00
3d " " Hiram Taylor, Morrell's American Shepherd.
Best merino buck 1 year old, Jacob Scrambling, ... 3 00
2d " " P. K. Leech, ... 1 00
3d " " N. Dickinson, Morrell's Am. Shepherd.
Best common buck, William Canfield, Clinton, ... 4 00
2d " J. H. Butterfield, Shelby, ... 2 00
Best 5 ewes, J. H. Butterfield, ... 4 00
2d " Hiram Taylor, ... 2 00
3d " B. G. Whitney, Armada, Morrell's American Shepherd.
Best pen 5 ewe lambs, N. Dickinson, ... 3 00
2d " " 2 00
Best pen 5 common ewes, William Canfield, ... 4 00
" 5 ewe lambs, G. W. Phillips, Armada, ... 3 00
" 5 buck lambs, N. Dickinson, ... 3 00
2d " Hiram Taylor, ... 2 00
3d " B. G. Whitney, ... Morrell's American Shepherd.

SWINE.

Best boar, Philip W. Price, Washington, ... 2 00
" litter of pigs, John Price, ... 2 00

FARM IMPLEMENTS.

Best double harness, J. E. & Geo. VanEpps, Mt. Clemens, ... 2 00
" " carriage harness, " " " ... 2 00
" single " J. Bogart, Romeo, ... 1 00
" 6 axes, Dexter Mussey, Romeo, ... 1 00
" grain cradle, Ayers & Sibbets, Romeo, ... 1 00
" half doz. hand rakes, C. C. Farrar, Romeo, ... 1 00
" plough, Ayers & Sibbets, Romeo, ... 3 00

There was exnibited by Mr. Waggoner, a patent well curb, which must be a valuable appendage to the farm, and to all who wish an easy and convenient method for drawing water. They are manufactured at Mt. Clemens.

FRUITS AND FLOWERS.

Best and greatest variety of table fruit, G. W. Patterson, Macomb, ... $2 00
2d do do do Ayalus Striber, Bruce, American Fruit Book.
Best 6 winter varities of table fruit, John Crittenden, Macomb, 2 00
2d do do do John Bates, Washington, American Fruit Book.
Best pears, D. C. Walker, Romeo, ... 1 00
2d do Dr. E. Hall, Clinton, ... American Fruit Book.
Best peaches, Dr. E. Hall, Clinton, ... 1 00
Best grapes, do do 1 00
Best quinces, C. Renmould, Clinton, ... 1 00
2d do J. L. Conger, Harrison, ... American Fruit Book.

DOMESTIC MANUFACTURES.

Best pair woolen blankets, Acelia S. Snover, ... $3 00
2d do cotton woolen blankets, Elizabeth Miller, ... 2 00
Best pair coverlids, Mrs. J. B. St. John, ... 3 00
2d do Mrs. N. Perry, ... 2 00
Best 10 yards white flannel, Acelia S. Snover, ... 3 00
2d do grey flannel, Samuel Ladd, ... 2 00
Best 10 yards woolen carpet, Mrs. Ira Philips, ... 3 00
2d do do Mrs. Wm. Hall, ... 2 00
Best 10 yards rag carpet, Mrs. S. Ladd, ... 2 00
2d do do do 1 00
Best pair woolen knit stockings, Mrs. Elizabeth Miller, ... 1 00
2d do do Mrs. M. Sellock, ... 50
Best pair cot. stockings, Mrs. E. Miller, ... 1 00
2d do A. N. Pierce, ... 50
Best pair silk stockings, E. M. Sterling, ... 1 00
Best pair wool stockings, Mrs. J. H. Butterfield, ... 1 00
2d do A. N. Pierce, ... 50

Best pair wool knit mittens, Mrs. A. W. Sterling, ... 75
Best 10 yards tow cloth, Celia Snover, ... 50
Best 10 yards woolen plaid, Jas. H. Porter, ... 2 00
The attention of the public is respectfully called to a superior cooking stove manufactured at Romeo by Ayers & Sibbet, as being a very valuable articles in that department, and worthy a trial by all house keepers.
Best specimen cabinet work, J. Batty, ... 2 00

NEEDLE, SHELL AND WAX WORK.

Lamp mat, on raised work, Mrs. A. Buell, ...$1 00
Ottoman cover, Mrs. S. Knight, ... 50
Best ottoman cover, single stitch work, Mary J. Stephens, ... 1 00
2d do do Mrs. G. Fletcher, ... 1 00
Best perforated vase, Miss E. M. Sterling, ... 1 00
Best do tapestry toilet cushion, Mrs. G. C. Fletcher, .. 1 00
" do pair embro. suspenders, Susan M. Stephens, ... 50
" do crotchet purse, Mrs. Magee, ... 1 00
" do do bag, do 50
" do wrought cap, Mrs. W. Loud, ... 1 00
Best monochromatic painting, Miss Augusta Allen, ... 1 00
2d do do Matilda S. Allen, ... 50
3d do do Miss Augusta Allen, ... 50
One do picture, Sarah Batty, ... 50
" crotchet bonnet, Miss J. O'Neil, ... 50
" toilet stand, Mrs. J. Stockton, ... 50
" quilt, Mrs. John Bates, ... 1 00
" do Mrs. Jas. McChesney, ... 1 00
" vase of wax flowers, Mrs. S. W. Allen, ... 1 00
Best made dress coat, Mrs. J. H. Butterfield, ... 2 00

FIELD CROPS.

Best red chaff bald wheat, 43 bushels per acre, weight 62 lbs, Thomas Mitchell, Ray, ... 4 00
2d best wheat, Peruvian, 45 bushels per acre, weight, 66 lbs, Antoine Morass, Clinton, ...Farmer's Dictionary.
Best yellow corn, 150 bushels of ears on an acre, L. Bennet, Armada, ... 4 00
2d do 78 bushels shelled per 140 rods of ground, Luther Shaw, Armada, ...Farmer's Dictionary.

Best potatoes, Antoine Morass,	$2 00

BUTTER, CHEESE, &C.

Best cheese, Argalus Streeter, Bruce,	$2 00
2d do do do	Rural Economy.
Best June butter, Marcus Nye, Shelby,	3 00
Best fall do Harry S. Courtes, Armada,	2 00
Best cap honey, Silas Dixon, Clinton,	1 00
Dest bread, Wiley Bancroft, Armada,	2 00

VEGETABLES.

Best pumpkins, B. G. Whitney, Armada,	$0 50
Best squash, Rev. G. W. Newcomb, Clinton,	50
Best 6 heads cabbage, T. L. Powers,	50
Best vegetables and flowers, C. Reumold,	2 00

W. CANFIELD, *Secretary*.

REPORT

Of the Monroe County Agricultural Society.

J. C. HOLMES, Esq., *Sec. Mich. State Ag. Society:*

The Monroe County Agricultural Society, by an amendment to their constitution, having made it the duty of the Secretary, (instead of the president, as heretofore,) to make the annual report to the State Society, I find a duty imposed upon me difficult to perform, not knowing what is required or expected. I, therefore, in as brief a manner as possible, report:

That the society has now been in existence three years, having held its third annual fair on the 1st and 2d days of October, 1851.

The society numbers two hundred and eight-six members. The fee for membership is, for the first year, fifty cents; and annually, thereafter, twenty-five cents. This gives to the society about one hundred and fifty dollars annually, including the amount received from visitors on the days of the fair. About $100 is received from the county, making available for the uses of the society about $250.

The following sums were awarded at the late annual fair, 1851:

On field crops,	$24 00
Working oxen and steers,	17 00
Other neat cattle,	31 50
Horses,	12 00
Other horses and colts,	16 00
Matched horses,	5 00
Sheep,	12 00
Swine,	8 50
Products of the dairy,	14 00
Grains, fruits and vegetables,	11 11

Agricultural implements,	$3 00
Domestic manufactures,	12 75
Other articles manufactured in the county,	5 75
Fancy articles,	6 10
Luxuries for use of committees, consisting of bread, butter, cheese, honey, currant wine, etc.,	1 75
Trains of working oxen,	5 00
Plowing with horses,	5 00
do oxen,	5 00
Total amount of premiums,	$195 46

A discretionary committee was appointed for the examination of articles not enumerated in the list of premiums, and said committee recommended the payment of the sum of twenty dollars upon the various articles named; but as it requires the action of the executive committee, and no meeting of the same having been held since the fair in October, the sum named remains in the treasury. The balance, making up the sum of $250, was or is nearly expended in the expenses of procuring proper grounds and fixtures for the annual fair.

From the first organization of the Society, it has steadily increased in numbers and interests, and its effects are clearly visible upon the farming interest of our county. A desire to improve not only every species of farm stock, but also the quality and quantity of every kind of grain, fruit and vegetables indigenous to our soil and climate, seems to pervade the entire farming community, exciting a degree of emulation that cannot but prove highly beneficial to the agricultural interest of our county.

WM. WADSWORTH, JR.,
Sec. Monroe County Agricultural Society.

Monroe, Oct. 8, 1851.

ANNUAL ADDRESS,

Delivered before the Monroe County Agricultnral Society, at its third Annual Fair, held in the city of Monroe, Oct. 1st and 2d, 1851.

BY HON. WILLIAM H. MONTGOMERY.

Ladies and Gentlemen:

A year has passed since we were assembled at our last county fair, bringing in its annual round the festivities and enjoyments of winter, the buoyancy and fragrance of spring, the rich and abundant harvests of summer, and the lucious and crowning fruits of autumn; all these with the attendant blessings of health, admonish us of our many obligations to our Creator and bountiful benefactor, who has cast our lot in a land and clime where, for change of season, for variety of the products of the soil, for value and choice varieties of fruit, the growth and perfection of domestic animals, the great development of the physical and intellectual powers of the body and mind, we may not be surpassed by any portion of the globe. Being thus situated, surrounded with these blessings, it behooves us to see to it, that we do not allow them to pass unimproved, but that we diligently make use of the talent committed to our charge, that in the end we may be able to render a satisfactory account of our participation in the great drama of life.

The great crowning object to which we all aspire, that goal on which we place our eye in the grand experiment of life, is the attainment of happiness; it is so wisely ordered that it may be enjoyed by every human being and not monopolized by any; it is accessible by the great ones of the earth equally with the lowly; the Emperor on his throne may enjoy it no more than the peasant in his cot. Those most eager to obtain it are most likely to drive it from their grasp, while those negligent in the pursuit are seldom blessed with its presence—it is the almost invariable accompaniment of right and virtuous actions, the reward of honest industry, the inhabitant of a well disciplined mind.

In the pursuit we may follow any of the lanes of life, and feel sure of finding it not only at the end, but scattered along our pathway to cheer and encourage us in our toilsome progress—not only present with us but looming up in the distance to stimulate a more vigorous exertion, always increasing in magnitude and degree as our capacities are more susceptible of enjoyment.

It is the incentive to application of the professor in his studies, the artisan in his cabinet, the mechanic in his shop, and the agriculturist on his farm; and how exhilerating the thought that when pursued with an honest intent and from a right motive, our efforts are ever crowned with success.

It is not enough for our present nor prospective happiness that we rest content with things as they exist; our march is onward, we cannot long remain stationary; the mind of man, like a sunken pool, if not ruffled by the busy scenes of life, soon becomes stagnant, and instead of the bright and sparkling beverage, infusing strength and vigor into those who partake, we have a turgid mass, overgrown with a polluting scum, throwing a deleterious miasmatic influence broadcast over the community, engendering the seeds of physical and moral dissolution.

The mind, not only, but the physical powers of man, must be brought into requisition to subserve the purposes for which they were designed by creative wisdom; for we find that when the "Lord God had created man, he put him into the midst of the garden of Eden, to dress it and to keep it," well knowing what we have all learned by experience and observation that the mind and body, to be healthy and useful, must be employed. As I design noticing briefly this subject hereafter, I will pass to the subject matter of our present organization, which is the promotion of one of the great industrial avocations of life, agriculture with its accessories, manufactures and the mechanic arts.

The science of agriculture is of very ancient origin, having been followed to some extent by the first progenitors of our race. We have no authentic knowledge of its having been reduced to a science till some 2,000 years after the creation, when the Egyptians are said to have practiced it to some extent, and probably carried the art to a high degree of perfection. A country so fertile as Egypt, in process of time, being somewhat limited in extent, would be overrun with inhabitants, and we find colonies of them emigrating to new locations, carrying with them their knowledge of the arts of agriculture, which continuing to extend with the increase and migration of the inhabitants, overspread the whole of Europe, and we, the descendants of European nations are now deriving advantages from the investigation and scientific pursuits of Ociris, Calumela, Virgil, Pliny, Cicero, and others, who in their time lent the powerful energies of their minds to the advancement of agricultural knowledge.

And there is no science which requires more close observation, greater research, or more rigid investigation than this. To become fully acquainted with the properties of the soil we cultivate, to learn its relative proportions, and the adaptation to the production of the various cereal, culmifferous, and leguminous crops, by analysis, we must necessarily have some knowledge of chemistry.

To inform yourselves of the habits, procreative powers and the means successfully to counteract the encroachments of the insect tribes upon the crops we cultivate, we have need of the science of entomology.

In adapting our crops to the condition of the soil, with reference to its capability to produce and bring to a state of perfection the various kinds, we have need to know of what those crops consist; this we must also learn by analysis; and in this branch of our agricultural labors we shall be very much aided in our classification of the various plants by a knowledge of botany.

In rearing our domestic animals, informing ourselves of their wants, capacities, natures, the diseases to which they are subject, with their proper treatment, we must have some acquaintance with the science of Zoology; and in the science of ornithology we must be sufficiently versed to raise our turkeys, geese and ducks, our dorkings, polands and bantums; lastly and not the least of our pursuits, is the study of physiology, by which we may learn some things with reference to our own structure, which it is important to ourselves not only but to our posterity, that we should know. How are we to know these things? Much of these sciences may be taught to children too young to comprehend many of the branches of a common English education. Where are the qualified teachers to be procured? From one of the wings of the State University, where we must place our young men and young ladies, and insist upon such course of instruction as will fit them for teachers; the fund with which that institution has been reared, and which is yearly applied to its support, belongs to the agricultural population in proportion to their numbers; and when we speak with a determination to be obeyed, our legislature will not dare treat lightly the mandate.

We have great need of an agricultural department, where our young men may obtain a knowledge of all the sciences having reference to a regular systematic course of agriculture. This once attained and infused into the whole agricultural community, we shall no longer hear the equivocal compliment of *he is a plodding farmer.* The primary object in the cultivation of the soil should be, to obtain the greatest possible amount of products with the least possible expense, at the same time to maintain or to increase the original fertelity of the land by a gradual deepening of the soil, and by the addition of fertilizing agencies, to increase its productiveness. To ensure success requires an intimate knowledge of its constituent properties, the various earths composing it, together with the minerals, phosphates, and alkalies, with which it is impregnated, what is deficient, and what is superabundant. This having been satisfactorily determined, the work of regeneration may be fairly begun, and by a proper rotation of cropping, with the timely additions of clover, plaster, lime, ashes and the prompt return of the offal of the fields, annually prepared in the barnyard, the farmer may rest assured that bounteous nature will not neglect to reciprocate his good offices, but will amply compensate his toil and attention with the rich harvest, and the abundant production of fruits in their season. Owing to the natural fertility of the soil, the farmers of this county have not paid that attention to the preparation and application of manures, which their true interest requires, and so long as that fertelity remains we shall not probably introduce so much system in its cultivation, nor be compelled to exercise the same skill in its management as is necessary in countries having been longer under improvement; but the time will come when our exhausting method of farming must be changed, and for every crop we take from our fields we must return an equivalent. Owing to the large tracts of land bordering our numerous streams, the difficulty of perfect drainage by individuals, and the fact that the openings and the prairie country in the interior of the State could be more easily brought into cultivation, the resources of Monroe county are much longer in being developed than many of the interior counties; but the time is approaching, when the true sources of wealth from the products of the soil, the dairy, wool growing, and the raising of superior stock, she will not be excelled by any section of Michigan. From our locality and the properties of the soil, this is essentially a grazing county, which business, together with wool growing, is always safe and generally remunerative. I have been apprehensive that our atmosphere was too damp, our grass was too luxuriant, and climate too variable to permit a healthy growth of fine wooled sheep, not subject to disease nor to deteriorate in the quality of fiber; and the experiment of persons long in the business assures us that the larger frames and long wooled sheep are more hardy and better adapted to countries similar to ours, and for fattening purposes much superior.

Although slow in our improvements, we have yet in our timbered lands sure guaranties of wealth, as evinced by the piles of lumber, timber and staves strewn along the railroad, and at the wharves, waiting for an eastern market. We have also in our numerous stone quarries a source of wealth inexhaustable in supply, to which many of the interior counties are tributary for their stone and lime; these, connected with the advantages of an easy access by water or railroad to an eastern market, gives to the farmers of this county a position at least equal to any part of Michigan or other of the north-western States.

I have before remarked that the farmer who thoroughly understands his business and pursues it systematically, must necessarily be somewhat versed in chemistry. The inquiry may arise where and what books will he require, and how procure a laboratory; a good agricultural paper, the Michigan Farmer, is one of the sources of knowledge. If more is necessary, they can be procured at any of the book stores; but with so simple a laboratory, a good agricultural paper will give him all the rudiments; and every man who owns a barn yard has a laboratory which will for the present answer all practical purposes. There should be collected the various products of the farm not destined for market, such as straw, cornstocks, pea vines, refuse hay, &c., to be converted by partial decomposition into the aliment of future crops. And here I would remark that those who thresh their crops in the field, leaving piles of straw to be bleached by the snow and rains without deriving any material benefit from them until obliged to remove them to make room for future cultivation, are pursuing a ruinous and mistaken policy; this straw, if properly deposited in the barn yard from time to time during the winter, would add much to the cleanliness, comfort and health of your stock. And by absorbing the liquid manure of the yard, and being intermixed by the treading of your cattle with the offal of the stables will be in good condition on the opening of spring to contribute to a luxuriant growth of corn or to any of the leguminous crops, such as peas, potatees, beans, &c., which require a nutriment just such as is supplied during the early stages of decomposition of vegetable substances, and which, if not so appropriated passes off in the atmosphere, a floating tax upon the farm and the health of the neighborhood.

After being thoroughly decomposed it is then in the right state to give sustenance to the wheat or other culmifferous crops. I have wandered from the laboratory to the field; and this should be the most frequented path of every farmer. But a few words more about the barn yard. This I consider the savings bank of the farmer, from which, if he deposits freely during the fall and winter, he may safely draw in the spring, without endorsers, or the fear of protests at the end of ninety days, in the form of meagre crops or barren fields.

No man of sense would think of storing his wheat or corn in the highway with the hope of finding it there when wanted; yet some who are tolerably well informed and passable farmers too, are in the practice of feeding ther stock in the highway, or in such a loose and slovenly manner about the fields, as to lose most of the valuable properties of the manure that would accumulate were their stock properly guarded and fed in racks secured from storms and cold winds.

In a climate like this, good sheds for the protection and feeding of stock are preferable to close stabling.

Pure air is as essential to the health of animals as to man; and without more labor and attention than is usually bestowed on our stock they cannot always have it when stabled; an animal breathing the close air of the stable is poorly qualified to elaborate the nourishing and wholesome properties of milk or good butter. Nature evidently designed them for a greater range, and it is a matter of nice discrimination, calling for the exercise of the judgment, the wisdom of experience, results of long and close observation, to determine how far their natural propensities may be restrained without interfering with their most perfect healthy development of carcase or causing a lasting and serious injury to their constitutions. In the choice of stock we all have our preferences; some choose the Durhams, while others prefer the Devons, and others again the mixed, or a cross with either and our common stock, much depending on the uses for which they are designed. As milkers the Durhams stand deservedly high; and for draught, the Devons are equally so; neither are so hardy as our common stock, and as our cattle are usually treated, a cross with our more hardy breeds is preferable to the full bloods, which must have high keeping and much care during our winters, and full feed during summer, to perfect their growth.

Although we are awake in this county to the importance of raising good fruit, we are far behind some of our neighbor counties, having a supply from the old orchards, set by those who preceded us in the settlement of the eastern part of the county, we have not felt the necessity of raising our own, except to a limited extent. Notwithstanding that is passably good, and the best perhaps which could be obtained at the time of its growth, it is far inferior to the many choice varieties more recently introduced, and every farmer, business man, and mechanic, ought to devote at least time enough to horticultural pursuits to raise his own fruit, and an abundance of it. Labors of this kind have their reward; they add much to the health and comfort of himself and family; they give his grounds a cheering and inviting aspect, making home pleasant, and furnishing employment for a leisure hour, from the labors of the field, the shop, or the turmoil of business; they secure an abundant supply, and with care in the selection and cultivation of the choicest kinds, which are not always to be had in the market, and the gratification of eating the products of our own labor in our own independent way, and I trust that we all know enough of human nature to know that what we raise with much care and attention, by a little stretch of the imagination, can always be the best.

Our fruit trees, like our stock, must be well fed, with their appropriate aliment and they will give us generous crops. In some, and I believe in all the adjoining counties in this State, they have horticultural societies separate from the agricultural societies, which hold their weekly, semi-monthly, or monthly meetings during the ripening of the various fruits in their season. These societies are doing much good by the introduction and dissemination of choice varieties of fruit of the various kinds grown in this climate. Ought not we of Monroe county to form a society for the same purpose?—the good results are incalculable. What can be more pleasant and agreeable than for the farmers of the county, their wives and daughters, to meet the mechanics, the business and professional men in the community, with their wives and daughters, and emulate each other in the production of splendid flowers or choice fruits, and by a mutual interchange of friendly greetings open a broad field of acquaintance? I will recommend this subject to your favorable consideration.

There is one subject to which I wish to call the attention of my brother farmers, which, though not attended with immediate pecuniary results, will nevertheless pay for the outlay and labor in after years. It is the setting of shade trees. No outward appendage to a well built and neatly arranged dwelling is more pleasant or desirable than a growth of well selected and flourishing shade trees. They add much to the appearance, and very much to the comfort of the dwellers within, furnishing father, sons and workmen a convenient and pleasant resort to while away an hour in the middle of the day with a book or newspaper, to keep posted up with the passing events and improvements of the age, and where the children may be protected in their out-door sports from the heat of the sun. Here also is an inviting retreat where father and sons may hold council over the multifarious employments on the farm. They also in a measure protect our dwellings from the vertical rays of the sun, preserving a purity and freshness in the air around them, making it not only more pleasant and healthy for our wives and daughters, but altogether more desirable as a place of resort for ourselves and sons than the awning of some village store, or those recesses of vice and immorality, steaming with the noxious vapors of empty casks, and the repulsive breath and more repulsive and idiotic expressions of the inebriate. By the cultivation of fruit and shade trees a salutary moral influence may be exercised throughout the whole community; no persons, however negligent in this matter, but admire the appearance, approve the practice, and resolve in their imaginations, how nice and desirable they will have all these things, sometime in the distant future. Now, my friends, is the time to put into execution these resolutions, while in the busy scenes of life, it furnishes a recreation and a pastime to more laborious pursuits, and is an important and desirable adornment to our dwellings, helping to make them indeed, what the finer sensibilities of our nature would have them, a home, sweet home.

Another department in horticulture which contributes in no inconsiderable degree to those pleasures and enjoyments which help to make home pleasant and desirable, is the cultivation of flowers. In a country like ours where no branch of labor is overstocked, and the avenues to wealth, fame and distinction, so broad and inviting, it is not to be expected that the adult male population will devote much attention to this business. Therefore, the development of flora's treasures is assigned to the ladies—that they have not been unmindful of the responsibility devolving upon them in this healthy and delightful exercise, the display of bouquets, flowering shrubs and exotics, most fully demonstrates. In the display of articles of domestic manufacture, we also have assurance they have not neglected the use of the spindle, the loom and the needle.

In a country so full of inventive genius, where almost every branch of mechanism is being performed by machinery, we do not expect to see much of domestic manufacture—the low price for which many articles can be furnished from the shops precludes the idea of their being constructed by hand, from pecuniary considerations.

There is one branch of female industry, however, which cannot and ought not to be supplanted by machinery, and that is good house keeping. It matters not what may be their situation in life, no young or old lady is fully qualified to discharge the duties incumbent upon them as good wives, (and to that estate I believe they all aspire,) without having obtained a thorough practical education in its various routine of duties. Without this, they are very liable to be imposed upon by their help, their good tempers are often sorely tried by vexatious disappointments when relying upon theoretical knowledge, and they at length become resigned to heavy bread, with heavy sighs from heavy hearts, whereas by a little industry and application, and the exercise of those female virtues, amiability and perseverance, they may have light bread, light hearts, and light hearted and true husbands. Be assured, ladies, nothing will so effectually drive a man from home to seek a solace from the cares and perplexities of life, as the want of a correct knowledge in the various duties of good house keeping. What are the so called polite accomplishments, to a man in the agony of swallowing a cold wishewashy dish of coffee, or when masticating a burned up beef steak? The sound of the piano adds but to the jargon of children crying for the mothers protection, and soothing influences. I would not deprecate these accomplishments; on the contrary I admire and much approve them, but let them never interfere with the more useful and necessary avocations of female industry. In the display of farming implements, household furniture, wheel carriages, &c., we are assured that we need not go abroad to procure either articles of convenience or luxury. Persons so fastidious in their tastes as not to be suited with the handi-work of our mechanics are not wholesome citizens in the community; their extravagance, example and desire for display has a most pernicious tendency in leading many who are weak in mind as well as means into habits of luxury and dissipation, unsuited to their condition in life, on the genius and republican simplicity of our free institutions.

There are certain fixed principles co-existent with man which must be respected and observed as a means of arriving to that condition we were designed to occupy in the sphere of creative wisdom, among which are life, liberty and the pursuit of happiness, in the promotion of these rights, inalienable in their nature, the great desideratum with us should be the maturity and perfection of our species, the most perfect development of body and mind of which our race is susceptible; in pursuance of this object we necessarily call into requisition the various sciences and arts known among men, making them all subservient to one grand purpose. Any of the pursuits followed by men, whether for scientific or pecuniary objects, which have not in their tendency a direct or remote promotion of this desirable end, are of doubtful utility, and should be discountenanced in the community. Most prominent among these is the conversion of the products of agriculture into alcoholic beverages. I do not design going into the subject at length, but would call the attention of those engaged in the cultivation of the soil to the increased amount of taxation to which they are annually subjected by this most miserable business; more than three-fourths of all our expenditures for the promotion of justice and the support of pauperism result from this calling, and those who cultivate the soil with a view to a sale for such purposes are indirectly accessory to the evils resulting from its use in the community. Experience has long since taught us that it is not a promoter of life; on the contrary in very many instances it is the active cause of a speedy dissolution. Neither does it contribute in any form or in any of its appliances to an increased amount of liberty; the very reverse of this is the inevitable consequence resulting from the use of it, and a resort to its delirious influences on the brain with a view to drowning care or for the promotion of happiness is placing a slow match to the magazine of life which will finaly terminate in its destruction.

It is a principle well established by experience, that the faculties of the mind can only be exercised to their fullest extent through the medium of a perfect organization of the body, and that organization can only be attained by the fostering care, moral and physical training of the body and mind from infancy up to manhood. On the mothers in our country rests a great responsibility in this matter.

To them is committed the task of watching over their infancy, providing their nourishment and giving a right inclination to their wants and desires, protecting them from evil influences, instilling into their pliant minds the first principles of moral rectitude, and with great solicitude for their future well being to exercise unwearied patience in ministering to their numerous wants and restraining their volatile dispositions.

Mothers, do you fully appreciate the vast responsibility resting upon you in contributing to maintain the supremacy, and elevate the condition of our race. To you we look for the moral discipline and healthy physical development of body and mind of the rising generation.

On you essentially depends the future prosperity of our growing republic. By a wise provision of Him who rules the destinies of nations and of men, you are peculiarly fitted by your fortitude and perseverance, your virtuous minds and winning ways to draw the pliant minds of youth along those pathways of life which lead to usefulness and future greatness.

Daughters, young ladies of Monroe, on you also devolves a responsible and important duty in the experiment of life; you possess an influence which properly exercised controls the turbulent passions of the sterner sex, subdues and gives them a right inclination; you possess a charm which, exercised with discretion, will drive away many of the evil spirits by which many of the youth of our country are led estray.

No guard so effectual in leading innocent youth to virtuous manhood as a mother's precept, aided by a sister's love and example. From the many advantages you possess, we have reason to expect much good will result to this community from your having lived in it.

There is a period when the father's influence, by precept or example, begins to tell plainly whether it is for evil or for good, and as the vigor and health of body and mind are cared for or neglected, so will the yonng man come upon the theatre of active life, a useful citizen and an ornament to society, or merely the form without the principles of a man.

It is a duty every parent owes to the community in which he resides, to contribute by his precept and example to make his sons and daughters conspicuous for their many virtues. In so doing you not only promote your own happiness but their's also, besides adding to the store of life a fund on which more unfortunates are permitted to draw. Young men of Monroe county, notwithstanding all the care and solicitude manifested for you by your parents, friends, and the community generally, you are to be the artificers of your own fortunes. Your growth of body and mind may have been anxiously watched from day to day; every pains may have been taken with your education which the circumstances and situation of parents would permit, and yet a pernicious example, if followed, a vicious habit, if indulged, may prostrate all the fond hopes of parents and friends, and turn you an outcast upon the cold charities of the world, to end prematurely a miserable existence. How important then that you should be very circumspect in your manner of life, guarding against every vicious indulgence as the opening of a pathway to infamy, disease and death. Upon you we rely as recruits about to enter upon the great battle field of actual life. In a few more years we shall have resigned our commissions and retired from the service. Our places must be filled by well disciplined, efficient, active men, and we trust you will be prepared at any period to assume the responsibility which prospectively devolves upon you. Your knowledge, like your pursuits, must necessarily be various. You cannot all be farmers, nor artisans, nor professors; but you can all be equally honored and respected, whatever may be your calling, if faithful and efficient in the discharge of the duties incumbent on you as men and good citizens. "There is a tide in the affairs of men, which taken at the flood leads on to fortune; omitted, all the voyage of their lives is spent in shallows and in misery."

Shakspeare well understood human nature. You no doubt have, with your little experience in life, seen it verified. A young man over indulged is poorly prepared to guide his bark up and down the stream of life. His chances are great of making shipwreck on the first breakers he endeavors to pass, and even some will founder on a smooth sea—such are ever repairing but seldom get under way. Prudence and economy are cardinal virtues with young men, which if practiced in the setting out in life become household deities whom we feel it no desecration to worship; if neglected they compel us to do them homage or suffer for our disobedience. A very important requisite to the attainment of the object of pursuit in all the avocations of life, is the formation of habits of industry and perseverance; these are acquired not by forming a resolution merely, but that resolution must be carried into execution from day to day, until by the continued practice the habit becomes formed, and what was at one time deemed impracticable for the intellectual or physical powers to accomplish, at length becomes an easy task, or a recreative amusement.

We live in a community where stations of honor, distinction and emolument are accessible to all; industry, application and perseverance are sure guarantees of success when guided by integrity and honesty of purpose. We have no dignitaries nor favored ones by divine right. Wisdom, virtue and intelligence, are the ensigns of nobility with a free people. A broad field is open before you; many of its scenes are inviting, alluring, deceptive. Prudence and discretion will guide you safely through to that haven of rest where we all finally expect to arrive at a never ending jubilee.

REPORT

Of the Oakland County Agricultural Society.

J. C. HOLMES, *Sec. Mich. State Ag. Society:*

At the annual meeting of the Oakland County Agricultural Society, held at the Court House in Pontiac, on the first Tuesday in January, 1851, the following amendments to the constitution were voted:

That the constitution be so amended that the president and recording secretary be ex-officio members of the executive committee. Also that part which authorized the electing of one person to the office of secretary and treasurer, be repealed, and that the president preside at all meetings of the executive committee, and the secretary keep a record of their proceedings.

The following persons were elected officers of the Society for the ensuing year:

President—Linus Cone, Troy.

VICE PRESIDENTS.

Avon—Calvin A. Green.
Farmington—Nathan Power.
Waterford—Isaac J. Voorhies.
Independence—George Miller.
Brandon—N. D. Bingham.
Bloomfield—John Chamberlain.
Novi—Samuel Rogers.
Troy—Clark Beardslee.
White Lake—Edward Baughman.
Royal Oak—David Chase.

Oakland—Wm. M. Axford.

Southfield—David Goss.

Oxford—Eldredge D. Deming.

West Bloomfield—Murrin J. Irish.

Pontiac—Benjamin Phelps.

Springfield—James G. Simonson.

Commerce—Amasa Andrews.

Highland—O. P. Davidson.

Milford—Henry K. Foote.

Rose—Delevan Burrows.

Holley————Patterson.

Addison—Seymour Arnold.

Orion—Elijah B. Clark.

Lyon—Samuel Sulvy.

Recording Secretary—H. N. Howard.

Treasurer—A. W. Hovey.

Corresponding Secretary—Wm. W. Phelps.

The following persons were elected as executive committee:

Oxford—John Thomas.

Avon—Charles Baldwin.

Waterford—A. Whitehead.

Troy—James Bailey.

White Lake—Robert Garrin.

Pontiac—Henry Waldron.

West Bloomfield—Wm. S. Henderson, Asa B. Hudson.

The meeting of the executive committee was fixed to be held on the 26th day of May at 10 o'clock A. M.

The meeting then adjourned.

On the 26th day of May, the executive committee met at the Court House, in Pontiac, and there not being a quorum of members present, adjourned to meet on Friday the 6th day of June at same place.

On the 6th day of June committee met pursuant to adjournment, and resolved to hold a fair this fall, and also voted to appropriate the sum of five hundred dollars, exclusive of any discretionary premiums that might be awarded.

A committee, consisting of Andrew C. Walker and Charles Baldwin, was appOinted to make out a premium list.

The committee reported and report was accepted and adopted.

The president and secretary were appointed a committee to procure the printing of bills, &c.

A. B. Hadsell was appointed a committee to solicit funds for the purpose of aiding in erecting buildings and preparing grounds for the fair, &c.

The following persons were appointed examining committees, to wit:

On Farms—Messrs. Linus Cone, Andrew C. Walker and Harrison Voorhies.

On Cattle—Messrs. Eber Durham, Aaron Burley, H. V. D. Bogart, George F. Thurston and Jacob Hadley.

On Horses—Messrs. Orville C. Morris, John Sprague, Charles Grant, Philip Frisbee and Thomas Gerles.

On Sheep—Messrs. Moses Wisner, Rollin Sprague, Washington Stanley, Harrison Voorhies and Isaiah B. Ward.

On Swine—Messrs. John Springer, Wm. Whitfield and Wm. F. Bingham.

On Farming Implements—Messrs. Wm. Besley, Alexander Wattles, Jessee Tenney, Theron W. Barber, and Edward C. Baughman.

On Plowing—Messrs. Riley Cone, Joseph M. Irish, John Garner, Zebrin Voorheis and George Tibbets.

On Butter and Cheese—Messrs. Wm. Draper, R. P. Bateman and Mason J. James.

On Domestic Manufactures—Messrs. B. B. Morris, O. D. Richardson, Mrs. John Sprague, Mrs. George Tibbets and Mrs. A. B. Mathews.

On Needle, Shell and Wax Work, Paintings and Drawings—A. B. Mathews, Mrs. A. A. Lull, Mrs. Geo. W. Collins, Mrs. E. T. Harris and Mrs. W. S. Henderson.

On Flowers—Thomas J. Drake, Miss Sarah Beach, Miss Jane Hanscom, Miss Frances A. Hunt and Miss Lucinda Martin.

On Fruit—Messrs. Francis Darrow, Wm. G. Stone, Hiram A. Rood, Cyrus A. Chipman and Samuel Rogers.

On Vegetables—Messrs. Clark Beardslee, Elnathan Phelps, Horatio Scott, Cyrus Smith and Embury Ferguson.

On Grain—Edward Barkham, Joseph Jones, John P. Wycoff, Ezra Carpenter and Alfred Phelps.

On Poultry—Messrs. Hester L. Stevens, Peter Hogan and George W. Phelps.

Committee adjournnd to meet at the court house, in Pontiac, on the first Saturday in September, at one o'clock P. M.

Committee met pursuant to adjournment at the court house in Pontiac, on the first Saturday in September, at one o'clock P. M.

J. L. Brownell was appointed a committee to wait upon Judge Hunt and solicit said Hunt to deliver the address at the annual fair.

Committee reported that Judge Hunt would accept of the invitation, provided no other person could be procured.

It was determined that the fair should be held at the village of Pontiac, and John P. Le Roy, Esq., was appointed a committee to solicit funds from the citizens to aid in fitting up the grounds.

A committee of three was appointed, consisting of James Bailey, John Thomas and Wm. S. Henderson, to locate the grounds, erect suitable buildings, and make the necessary arrangements for the fair.

Moses G. Spear, the sheriff of the county, was appointed marshal for the two days of the fair.

Committee adjourned to meet at the court house on Saturday the 4th day of October.

Saturday, October 4th—Committee met at the court house and voted that season tickets be sold to old members for fifty cents and ten cent tickets to admit members for one day. It was also agreed that the time for plowing one-fourth of an acre be one hour and ten minutes.

A. B. Hudsell was appointed a committee to obtain suitable persons for door-keepers at the fair.

Committee adjourned to meet at the fair grounds on the 7th at 10 o'clock A. M.

Tuesday, October 7th—Committee met pursuant to adjournment, at the fair grounds—completed the arrangements for the fair by filling vacancies in the examining committees, &c. Fair held on the same and the following day. On the second day an address was delivered by the Hon James B. Hunt, which was listened to with attention by the vast concourse present; after which, the reports of the various committees were read, and the meeting adjourned.

The executive committee met in the evening at the office of J. B. Hunt, Esq., to review the reports, and complete the nnfinished business. Committee adjourned to meet on the first Tuesday in January, 1852.

January 6th, 1852--Committee met pursuant to adjournment, at the Court House in Pontiac, and the committee on farms reported that no application had been made for premiums; also the same on field crops.

Committee on printing made a report, which was accepted and adopted.

The report of the treasurer was also accepted and adopted.

REPORT OF TREASURER.

Oakland County Agricultural Society in account with A. W. Hovey.

1851.	DR	
June 16.	To paid H. N. Howard,	$9 00
23.	C. M. Eldredge,	3 00
Sept. 25.	C. W. Tuthill,	130 00
Oct. 4.	Phelps & Stevens,	5 00
9.	A. B. Hudsell, (three door-keepers,)	9 00
9.	Whitney, (watchman,)	50
9.	Expenses at Fair,	83 94
10.	J. B. Hunt,	5 00
11.	W. Tilden, (premium 1849,)	2 00
11.	O. C. & R. B. Morris,	1 02
11.	George Dorr, (premium 1849,)	1 00
Nov. 5.	Thompson, printing address,	8 00
5.	Premiums for 1851,	438 50
5.	Balance cash in hand,	103 36
Total		$801 32

1851.	CR.	
June 16.	By cash from County Treasurer,	30 00
Sept. 25.	" " "	160 48
Oct. 9.	citizens of Pontiac,	106 00
9.	membership tickets,	208 00
9.	admission fees at fair,	288 32
9.	sale of fruit,	8 52
Total		$801 32

The undersigned, treasurer of Oakland County Agricultural Society, respectfully submits the foregoing exhibit of the financial condition of the Society, showing a balance of cash in my hands of one hundred and three dollars and thirty-six cents. There is also in the hands of the county treasurer, as I am informed by that officer, the sum of about ninety-five dollars, belonging to the Agricultural Society.

All of which is respectfully submitted.

A. W. HOVEY,

Treasurer of Oakland County Ag. Society.

LIST OF PREMIUMS.

CATTLE—SHORT HORNS.

1st premium,	bull, 14 months old, A. S. Brooks,		$5 00
1st do	3 year old cow,	do	5 00
1st do	yearling heifer,	do	3 00

DEVONS.

1st premium,	4 year old bull, John L. Brownell,		$5 00
2d do	one bull 17 months old, Charles Derbyshire,		3 00
1st do	one cow 6 years old, John Thomas,		5 00
2d do	6 year old cow,	do	3 00

GRADE.

1st premium,	1 bull 2 years old, Birdseye Dewey,	$4 00
2d do	1 bull 2 years old, Ira Clark Seeley,	2 00
1st do	1 yearling bull, N. G. Terry,	2 00
1st do	1 bull calf, Samuel Blackwood,	2 00
2d do	1 do Rial Irish,	1 00

1st premium, 1 milch cow, H. V. D. Bogart, ... $5 00
2d do 1 do John H. Button, ... 3 00
1st do 1 two year old heifer, Wm. H. Pier, ... 3 00
2d do 1 do do John H. Button, ... 2 00
1st do 1 yearling heifer, Ira Clark Seeley, ... 2 00
2d do 1 do John W. Leonard, ... 1 00
1st do 1 heifer calf, A. S. Brooks, ... 2 00
2d do 1 do H. V. D. Bogart, ... 1 00

WORKING OXEN.

1st premium, 1 yoke 4 year old oxen, B. D. Coonley, ... $4 00
2d do 1 do do do 3 00
3d do 1 do do Edward Martin, ... 2 00
1st do 3 year old steers, John Norton, ... 4 00
2d do 3 do George Tibbetts, ... 3 00
1st do 2 do John H. Button, ... 3 00
2d do 2 do J. G. Simonson, ... 2 00
1st do yearling steers, Samuel Blackwood, ... 3 00
2d do do John L. Brownell, ... 2 00

FAT CATTLE.

1st premium, 1 yoke oxen, Owen B. L. Trowbridge, ... $4 00
1st do 1 fat cow, Samuel Blackwood, ... 3 00

AYRSHIRE.

1st premium, cow and calf, John Thomas, ... $5 00

DISCRETIONARY PREMIUM.

1 durham bull, Andrew C. Walker, ... $4 00
1 two year old heifer, Isaac I. Voorhies, ... 2 00

HORSES.

1st premium, 1 four year old stallion, Calvin A. Green, ... $6 00
2d " 1 " " Wm. Whitfield, ... 4 00
3d " 1 " " Samuel Batchelor, ... 3 00
1st " brood mare, with foal on foot, C. W. Green, ... 6 00
2d " " " M. Fuller, ... 4 00
3d " " " Wm. Green, ... 3 00
1st " 3 year old horse, mare or gelding, J. D. Swan, . 4 00
2d " " " Jacob Hiller, . 3 00
1st " 2 year old horse or mare, C. A. Durfee, ... 3 00
2d ' " " Alva Butler, ... 2 00
3d " " " R. P. Hosner, ... 1 00

1st premium, best yearling colt, Willard White, ... $2 00
2d " " E. H. Cressey, ... 1 00
1st " best pair of horses over 4 years old, J. Thomas, . 6 00
2d " " " A. Wattles, Jr., 4 00
1st " best gelding over 4 years old, H. V. D. Bogart, . . 4 00
2d " " " H. N. Howard, . . 3 00

DISCRETIONARY PREMIUMS.

1st premium, best 3 year old stallion, F. A. Williams, ... $1 00
2d " " " John Renwick, ... 1 00
3d " " " A. M. Ellenwood, ... 1 00
1st " best 2 year old stallion, William Tilden, ... 1 00
1st " best 4 year old gelding, John Chamberlain, ... 1 00
1st " best yearlings, Samuel Pearsall, ... 50
2d " " Matthew Stanley, ... 50
Best 4 year old mare, B. B. Mosier, ... 2 00

SHEEP—MERINOS.

1st premium, 1 merino buck, Linus Cone, ... $4 00
2d " 1 " H. N. Howard, ... 3 00
3d " 1 " J. A. Peck, ... 3 00
1st " best pen 5 merino ewes, H. N. Howard, ... 5 00
2d " " " C. A. Chipman, ... 3 00
1st " long wool buck, Leicester, John Thomas, ... 4 00
2d " " " " 3 00
1st premium, Southdown buck, William Whitfield, ... 4 00
2d " " " " 2 00
1st " best lot 5 Southdown ewes, " 4 00
2d " " " Daniel Whitfield, ... 2 00
1st " best fat sheep, 3 wethers, " 2 00
1st " grade sheep, French and Spanish merinos, Harris Newton, ... 4 00
2d " grade sheep, French and Spanish merinos, N. S. Schuyler 3 00
1st " 5 ewes, French and Spanish, Harris Newton, ... 5 00

SWINE.

1st premium, boar, Samuel Rood, ... 4 00
2d " " Erasmus E. Sherwood, ... 3 00

1st premium, breeding sow, Samuel Rood,................$4 00
2d " " Horatio N. Howard,........... 3 00
1st " lot of 5 pigs, Samuel Rood,................ 4 00
2d " " L. G. Burch,................ 3 00

DISCRETIONARY PREMIUMS.

1st premium, breeding sow and 3 barrows J. Durkee,........ 2 00
2d " " " A. Dennison,..... 1 00
1st " sow and six pigs, John Chamberlin,........... 2 00

FARMING IMPLEMENTS.

1st premium, plow for general purposes, Aaron Smith,...... 4 00
1st " " stiff soil, Aaron Smith,............ 4 00
1st " roller for general use, " 3 00
1st " farm wagon, B. & J. E. Norton,............ 5 00
1st " cultivator for general use, O. C. Morris,....... 2 00
1st " straw cutter, Amasa Andrews,.............. 2 00
1st " best horse cart for farms, Moses Wisner,...... 3 00
1st " " rake, Joseph M. Irish,............ 2 00
1st " single harness, Alvin C. High,.............. 2 00
1st " corn brooms, A. Wattles, Jr.,................ 1 00
1st " threshing machine, George W. Merrill,....... 5 00
1st " parlor stove, George W. Merrill,............. 2 00
1st " cook stove, " " 3 00

DISCRETIONARY PREMIUMS.

D. O. & W. S. Penfield, for a lot of farming implements and horticultural tools,.................................... 5 00
1 double harness, C. Webster,.......................... 2 00

PLOWING.

1st premium, with horses, Isaac Osmun,................. 5 00
2d " " A. Wattles,................... 3 00
3d " " Samuel Hubbell,............... 2 00

BOY'S PLOWING.

1st premium, Thomas W. Whitfield, 13 years old,.......... 3 00
William S. Henderson, treble team, (dis. prem.,)........... 5 00

BUTTER AND CHEESE.

1st premium, 10 ℔s butter in rolls, W. Whitfield,.......... 3 00
2d " " " Miss M. Beardslee,...... 2 00
3d " " " James Bailey,.......... 1 00

1st premium, best 50 lbs of butter, O. P. Davison, ...$3 00
2d " " " John P. Wycoff, ... 2 00
3d " " " John Windiate, ... 1 00

DISCRETIONARY PREMIUMS.

C. D. Woolcot, ... 50
J. P. Wycoff, ... 50
Charles Elliot, ... 50
O. P. Davison, ... 50
Jacob Hiller, ... 50
L. G. Burch, ... 50
Mrs. A. Whitehead, ... 50
C. Smith, ... 50
3 fine loaves of bread, made of flour from Paddock's Mills, by Mrs. George W. Wisner, ... 50
Best old cheese, Luther Lapham, ... 3 00
2d " J. H. Murray, ... 2 00
Best new cheese, Luther Lapham, ... 3 00
2d " J. H. Murray, ... 2 00

CHEESE—(DISCRETIONARY.)

Clark Beardsley, ... 1 00
Samuel Hubbell, ... 1 00
1st premium, 10 lbs honey, Samuel Rood, ... 1 50
2d " " " Edward Turbush, ... 75
Discretionary premiums, Henry Waldron, ... 50
Best bee hive, J. H. Murray, ... 1 00
" 10 lbs of mople sugar, B. Coonley, ... 1 00

DOMESTIC MANUFACTURES.

1st premium, 10 yds flannel, Mrs. Benjamin Phelps, ... 3 00
2d " " " Mrs. George Tibbetts, ... 2 00
1st " 10 yards rag carpet, Mrs. Stephen Reeves, ... 3 00
2d " " " Mrs. Levi Burlingham, ... 2 50
1st " silk knit stockings, Mrs. G. W. Collins, ... 1 00
1st " wool " " 1 00
2d " " " Mrs. H. M. Knapp, ... 50
1st " wool mittens, Miss M. Beardslee, ... 1 00
1st " best 10 yards linen cloth, Mrs. D. Hammond, ... 1 50

1st premium, 1 pair cotton knit stockings, Mrs. C. Hadsell,..$1 00
2d " " " Mrs. G. W. Collins, 50
1st " best woolen shawl, Mrs. John Richardson,..... 2 00
2d " " " Mrs. E. Chase,........... 1 50

DISCRETIONARY PREMIUMS.

1st premium, woolen and cotton coverlid, Mrs. Geo. Miller,..$1 00
2d " " " " Mrs. J. Richardson, 1 75
3d " " " " Mrs. James Bailey,. 50
4th " " " " Mrs. C. D. Woolcot, 50
Best pair linen knit stockings, Mrs. G. W. Collins,.......... 50
" worsted " Miss M. Beardslee,........... 25
" woolen socks, Mrs. H. M. Knapp,............... 25
1 pair, nearly all wool, blankets, heavy and good, wove in kersey, James Bailey,................................ 1 00

NEEDLE, SHELL AND WAX WORK, PAINTINGS AND DRAWINGS.

Best piece worsted work, Mrs. A. B. Mathews,............. 1 00
" worked quilt, Miss C. A. Barnum,.................. 1 00
" white quilt, Mrs. Henry Mead,..................... 1 00
" silk bonnet, Mrs. Markham,...................... 1 00
" knit crochet hat, Miss A. Andrews,................. 1 00
" 2 lamp mats, Mrs. George W. Wisner,............... 1 00
" ornamented shell work, Mrs. S. Brotherton,............ 1 00
" oil or water colored paintings, Mrs. G. W. Wisner,..... 1 00

DISCRETIONARY PREMIUMS.

1 worsted rug or mat, Mrs. E. P. Harris,................. 50
Specimens of down sewings, Mrs. O. Taylor,............. 50
1 bead purse and port folio, Miss S. Hubbell,............. 1 00
3 pieces fine chromatic paintings, George Rogers,.......... 1 00
1 piece " " Miss Amelia Peck,........ 50
1 port folio, Miss A. Andrews,....................... 50
1 crochet napkin, Mrs. H. A. Rood,.................... 50
1 bunch domestic sewing silk, Mrs. G. W. Collins,.......... 50
1 pr silk mitts, Mrs. G. W. Collins,..................... 50
1 pr knit slippers, Mrs. H. A. Rood,.................... 25
1 crochet collar, Miss A. E. Chipman,.................. 25

1 crochet collar, Miss W. Wilson,	50
1 crochet tida, Mrs. G. W. Wisner,	$ 50
1 " " "	50
1 chromatic painting, Miss N. D. Bingham,	25
1 bed quilt, Mrs. Linus Cone,	50
2 " Mrs. Samuel Brotherton,	50
1 " Luman Brownson,	50
1 " John Barge,	50
1 " Henry Waldron,	50
1 work stand, G. W. Wisner,	50
1st best bottle currant wine, A. B. Matthews,	1 00
2d " " E. C. Beach,	50
3d " " John Dains,	50

FLOWERS.

Greatest variety and quality, Mrs. A. B. Mathews,	1 50
Square box containing 30 varieties of dahlias, M. Wisner,	1 50
Best eight dissimilar blooms in a vase, M. Wisner,	50
Discretionary premium, eight dissimilar blooms, Mrs. A. B. Mathews	50
Best collection of green house plants, Mrs. H. A. Rood,	2 00
" floral design, Miss Elizabeth Ruggless,	1 00
" hand boquet, flat, Mrs. O. D. Richardson,	50
" " round, Mrs. A. H. Hanscom,	50
" arranged basket of flowers, Miss Maria T. Hunt,	1 00
Discretionary premium, 1 basket of flowers, Mrs. H. A. Rood,	50
A beautiful gerranium, studded with flowers, Mrs. Ackerman,	50

FRUIT.

Best and greatest variety of table appless, A. Terry,	1 00
" seedling apples, John Chamberlain,	2 00
" six varieties of winter appless, 6 of each, A. Terry,	3 00
" specimen of winter fruit, not less than 6 varieties, C. A. Chipman,	1 00
2d best, Samuel Rood,	50
Best variety of pears, Samuel Rood,	2 00
" twelve peaches, Wm. Fisher,	1 00

Best variety of seedlings, John Chamberlin, ... $1 00
" 12 quinces, A. Barber, ... 1 00
2d " G. W. Collins, ... 50
Best collection of grapes, John Frank, ... 1 00
" single bunch " " 50
" lot of watermelons, E. H. Bristol, ... 1 00

DISCRETIONARY PREMIUMS—WITHOUT DISTINCTION.

Apples, Dinah Smith, ... 1 00
" Lewis W. Adams, ... 1 00
" James Bailey, ... 1 00
" O. P. Davison, ... 1 00
" Isaac I. Voorhies, ... 1 00
" Isaac Osmun, ... 1 00
" Eddy Morrison, ... 1 00
" Linus Cone, ... 1 00
" George Tibbets, ... 1 00
" R. Adams, ... 1 00
" R. Brownson, ... 1 00
" James T. Allen, ... 1 00
" William Fisher, ... 1 00
" Rev. I. W. Ruggles, ... 1 00

PEARS—(DIS. PREMIUMS.)

Cyrus A. Chipman, ... 50
G. W. Collins, ... 50
Quinces, Wm. Fisher, ... 50
" A. Terry, ... 50
" A. Spear, ... 50
Grapes, A. Terry, ... 50
Peaches, N. D. Bingham, ... 50

VEGETABLES.

Best 12 carrots, Harvey Howard, ... 50
" ½ peck Lima beans, Mrs. Geo. W. Wisner, ... 50
" 6 heads of cabbage, O. D. Richardson, ... 50
" 6 beets, T. Gerralds, ... 50
" tomatoes, " 50

Best 12 onions, M. Cross, ... $ 50
" 6 pumpkins, N. D. Bingham, ... 50
" ½ bushel potatoes, A. Terry, ... 50
" flat turnips, G. W. Collins, ... 50
" 12 squashes, S. Stevens, ... 50
" 12 parsnips, S. Stevens, ... 50
" and greatest variety raised by one exhibitor, S. Stevens, ... 1 00

DISCRETIONARY PREMIUMS.

Lot of beets, Miss Dinah Smith, ... 50
" Harvey Howard, ... 50
Lot of carrots, Avery Jones, ... 50
" John Crombie, ... 50
Lot of potatoes, Clark Beardslee, ... 50
" watermelons, R. Adams, ... 50
1 bushel merino potatoes, M. Cross, ... 50

GRAIN.

Best sample of wheat, N. D. Bingham, ... 1 00
Discretionary, 1 sample of blue stem, G. Persoll, ... 50
Best sample of corn in the ear, L. W. Adams, ... 1 00
Discretionary, " Jedediah Durkee, ... 50
Best sample oats, Lyman Fuller, ... 1 00
" barrel of flour, M. Walters, ... 2 00
Sample of Egyptian millet, O. D. Richardson, ... 25

POULTRY.

Best lot of large fowls, M. Wisner, ... 50
" Shanghe, J. Crombie, ... 50
" poultry, Walter Whitfield, ... 1 00
Discretionary, lot of Bantams, A. W. Hovey, ... 50
" " Polands, " 50

MISCELLANEOUS ARTICLES.

Printing, W. M. Thompson, ... 1 00
" Phelps & Stevens, ... 1 00
2 pieces of marble, W. L. Strong, ... 1 00
1 horse sleigh, C. A. Brownson, ... 1 00

1 fleece of merino wool, L. Cone, ... $1 00
1 tiling machine, John Daines, ... 10 00
1 lot of drain tile, John Daines, ... 1 00
1 apple-paring machine, Thom. Buel, ... 50

H. N. HOWARD,
Sec. Oakland Co Ag. Society.

ADDRESS

Delivered before the Oakland County Agricultural Society.

BY HON. J. B. HUNT.

Fellow Citizens:

We have again met together to hold the anniversary fair of the working men of the county; our assembling together is the best evidence that we take an interest in the progress of labor. The apparently stern sentence which was pronounced upon man in the garden of Eden, "that in the sweat of thy face shalt thou eat bread till thou return unto the ground," is turned to a blessing, when labor meets with its proper reward; life would soon become intolerable without some useful employment; and when the law inflicts its most severe penalty, the unhappy wretch is doomed to solitary confiement *without* labor. It is the best proof of the improvement of the age in which we live that men meet together for the purpose of bestowing rewards on the most successful and skillful of their class, and that that class is composed of the men who have the manhood to work for their own independence and standing in society. This is almost a new feature in the history of our agricultural population; in the olden time and in the ruder ages of the world, the army and the church were the well known paths which led to rewards and honors, the laborer was the oppressed and wretched serf of the soil. In those days, England, Saxon England, although governed by a milder rule than any of the continental nations, yet even there, the iron collar was worn by the field laborer as the badge both of his slavery and his degradation. And Blackstone says, 'the opprobrious name of villain,' embraced the whole agricultural population of the country. On the continent of Europe the same unhappy condition of the laborer existed, and if possible enforced with more rigor by the nobles. This was no new state of things; it had regularly descended from the breaking up of the Roman Empire; it had existed during the imperial government of Rome; through all these dreary years, the field laborer had been subject to the most cruel oppression, and if he ever arose from that abject condition it was by becoming a common soldier, and by daring and bloody feats, fighting himself up to distinction.

One of the greatest difficulties which the agriculturist had in changing his social position, was the manner in which the lands were held; in all those countries where the feudal system prevailed, the lands were held of the chief by military service, or they belonged to the religious houses; in both cases they were rented and under let from one dependent to another, subject to every charge that ingenuity could invent, until the laborer who done the work, with the worst kind of implements and in the most rude method of agriculture, received a pittance scarcely sufficient to sustain life. The violent oppression of the military baron was scarcely more onerous than the tender mercies of the church; 'they were ground between the upper and the nether mill stone.' They remained ignorant, for they were too much despised to be the associates of the better informed, and the great scarcity and expense of books in those days, made this means of knowledge beyond the reach of the poor.

Fortunately for us, a variety of great events running through the last 500 years has made a vast change in the condition of the agriculturist and the artisan. Whatever change conduced to the improvement of the human race ameliorated their social condition; for men gain intelligence as they acquire political power in the State. One of these great events was the reformation in the Romish church; the human mind seemed to break loose at once from the control of the Priesthood and from the superstitions by which it had been enfeebled, and human freedom made a rapid advance, because on this subject, at least, men began to think; the introduction of the art of printing about the same time, was another of these forward movements, for then men learned to read; and when men begin to think and to read, they *must* improve. These thinkers and readers soon commenced writing and printing for others these new thoughts, and so the faint light of morning soon assumed a mid day brilliancy. The discovery of America was another of these grent events, because it gave men room to act, and this room, this new field, was not encumbered by old land titles or governed by old laws and usages, made and sanctified by time to the interest of a particular and favored class—and by one of those wise and providential arrangements, which God in his infinite mercy bestows on the children of men, this new field was given over to those who had felt, and fled from the oppression of those bad laws; they had suffered upon a question which never should have been the subject of human laws, for the service which man owes to his Creator, is not to be circumscribed by his erring fellows. It was this great subject of human rights which compelled these men to occupy the wilderness of the new found continent. And although we can not admire the system of government first adopted by the colonists, yet it was an advance in both political and religious freedom, each expanded with the growth of the colonies, notwithstanding they were fettered and kept back by charters bestowed by English kings, and by laws enacted from time to time in conformity with the views of the mother country.

In due time came our glorious revolution, and the severance of the government from European domination; and this was another, and to us, the greatest of those events in the history of the world, which tend to the improvement of the condition of man; from this period all the impediments which had been thrown around the transfer of real estate were removed. Our English ancestry and English rule had bestowed upon us the spirit of the British laws, the richest legacy which one people could bestow upon another; yet these laws, by the wisdom of the great men of the revolution, were so modified, that lands were no longer tied up in families from one generation to another, the whole system of entails was swept away, land could no longer be let on long and sometimes never ending leases, and whenever a landed proprietor died intestate or became insolvent, the lands were divided among the heirs or sold on execution for the benefit of the creditors. Land monopoly, the monster which we know nothing of in our day, received its death blow. From that period the laboring man could acquire land at a reasonable price whenever he obtained the means; and by a provision of our statute laws, recently engrafted into the constitution, forty acres of land is forever exempted from sale on execution to the farmer, and a city or village lot not exceeding in value fifteen hundred dollars, to other householders. By these various improvements in the condition of the husbandman, the land is divided up into small parcels, the country becomes thickly populated with a hardy, industrious and intelligent class of citizens, each holding his land in his own right, and each being the equal and peer of his neighbor, so that a total and complete revolution has by these various causes been effected in the condition of the farmers. The laborers on the soil are the owners, instead of being the serfs of the bye gone days. The American farmer occupies the place of the feudal baron, and has taken a position in the first instead of the last class of citizens.

In this dignified position it behooves him to use proper care and discretion to sustain the place which the laws have assigned him. He must be a gentleman. We do not mean by that phrase, that he must not labor. All gentlemen work in these days. The loafer is the only one that lives without work. And we all know in what estimation he is held. But we mean by that epithet, intelligence, kindly feelings and that regard for others which is taught as one of our highest christian duties, to "do to others as you would have them do unto you." If we have not brought ourselves up to this standard, let us see to it that our children are better instructed. We have an excellent system of common schools, and these schools will continue to be the nurseries of our law-givers and statesmen, and the certain guarantees of our political independence, for they keep up that intelligence so essential to the preservation of democratic institutions. But these schools and our colleges and seminaries of learning are employed principally in instructing the mind, the head of the student, while the heart is neglected and allowed to run to waste. This last is the schooling which we most need, and which can be best learned from a father's example and the teaching of a mother's love; this branch of education has been too much overlooked and undervalued. Our sons are sent off to learn the dead languages, mathematics, &c.; and this we call educating them for the business of life. Our daughters are sometimes sent off on the same vain errand. This kind of education is not sufficient to enable them to discharge the duties and resist the temptations which come upon them in the avocations and duties which surround them in every day transactions; they have not had that moral training which teaches men to be true to themselves under all circumstances and to be honest in the worst of times. For woe be to the weak minded, unprincipled man or woman when the hour of trial or temptation comes upon them. God, in his infinite mercy, may have compassion upon them, but men never have. They become a bye word and reproach in the world.

We have associated together and formed this society for mutual improvement and for the benefit of our children. By thus associating we say to the world that we labor—that we take pride in this labor—we exhibit the fruits of it as the best evidence of our moral worth and social condition. We do it also as an example to all, that they may understand that there is no degradation in labor; that the most skillful agriculturist and mechanic, like the victor at the olympic games, is to receive the badge of distinction at the hands of his countrymen. There is something worthy of admiration in the hard struggles of an honest and laborious man for a competency, and when that is obtained the combat is continued for sufficient time to insure an independence. It is these efforts which have cleared the forests from the sea-board to the Mississippi, which will clear them to the Rocky Mountains, and from the Pacific back to the mountains again. Has the victor upon an hundred battle fields ever done any thing like this for the human race? and yet it is to this class of men that history has been devoted, and been made so barren of interest for all useful purposes. While the working man who has been turning the wilderness into a garden, building up cities and furnishing products for manufacture and commerce, is only known to a small circle of friends, and in a few years his name passes from the memory of the world forever. But however it may have been neglected, this industry is the mainspring to success, and you can not be too earnestly entreated to be diligent in labor. Work cheerfully, for it is for the health both of the body and the soul; it is the distinctive mark of a good citizen; for he who labors is not apt to squander, and the industrious child is almost certain to be a respectable man or woman. Look upon labor then as one of the cardinal virtues; teach it to your children as the true road to honor and worth and high estimation among men, as it is also the fulfilling of that precept which directs us to do with all our might whatsoever our hands find to do.

It is but recently, fellow citizens, that we were engaged in the melancholy task of assisting at the funeral ceremonies of one of these sons of labor. He came to this State many years ago, a young man, without friends to give him business or to aid and assist him by their counsel. His only capital was his surveyor's compass and his own brave heart; at first he encountered difficulties as most men do who commence the battle of life under such circumstances, but he was a man whom no ordinary difficulties could daunt. He persevered in his laborious duties, spending months in the woods, frequently short of provisions, sometimes living days together on wild honey alone, and barely sufficient of that. His energy and perseverance soon attracted public attention; he rose in the estimation of those who knew him best, and in a few years he was elected delegate to Congress from the then Territory; he discharged the duties of his office so satisfactory to his constituents, that he was again re-elected, and when Michigan became a State, Lucius Lyon was chosen one of her first Senators to represent her interests in the councils of the nation. In these few years his labor and industry had given him one of the highest positions in the country, and the good and wise everywhere were glad to claim him as an associate and friend. After his senatorial term had expired, and as soon as the State was districted for congressional representation, he was elected to the House of Representatives; afterwards appointed Surveyor General of the District of Ohio, Indiana and Michigan. At the expiration of the term of this last office, we find him again in the woods with his compass, still laboring as when he commenced life, a poor and friendless boy. He had not been spoiled by the favors of the world. It is such men as Lucius Lyon who give dignity to labor. Michigan may well be proud of him, and his example should be held up to the youth of the State not only as worthy of imitation, but that it may serve to cheer on the desponding sons of toil in their efforts to success. It affords us a melancholy satisfaction to refer to this pure minded, honest and laborious man, as one who had the best interests of Michigan at heart, but whose patriotism was sufficient to embrace the whole Union.

In addition to industry and moral training, intelligence is essential to the perfection of the character of a farmer. It has been said that there is nothing new under the sun and it is equally true, that whatever is known may be found in the books. Go then to the books and learn whatever may be taught of science and skill in the arts, in manufactures, agriculture, and the history of the past. The world is filled with the spirit of improvement; be not behind the age in which you live; select your books with care and read them for the purpose of learning, not for amusement or as the means of killing time. If you read only for these latter objects, no benefit is derived. In this way men become dreamers instead of thinkers and actors. And what is more injurious, it sometimes creates a distaste for the business and duties which their situation and condition have imposed upon them. A good citizen should discharge these requirements first before he indulges these tastes or generates morbid feelings in reading unprofitable books. It had become a saying almost as early as in the Augustine age of Rome, "that it is as easy to draw water in a sieve, as to learn without a book;" and in those days books were expensive, because there was no printing, and writers were scarce because the great body of the people were uneducated. Although reading is one of the best and readiest means of acquiring knowledge, yet we would not advise farmers to depend too much on book farming. We may gather from it the experience of others, and it serves to draw the attention of the reader to the subject, so that he may compare his own knowledge with that of the author. From information thus obtained a judgment can be formed from which to act in similar cases. We frequently find farmers who have no faith in this book knowledge, merely because it has been written or printed, while they probably would have believed the whole of it, if it had been stated to them in conversation. These err as much on the one side, as those do on the other, in trying experiments on every subject which is brought to their observation.

There is no doubt but that as a people we are too apt to be led off by theories; we cannot be too careful on this subject. A few years ago we had a silk fever, and large prices were paid for the

Morus Multicaulis. Calculations were made upon paper, showing the enormous profits to be made from raising the mulberry to feed silk worms—books were published on the subject, and Congress in its wisdom, ordered a certain number for distribution for the purpose of introducing it as a national business; many excitable persons embarked in it, and very few succeeded. The subject for the last eight or ten years has almost passed from the public mind, a few enterprising and industrious Ladies alone continuing the cultivation, rather as an amusement for their leisure hours in the ornamental, than in the profitable business of housewifery. We would not discourage these attempts at improvement, if they are not carried on to the exclusion of the well known and more certain means of success. We Americans are very prone to seize upon some new road to wealth, and crowd the path thick with the votaries of fortune. We tried it in 1835 and '36, in buying and selling wild lands and corner lots, and during the next year we imported from foreign countries three and a half million bushels of wheat; this great agricultural country brought its bread from Europe, because its farmers had turned speculators, purchasing lands that they did not want and could not cultivate; each one appeared to buy upon the principle that he could sell to some greater fool than himself, and for a little more than he had paid, and that little was modestly put at an 100 per cent on the cost; wheat was worth two dollars per bushels, and flour ten or twelve dollars per barrel. And this too, when our crops had not been injured; when no destroyer had been abroad, except this mania to become rich without labor. To escape from these difficulties, we tried another experiment, which was to make money plenty—and in this we succeeded. Our little village could boast of its five banking houses; we were going through a new problem in mathematics. In former times the money lenders associated together and established banks; this time the borrowers associated together, and made promissory notes, and called them money, and then scattered them abroad "thick as the autumnal leaves which strew the plains of Vallambrosia." And then, fellow-citizens, they failed. They could say, "Othello's occupation is gone." The courts came in aid of the banks, and decided that the whole farce was unconstitutional. There is no doubt that it was as bad and as foolish an operation as ever sane men went into; but it was unkind, to say the least of it, to throw the whole loss upon the poor bill-holders. Our ideal wealth was gone—the spell was broken—corner lots came down, and wild lands were sold for taxes; we were in fact no poorer than we had been during all the excitement, but then, we appeared to obtain a realizing sense of our poverty. While these little private operations were going on, a larger and bolder game was being played by States and corporations; they issued millions of stock bearing interest, and this stock was chiefly sold in Europe, and the proceeds came back principally in silks, railroad iron, wine, and goods of every description. These sales amounted to nearly two hundred million of dollars—the country appeared to be in the most prosperous condition—we were flooded with all the commodities that skill could create or money could purchase. But in a short time the interest became due—we had no spare produce to send forward—we had, as we have before stated, been obliged to buy our bread from abroad—the silks had been worn to rags, the railroad iron laid on the roads, the champaigne drank up, and the other goods disposed of in an equally profitable manner; this interest was required to be paid in something which was money in London, in Thread Needle Street, at the bank of England; the dicker pay of our western States was not current in that quarter. The money market in America became what the bankers call, tight. In the fearful revulsion which took place, men and banks and property of all kinds, went down like the pins in a nine-pin alley, when bowled at by a skillful player. Legislatures were assembled in haste, to permit banks to become insolvent—respectable men spoke openly of repudiating State debts; bankrupt laws were passed to remove from individuals the weight of indebtedness under which they had been buried, and from which there was no other resurrection. In our own State, land became by legislative enactment, almost a legal tender, at two-thirds of its appraised value; and if the unfortunate creditor did not like that kind of pay, he was only required to wait six months, and then the law allowed him another appraisal. In our efforts to hinder, delay and prevent the collection of debts, we had been compelled to violate the constitution of the United States; the legislation of the State had in fact become dishonest. All this humiliation had been brought upon us, because we had been trying in various ways to become rich without labor; abandoning the sure and ordinary means of acquiring property, for the desperate chances of speculation. All classes of our citizens were engaged in it, from the pulpit down to the grave digger. It is a sad history, but "if history is philosophy teaching by example," then it is well to review these histories, and have these examples brought before us as a warning to the future, so that at least this generation may not pass through the same ordeal a second time.

The Agriculturist especially, should keep himself clear from all these extraneous operations if he intends to farm it well and systematically. But how few farmers have we, that do it systematically; there is no great branch of business carried on at the present day, which is managed with as little system as farming. On looking over the whole county, how few you find who conduct the whole operation right. We are happy to say, that we can find many successful farmers among them, but they fall infinitely short of what they should be. The whole business appears to be carried on without any certainty; not one farmer in ten knows the exact quantity of land he has in each field on his farm, and when he has sown his seed, he is not certain whether he has put on a bushel or a bushel and a half to the acre, of wheat—whether he has put six or ten quarts of grass seed to the acre; and when the grain is cut he cannot tell exactly how many bushels he has grown to the acre; it is guess-work all around. Another great defect is, we undertake the cultivation of too much land each season—we have not sufficient force to carry it out—we are compelled to hurry the work—leave one thing unfinished to sommence another which is in a more suffering condition—and in this way the whole is slighted and run over. We undertake to put in a large crop of wheat, and for that purpose the team is divided, if more than one; and if there is but one team, the plowing is done as shallow as possible to hasten the process, instead of taking time and plowing deep. We put in a large crop of corn when we have not manure for more than half of it, and not half hands enough to give it a thorough hoeing. Large meadows and little grass, because the ground has been worn out in raising wheat years before. The fences are allowed to rot down because there is such a long string of it. The bushes grow up over the fences, because there is so much other work to be attended to, and so the bushes are put off to a more convenient season—all goes on in this slovenly manner. We have a class of men who call themselves farmers, who make the street their cow-yard in summer, and frequently throughout the year; the hog-pen is made to grace the door-yard. With such men, manure has no value, deep plowing is all folly, and ditching does not pay.

We would not, if we were capable of doing so, give a system of farming, on an occasion like this; but there are certain leading features to which the attention of every farmer should be called, although they are perfectly well known to every man who has ever worked a farm, and almost as universally neglected as they are known; and yet they are the foundation of his success. In the first place the land must be kept clear from surplus water, if he is desirous of raising anything but wild and miserable grass. It must be made rich, or it will grow nothing of any value, and when it is thus made dry and rich, if it is allowed to be overgrown by weeds, the grain is deprived of all sustenance. These three things are absolutely necessary, and yet some one or two of them are constantly neglected on almost every farm in the county. If the ground is not kept clear of the surplus water, its value is gone. If it is not well manured the grain or grass will not grow, for they require food as well as your horses and cattle. If, when this food is furnished by the proper kind of manure, what avails it, if you allow the weeds to get in and eat it up from your plants. But if these three pre-requisites are faithfully attended to, then with a proper rotation of crops, deep plowing and clean seed sown in due season, with God's blessings on your labors, you will have a full and an abundant crop. If you fail in either of these, although the bow of the covenant may still span the Heavens to show that the seed time and the harvest are still allotted to us by our Heavenly Father, to you the harvest time will come in vain, or be shorn of its richest blessings.

The farmer should also pay particular attention to the breed of animals. It costs no more to raise a fine horse or a fine ox than it does to raise a poor one. The value is generally more than double, and it is sold more readily. It is a general remark "that a good horse will always sell," while a bad one can only be sold by fraud and deception, or at a price that will not pay for raising. With cattle, the great object is to raise those which are the most valuable, either for beef or milk, and breeds should be selected for these purposes, and although it is expensive in the commencement to purchase the best of these varieties, yet if you will do so, in a few years you will have acquired a valuable property, and be in a situation to sell and thus get back your money with interest. The same system should be adopted with sheep and hogs. The first are becoming the great staple of the county, and the latter will pay well if good breeds are selected. We are a little behind on all this stock business, and there is no good reason for it. We have a great variety of good fruit in the county, but not half as much of it as there should be. Good fruit is as much more profitable than poor and worthless fruit, as a good horse is better than a bad one; and there is the same difference as to its selling qualities. The one will always sell, and the other it is almost an insult to offer to give away. The county of Oakland should be the first in the State upon all these agricultural products, and it remains with you to make it so. This Society can effect a complete revolution, if each of its members will resolve that it shall be accomplished, and will carry that resolution into practice. Take a deep interest in the progress of this work. Rally round your Society, no matter how many discouragements or disappointments you may encounter. If you have not succeeded this year in all that you anticipated, make a greater effort for the next. If judges make mistakes in awarding premiums or overlook articles of merit, remember that it is one of the infirmities of human nature to be continually erring; try to get better judges, and perhaps it would also be well to keep in mind, that we are all apt to consider our own things the best, and that this over estimate on our part may be the real cause of our disappointment. Let each member of the Society make it a point to bring something to the fair, and thus add to the annual exhibition. Recollect that the widow's mite was more acceptable than the offerings of those who gave of their abundance.

When we shall have succeeded in raising good crops and good stock, we must improve our means of communication with the best markets. Happily for us, our peninsular situation affords us the readiest means, when we have got our articles to the water. And to get them there, plank roads and rail roads are the best facilities. To build these plank roads, should be the business of the farmer, and if they would associate together for that purpose, every principal road in the county would be planked in five years, the expense of carrying off the surplus produce would be reduced three fourths, without including the difference in the wear of the carriages and teams, and a reasonable toll might be collected to keep the road in repair. When our produce is once on the water, we can send it to any part of the world for a market. Steam power has brought Liverpool within ten days sail of New York, and the telegraph wires daily announce the state of the markets in our cities on the sea board; our commerce is not fettered with heavy imports and duties in our trade with foreign nations, and a few million of dollars on one side or the other generally settles the balance of trade with the other nations of the world. Sometimes we pay this balance, and sometimes it is paid to us. But even when our Custom House Books show a balance against us, it is no evidence that we are indebted to foreign nations. If the balance is not excessively large, it may rather be considered the evidence of a healthy trade. For instance, a merchant exports from New York 100 bbls of flour, worth at that place $500; it is entered in the Custom House Books as an export of $500. This flour is sold at Liverpool for $700, and the money is invested in cloths. These cloths are returned to New York, and are worth in that market, $1,000 to the merchant. In this operation he has turned his $500 into a $1,000. The books at the Custom House show an import of $700, the value of the cloth at Liverpool, against an export of $500, the value of the flour at New York, exhibiting a balance against us on the books of $200, and yet the merchant has made $500; our imports have exceeded our exports. And still we have made money. An alarm is sometimes sought to be created on this subject, and the more unreasonable because the blame is attached to the government, while in fact it is the operation of individuals. If the merchants in our cities choose to purchase one hundred millions worth of foreign fabrics, what power has the government of controlling them; they are supposed to know how they will pay for these goods without official interference, and if they should not be able to pay for them, the loss falls on the vender, the person who sold them. The purchaser may become insolvent, but whether he gains or looses, it is no affair of the government or of any one else but his customer. Our merchants are generally sagacious enough to dispense with the advice of the politicians, and any interference on the part of the government, except to collect their tax on the goods, is a matter in which they have no concern. These goods so purchased are in ordinary times principally paid for in produce, and in such articles as we manufacture, and which may be saleable where these goods were bought. These are sent on to meet the bills as they become due; if these articles are not to be had or will not pay transportation and charges, they buy drafts on the places where the money is to be paid, or remit coin to settle the balance.

It would undoubtedly be better if we could pay for all the goods in produce, because the raising of produce is the great business of the country, and it would enhance the price at home and afford encouragement to raise greater crops. Yet we need not be alarmed to see the coin go out of the country for the purpose of paying our debts; there certainly can be no better use that the money can be put to, than the payment of debts. We can always get this money back if we have what will buy it, and if we can raise produce cheaper and better than other nations it will return. If the markets of the world are left open to us, American industry and enterprise will do the rest. The less restrictions there are upon trade, and the less governments interfere, the better it is for the merchant, manufacturer, farmer and consumer. Men soon ascertain what articles they can best manufacture or raise, or what pursuits it is most proper to follow without special legislation. Some fifteen years ago, we had a law on our statute book exempting sheep from taxation, few took advantage of it, because the country was not in a situation to go into sheep raising; the law was repealed, and as the country improved the farmers found the benefit of keeping this kind of stock, and now we have them in such abundance that thousands have been recently slaughtered merely for their pelts and tallow. But while the law existed, the farmer who raised horses or cattle, was obliged not only to pay his own tax, but also that part of the tax which the sheep owner should have paid; the law was therefore both unwise and unjust. And when these things are carried to any extent, they become oppressive. A few years since our legislature offered a premium of two cents on the pound on beet sugar, manufactured in the State; but even with this government premium, the business could not be sustained. Those who engaged in it lost money, and the farmers soon found that they could raise wheat and sheep to more advantage than they could beets; here was the same folly and injustice again, for the two cents a pound that was paid to the sugar man, must be raised by taxation on the property or articles raised by others. Governments are pretty certain of travelling out of the sphere of their legitimate duties when they hold out temptations to induce their citizens to go into particular kinds of business or undertake to foster the interest of any particular class, especially when the money thus bestowed is to be raised by taxing some one else, either directly or by putting the tax on articles which go into the general consumption of the whole community. This forcing the growth or manufacture of articles is not the way to enrich a country; for articles thus forced seldom pay the expense of the investment. The late Mr. Biddle, of Philadelphia, it is said, raised elegant pine apples at Andelusia, his country seat near that city, but it would have been ridiculous if all the farmers of Pennsylvania had gone to raising pine apples, because Mr. Biddle had been so successful, for the expense must have been enormous to a common farmer. We must raise articles adapted to our climate; to farm well we must learn to farm profitably. There is no advantage in raising a crop or an animal for market, if it cost more than the grain is worth or the beef or pork of the animal when disposed of. Economy must be combined with skill and industry in farming as well as any other business, or it had better be abandoned; amateur farming may do for men of capital and leisure, and when it is pursued as a pleasant avocation to amuse and while away the time. But when profit is the object, a close and strict account should be kept of the expense of each crop and of the whole farm. This is also one of the subjects which have been neglected by our farmers, and to which we would call their particular attention; indeed, we do not know of but one member of our society, who is in the habit of keeping such an account; and that fact itself should have raised him to the office of President of the society, had he not already arrived at that distinction by the great interest which he had taken in agriculture generally. His excellent example should be followed by every farmer in the county. What would be thought of a merchant who kept no profit and loss account; who could not tell by his books at the end of the year, whether he had been making or losing money? A farmer may tell you that his sheep have been profitable the last year on account of the high price of wool, but if you should ask for the exact amount of profits, he would be unable to state it, or to calculate it, because he has no known data upon which to go; he has kept no record of the quantity of hay and grain that they have eaten, or the value of those he had lost during the year. So it is with the raising of his horses, cattle and crops; he knows no more about it than his next door neighbor. But if an account was kept, he would soon know the exact measure of each field, the number of days spent in plowing each, the quantity of seed required, what the produce has been, and in this way ascertain the profit and loss on the crop, and so of each crop, and in a few years he can tell whether it is most advantageous to raise wheat or corn, cattle or sheep or horses; this one step would be a great advance in the system so much needed in husbandry; it would teach exactness, it would put an end to a good deal of slovenly farming. We should begin to know what we are about, and how we were to come out at the end of the year.

If in these remarks we have been unable to say much in commendation of the husbandry of the county, it is because that husbandry is at fault; but we do not intend to say that the farmers in a new country like this are so very much to blame; in the settlement of a country the first efforts must be for subsistence. Then the grounds are to be cleaned, the orchards planted, buildings erected and roads made; all these must be attended to before they can turn their attention to the improvements which are going on in the older States. But we have now accomplished most of this. We can begin to look forward to greater improvements and hope for better results. We are placed in one of the best grain growing and fruit raising countries in the world. The motto on our State coat of arms, is peculiarly appropriate to the location of our State. "If any one is desirous of seeing a beautiful Peninsular, let him look around." Our soil and climate are equal, if they do not surpass those of the most favored of the States in the Union. Let us unite to make Michigan the first in agriculture and our county the first in our State. By a return made by the State officers more than a year since, Oakland stood first in the great staples of wheat and wool, and in horses, cattle and sheep, and what is still more to our credit, we sent more children to school than any other county in the State. Let us keep in advance, not because our territory may be somewhat larger, but because we have the best farmers. Raise the character of the farming interest by raising the character of the farmer, and this must be left to your own good sense. It is now the leading interest of the county and the State. In the hands of the farmers are placed the control of the institutions of the country, and they must take the responsibility of seeing them sustained. Our present exhibition shows that we are progressing rapidly from our first effort two years since. Our stock of all kinds has improved almost miraculously. Our manufactures are multiplying. In fruit and flowers, butter, grain and swine, we have excelled the recent State fair, and still vast improvements can and must be made before we have discharged our duty. Not yet are we entitled to the meed of praise. But on this subject a distinguished Frenchman has said that our self-complacency is such, that it is almost impossible to flatter an American. If, says he, I should say to an American, you have an excellent country, the reply is, there never was such a country in the world; the sun never shone upon a better—we have every variety of climate and soil, &c. There is no end of our laudation. If says he, I advert to their government, I am met in the same strain. The answer is, it is the best government ever instituted by man. Here every man is free. We have no nobles to lord it over us. We do not give his words, but the substance of his remarks. Although this feeling of ours in regard to the country and its institutions is perhaps too often exhibited and may raise a smile or excite the indignation of a foreigner; yet it is the feeling which should pervade every American heart. It should be above and beyond all others in his affections; he should make every other sentiment subservient to this love of country, and it should extend to the whole country; it should be as broad as the American flag, and wherever in this wide land that flag is unfurled and given to the breeze, there should be our country with all its memories, as much so as the soil beneath our feet; whether on the mountains of New Hampshire or on the plains of Lousiana, or on the Atlantic or on the Pacific coast, everywhere and at all times this feeling should be uppermost—it should be for each State in the Union and for the whole Union—an American patriot on every part of the American soil.

We cannot close our remarks without expressing our thanks to the ladies of the society for the kind interest which they have taken in advancing the business and enhancing the pleasure of our annual exhibitions. They have not been backward in displaying that industry and skill which adds so much to all the comforts and conveniences of life, while it beautifies and adorns the homes in which are centered the chief attractions of man when the business and toils of the day are past. Our society would have been almost a failure, if they had not come to our aid and formed the most interesting part of the exhibition. We cannot express our admiration better than by again referring to our Frenchman, M. de Tocqueville, now or recently the minister of foreign affairs in the French republic. He visited this country for the purpose of enquiring into our democratic institutions, and wrote a book on the subject, better than any heretofore written by a foreigner on that subject. He subsequently wrote another book on the manners and customs of the American People, and in that he says: "Now when I am about to close my labors, if I should be asked to what cause in particular, I attribute the great success of the Americans, I should say, to the superiority of their women."

This gallant Frenchman attributes our success to your superiority, and we have no disposition to reverse that judgment. But if the destinies of this great country are in your hands, if you can wield it for good or for evil, see that you discharge the great trust in such a way that you leave it in a better condition than it was when it was placed in your charge. Emulate the character of the women of the revolution, so that your brothers, sons and daughters may look back with reverence and love to the pure teachings of your life and example.

REPORT

Of the Shiawassee County Agricultural Society.

J. C. Holmes, *Sec. Mich. State Ag. Society:*

In conformity with the 11th article of the constitution of the Michigan State Agricultural Society, the following is respectfully submitted as the second annual report of the Shiawassee County Agricultural Society and for the year 1851.

The holding of the county fair occupied the 3d and 4th days of October, favored with as beautiful weather as could be desired.

The executive committee adopted the plan of the State fair so far as enclosing a lot of nearly three acres, with a tight board fence and not less than seven feet high, with suitable stalls and pens for the exhibition of stock and articles that would be probably entered for exhibition.

The new court house (it being unfinished) was fitted up with lines and tables suitable for the exhibition of articles of household manufactures, machinery, specimens of field crops, garden vegetables and unenumerated articles.

Murmurs and dissatisfactions were heard before the anniversary meeting of the society, growing out of the distribution of the premium moneys for the year 1850; it being the first meeting, a systematic plan of operations had not yet been matured to act upon.

The plan now adopted is similar to the State society, and the past year proves conclusively that that is the true and only available way to raise the means and to keep the finances of the society in a flourishing condition.

The supervisors have levied the one-fifth mill tax in this county, which amounts to nearly eighty dollars annually for the benefit of the county society; and it is with pride and pleasure that we can point to the benefits already growing up in our midst. Stock of all kind has improved and is still improving under the progressive order of things. Where we were wont to see the native, stinted breed, we now can see the noble Devon and the stately Durham.

In sheep there has been more improvement made than in any other one branch of stock, but the hogs are trying to keep pace with them.

At the annual meeting of this society, Isaac Gale, of Bennington, had awarded to him the first premium on corn. He raised on three and half (3½) acres four hundred and forty-two bushels of good shelled corn, it being at the rate of one hundred and twenty-six bushels to the acre.

One other specimen was exhibited that produced one hundred bushels to the acre. Although the past season has been universally acknowledged to have been one of the poorest corn raising years for a long time, yet we see enough not to be discouraged, but every thing to encourage.

The above is a short sketch of the situation of our society, but we are looking forward to something yet to come that will be far more interesting and profitable.

The following is a list of officers for the year 1852:

President—Isaac Gale.

Secretary—P. S. Lyman.

Treasurer—James Cummins.

Executive Committee—C. S. Johnson, I. B. Barnes, Roger Haverland, Ezekiel Cook and Ira Stimpson.

The premiums awarded this year were about the same in amount as last, and were all promptly paid as soon as the certificates could be made out.

The society is now in a flourishing condition, with some funds in the treasury. The society numbers about 134 working members, and last October they entered for exhibition about 309 articles, averaging nearly three articles each; and now, when such a state of feeling exists, who can tell what cannot be done or what can check such gatherings that eventually must result in a movement towards a better condition.

Respectfully yours,

PLINY S. LYMAN,

Secretary.

Corunna, March 3, 1852.

ADDRESS

Delivered at the Second Annual Fair of the Shiawassee County Agricultnral Society, October 4th, 1851.

BY ANDREW PARSONS.

Until yesterday (by some mistake) I did not know that the Society expected me to deliver an address. I am therefore compelled by want of time to omit much of detail, and to confine myself mainly to the outline.

We are taught by revelation that the heavens and the earth, and all the host of them, are the *work* of God's hand. That when the Creator had finished this work, he *planted* a *garden*, where he caused to grow every tree that was pleasant to the sight, and good for food. That after the fall of man by disobedience and crime, when he determined to destroy all flesh with the earth, he commanded righteous Noah to make an ark, and gave him a particular fashion after which he was to construct the first ship that ever sailed the sea—that after the waters of the flood had subsided, God gave to Noah the promise that while the earth remaineth, seed time and harvest should not cease, and commanded him to bring forth abundently in the earth, and that Noah began to be a husbandman, and that he planted a vineyard. Thus do we learn that he who would desire to look down upon the husbandman and the mechanic, or despise him who labors to procure an honest livelihood, would look down upon God, his own Creator, who was the *first mechanic* and *tiller of the soil*, and would despise the employment and character of Noah, whom God saved alone with his family, for his righteousness and purity, that the earth might in time bear its teeming millions of the human race, and its productions necessary for their comfortable subsistence. The existence of mankind depends as much upon their labor, as did the world and all the creation upon the work of God's hand. All the comforts of life, all the wealth of individuals, of States and of Empires, the immensity and grandeur of towns and cities, the costly mansions and edifices which adorn and beautify them, are the result of *man's labor.*

When we look out upon Nature and behold this world with its green forests, its hills and valleys, its winding rivers, crystal lakes and broad seas—when we behold the star-bespangled firmament, and observe the working of that system which brings to us day and night, changes of seasons, with seed time and harvest, we are ready to exclaim, how wonderful are the works of the Almighty! and how honorable and glorious must be that Being who by His power has created this immensity!

To a great extent do we in like manner look with wondrous pleasure upon the extensive possessions of the man of affluence, the splendid and costly silks and fine fabrics of the rich merchants—the magnificent superstructures which give character to great cities of wealth and grandeur. But how different is the principle upon which many award the honor, where the works of man adorn and beautify objects of admiration, and afford comforts and luxuries for man's happiness. It is indeed to be lamented that for the want of a proper exercise of reason with some, they will honor that class of mankind who possess the fruit of man's works, rather than that class whom God has endowed with faculties and chosen as real vessels of honor to imitate His example as the first tiller of the soil, and the example of righteous Noah, as one of the first mechanics, and to adorn and beautify the earth.

It is, however, agreeable to know, that under the light of revelation, science and experience, among enlightened nations, the exercise of reason is becoming more general, and the laborer, as well as labor, which is essential to the support of human existence, has already become not only respectable, but highly *honorable.*

But a few years ago, a fair and a cattle show which now so nobly excites a just spirit of emulation would have called together but few of the classes besides the practical farmer or the mechanic; but with the increase of knowledge, exhibitions of agricultural and mechanical skill, industry and usefulness, have become the pride of States, of nations, and of the *world.*

With the spread of knowledge man learns to *consider* that all wealth is produced by labor, and that all classes depend absolutely upon the agriculturist, mechanic and laboring man for food, raiment, the comforts and luxuries of life. But a few years ago it was thought by many that a man could not enjoy the reputation of a *gentleman* if he depended upon his labor for a livelihood, and indeed that it was next to impossible that a woman who practiced cooking her own dinners and washing her own clothes, should be entitled to the rank of a *lady;* but the light which gives a true character to the employments of mankind, is now shining around us, and there are but a few at the present day who would not look with an eye of *pity* upon that benighted mind that would despise the man or the woman, or the son or the daughter, that labors for an honest living.

He who tills the soil receives his bread and the rewards of his labor, as it were, directly from the hand of God, while he who laboreth not, receives his bread from the man who earns it by the sweat of his brow, and hence is placed one degree lower and farther from that exalted source of all the blessings of life, the substantial happiness of man—one degree farther from that Being who by his power bringeth about seed time and harvest, and cold and heat, and summer and winter, and day and night, without which all flesh would perish.

The first means to be used in obtaining the necessaries of life, and of acquiring that degree of wealth essential to the enjoyments of the comforts and pleasures of plenty, is the tilling of the ground. Let the cultivation of the earth cease, and there would arise a famine which would consume the human race as a fire that sweepeth over the dry land consumeth the grass—our villages, our towns and our cities would be left desolate as the howling wilderness—our rail roads and canals would be useless, our commerce would die away, and there would be nothing left to make the earth an agreeable habitation for man. When it is seen that the cultivation of the ground is the foundation of and bringeth with it all the enjoyments and treasures of this world, it is evident that too much cannot be done to ensure the success of the agriculturist. It may be true that pride goeth before destruction, and a haughty spirit before a fall; but a just spirit of emulation is but "an ardor kindled by the praiseworthy examples of others to imitate them or to equal or to excel them;" hence arises the benefits resulting from fairs and exhibitions of this kind, rightly conducted. The farmer is too noble minded to be excited by a spirit of jealousy because another farmer exhibits better stock or better crops—instead of being jealous he is anxious to profit by any improvement he can to facilitate the labor of his own hands or increase the profits of his own labor—this is for his *interest* as well as his future good name. As with the farmer in this respect, so it is with the mechanic. Although the farmer goes before the mechanic, yet it is necessary that the mechanic should follow after, and finally go hand in hand with the farmer, as that the farmer should hold by his hand in the process of cultivation the plow which the mechanic has made —their interests are identified and inseparable.

Take from the farmer the improved implements and machinery which now enables him to carry on with such facility the cultivation of the soil, and place in their stead the implements in use but 30 years ago, and he would be strongly inclined to think that he could scarcely escape a famine before him or starvation behind him—give him the old wooden plow, the three cornered drag, the sickle and the flail, and take from him the cast iron plow, the improved harrows and cultivators, the grain cradle and the threshing machine—particularly in this country on the openings and prairies, take from the farmer the breaking-up plow, and you would take from him one-half of his ability to accomplish his work, or add one-half to his labor. The manufacturing establishment, with its complicated machinery, turning out its hundreds of thousands of yards of cloth yearly, is the work of the ingenious mechanic, and saves mankind a vast amount of labor. The implements and machinery used by the farmer are constructed by the mechanic—the work of the farmer and the mechanic indend goes hand in hand, and consequently both have a deep interest in the success of each other.

Exhibitions of this character call them both together, and are calculated not only to exhibit their numerical and political strength and importance, but to encourage and stimulate each other in the progress of their avocations, and to form those attachments which bind society together in harmony and happiness.

A few suggestions which we offer apply alike to the agriculturist, mechanic and manufacturer, and indeed to all classes of community, and if adhered to, will, under ordinary circumstances, ensure prosperity. Let every man be industrious and attend faithfully to his own business—never spend money for that which is not necessary to the substantial enjoyments of life; and above all, keep out of debt. The Divine command, "owe no man anything," is as necessary to be obeyed to secure enjoyment and exemption from troubles and harrassings, which destroy ones comfort and peace, as health is to exempt the body from pain and distress.

A want of industry is one material thing which makes the savage depend upon the fortune of the chase for meat, and indeed that makes him what he is—a savage. A want of economy is a degree of profligacy which eats out ones substance, or prevents him from accumulating any. We frequently hear of *lucky* farmers, who always have fine crops, good cattle, horses, sheep and hogs; but if we will observe we shall see that their *good luck* consists mainly in their industry, economy, and the careful management of their farms. Man is to a great extent the arbiter of his own fortune or luck. God sendeth rain, and his sun shines upon all alike; and if all farmers are alike industrious, economical and pains taking, their labors will generally be rewarded with like results. We hear many talk against what they call "book farming." Such have no relish for reading agricultural papers, and are ignorant of the fact that farming is a science, which, to *fully understand*, requires a knowledge of botany and chemistry, by which he learns the "phylosophy of his plants, the nourishment and treatment they require, and by analyzing his soil he discovers what is necessary to maintain and increase its fertility. Zoology and natural history teach him the character and constitution of his animals, and mechanics, the structure and use of his implements. Without *some* knowledge of this science, obtained either by study, observation or experience, he knows no more of farming than the untutored savage that is opposed to all books, and is even unwilling to associate with them whose knowledge has been increased by reading them. It is true that it is not necessary that we should learn *every* thing from books or agricultural papers.

Our own observation and experience should and does teach us much, but without the aid of others' experience and knowledge, our own would be as circumscribed and limited as that of the poor Indian, whose associations are only with them like himself. Strike from existence all books and periodicals—take away from us a knowledge of reading and writing, and communicating our thoughts and desires to one another, except by speech, and but a brief period would roll on before we should slide back into the dark ages of savage life, and have to depend principally upon *hunting* and *fishing* for subsistence.

It is by getting from one another what has been learned by study and experience, and by retaining it, and passing it to others, that we have emerged from heathen darkness; and it is evident that without books and newspaper or periodicals, what men learn could not be communicated as it is among us, but would, like man, pass into forgetfulness. Let no man scorn "book knowledge," for it is but *man's experience* and *knowledge* communicated as one friend communicates his thoughts and desires to another by letter, when it is not in his power to meet him face to face. We have many valuable agricultural periodicals published in our country, and among them one of the best is the Michigan Farmer, published in our own State. Every man should take it and *read* it. He will there find many valuable statistics, the experience of the best practical farmers in different portions of this State as well as some other States, touching every branch of agriculture, with much more valuable information than could be communicated to you by any person in any address. Some farmers may perhaps say, "we can raise *wheat* as well as others who read papers; all we have to do is to plow and till the ground well and sow our wheat and drag it in." But our farmers are beginning to learn that if a man depends upon raising wheat alone, he is depending upon a brittle thread; if his wheat fails, the labor of a year is gone. To ensure success, the farmer should generally raise some wheat, some corn, oats, potatoes, beans, some cattle, horses, sheep, hogs, make at least his own butter and cheese, (with some to sell if he can;) in fact he should live at home and out of his farm.

Our own county, though new in its agricultural developments, is calculated by its great variety and fertility of soil, its timber, its water and hydraulic privileges, to produce in abundance everything that grows in this latitude; and it is pleasing to observe by this exhibition that our farmers are beginning to turn their attention to the *varied* productions which tell of a rich agricultural country. We have good mills for manufacturing our flour, and all the machinery necessary for the accommodation of the people. Our plows and farming utensils are manufactured in our own county.

Our county produces as fine wheat, as good corn and vegetables and as delicate and rich fruits as any portion of the western country. Some of the stock exhibited here to-day may well challenge the emulation of much older counties of this State. The handiwork of the ladies in exhibition here shows that in their spirit, enterprise and skill they are determined not to be behind—that they will keep pace with the taste and improvements of the age. All this is encouraging to our farmers and mechanics, and indeed to classes, and should, as it doubtless will, excite that proper spirit of rivalry which is calculated to propel us onward in the path of prosperity until we arrive at that rank of wealth and greatness which some of our much older and rich counties now enjoy.

A few suggestions concerning the importance of education generly and the cultivation of the mind, and I shall have done. I cannot express more happily what I deem to be the true doctrine in regard to education as applicable to agriculture than to give you a brief extract from the Michigan Farmer, the paper above referred to. "Indeed, (says the Editor of that paper,) the connection between a highly cultivated mind and a highly cultivated soil is inseparable; for no cultivator of the soil can understand his business without being a cultivator of the mind, and the latter must precede the former—the mind of the farmer must be subject to an appropriate tillage before he is at all qualified to become a tiller of the soil. He may it is true plow and hoe and go through a set of mechanical round of duties just as he was set going by his good old father before him, who he guesses "knew how to farm as well as any body," but how much is he elevated in knowledge and skill above the ox he drives? And mother earth, how reluctantly and scantily does she pour her treasures into his lap, as though ashamed to own him as one of her lords! And how does she delight to crown with her richest abundance the man who brings to his task the knowledge and skill of a cultivated intellect, as though proud to own him as a son and honor him as a lord."

How forcible is the Scriptural saying that "wisdom is strength and knowledge is power." If we would be truly independent—if we would know how to support a republican government, great and rich in its agricultural products and mechanic arts, powerful in numbers and able to support its institutions, we must encourage and support schools and institutions of learning. Our common schools should in particular be fostered with great care; they are truly said to be (in this country) the cradle of liberty. Let us see that our children are thoroughly *rocked* in them, for while reposing there they rest from those vices and habits which follow the indolent and ignorant. And they rise up in manhood an honor to their fathers and mothers, possessing themselves a knowledge of the way of prosperity and happiness, and to be useful members in society, and strong pillars in the support of the government; and upon the principle that "train up a child in the way he should go, and when he is old he will not depart from it," when he shall have finished his course here he will be prepared to enter upon that life where misfortune is never known—where storms of adversity never come, and where labor and toil ceases; and where, instead of treasures of corn and of cattle and of the things of earth, which pass away with all flesh, are provided the rich treasures of unalloyed happiness, and where the storehouse of joyous plenty is forever full.

REPORT

Of the Van Buren County Agricultural Society.

The following call, signed by the board of supervisors and several other citizens of the county of Van Buren, was published in the "Paw Paw Free Press" and "Paw Paw Journal."

The undersigned, believing that the interests of the people of this county, and especially the farmers, would be promoted by an agricultural society, recommend that the farmers, and all others interested, meet at the court house in the village of Paw Paw, on Saturday the 28th day of June, 1851, at one o'clock P. M., for the purpose of organizing said society.

Wm. H. Harrison,	Samuel N. Gantt,
E. Barnum,	Jno. Cole,
L. Crane,	M. M. Briggs,
Wm. N. Pardee,	Harmon Harwick,
James Crane,	John Lyle,
Homer Adams,	John K. Pugsley,
Wm. Murch,	Wm. Hill,
John McKinney,	Benj. F. Chadwick,
G. A. Bently,	A. Stewart,
Ashbel Herron,	Elias G. Kinne,
S. T. Conway,	A. J. Goodrich,
John Andrews,	I. W. Willard,
C. P. Sheldon,	G. W. Doughty,
A. S. Brown,	J. W. Sherman,
L. G. Hill,	S. C. Grimes,
F. C. Annable,	Albert Mears,
Charles, Selleck,	B. Hall,
E. Durkee,	W. R. Hawkins.

Pursuant to this notice a meeting of the citizens of Van Buren county was held at the court house in the village of Paw Paw, on the 28th day of June, 1851, for the purpose of forming an agricultural society.

On motion, J. R. Monroe was chosen president of the meeting, and E. Barnum and O. Sission were chosen vice presidents, and E. A. Park, Secretary, E. O. Briggs, Assistant Secretary.

On motion, it was

Resolved, That the meeting proceed to organize a county agricultural society, and that the following constitution for such society be adopted:

CONSTITUTION OF THE VAN BUREN COUNTY AGRICULTURAL SOCIETY.

Article 1. This association shall be called the "Van Buren County Agricultural Society," its object shall be to encourage the improvement of agriculture, manufactures and the arts connected therewith, throughout the county of Van Buren.

Art. 2. The officers of this society shall consist of a president, one vice president in each organized township in the county, a secretary, a treasurer, and an executive committee, consisting of the president and secretary and five other members chosen for that purpose. The officers shall be chosen by ballot and by a majority of the votes, at the annual meeting of the society, and shall hold their offices for one year, and until others are chosen in their places: *Provided,* That the officers chosen upon the organization of the society shall be deemed members and shall remain officers only until others shall be duly elected at the first annual meeting, and if a vacancy happen it may be filled by appointment of the executive committee.

Art. 3. The duties of Officers shall be such as usually pertain to their respective offices; and such also as may be prescribed by the special order of the executive committee, as hereinafter provided. The treasurer shall receive and keep an accurate account of all the moneys belonging to the society; he shall pay out its moneys only on the order of the executive committee, and at each annual meeting of the society he shall make a full report of the financial condition and transactions.

Art. 4. The executive committee shall determine the place of holding each annual meeting and fair of the Society, and it shall call that meeting and fair at such time as it shall judge best, during the month of October, giving at least sixty days public notice thereof.

Art. 5. The executive committee shall direct the money appropriations of the society, and have the control of its property; it shall make the necessary preparation for the annual fair, and issue all proper public notices and circulars in relation thereto or to the general object of the society, it shall prepare the necessary by-laws of the society, and may prescribe such duties to the other officers of the society as are not inconsistent with the usual business of their respective offices; it shall itself obey the instructions which may be given to it, at the annual meeting of the society, and at the expiration of its term of service it shall make a full report of its proceedings. It shall be competent for the executive committee or a majority of them to appoint the president and secretary, to transact all such business as they may be authorized to do by the committee; and the secretary shall sign and the president shall countersign all orders on the treasurer, for the payment of any money directed by the committee to be paid for any purpose, and the secretary shall keep an accurate account of all moneys so drawn.

Art. 6. It shall be the duty of the executive committee annually to regulate and award premiums on such articles, productions, improvements as they deem best calculated to promote the agricultural and household manufacturing interests of the county, having special reference to the most economical or profitable mode of compensation in raising the crop, or stock, or in the fabrication of the article offered; *Provided always,* That before any premium shall be delivered, the persons claiming the same, or to whom the same shall be awarded, shall deliver to the president of the society in writing, an accurate statement and description, verified in such manner as the executive committee may direct, of the character of the soil and the process in preparing it, including the quantity and quality of the manure applied in raising the crop, or of the kind and quantity of the food in feeding the animal, as the case may be; also the kind and cost of labor employed, and total products of the crop, or the increase in value of the animal, with a view of showing accurately the exact resulting products.

Art. 7. The executive committee shall meet annually at such place as it may itself choose, on or before the 1st Monday in January, and shall then immediately prepare a report and abstract of the transactions of the society during the preceding year, embracing such valuable reports from the committees, statements of experiments, cultivation and improvements, proceedings, correspondence, statistics and other matter, the publication of which will exhibit the condition of the agricultural interests of Van Buren county, and a diffusion of which will, in the judgment of the committee, add to the productiveness of agricultural and household labor, and therefore promote the general prosperity of the county.

Art. 8. No officer of this society shall receive any compensation for his services.

Art. 9. Any person may become a member with his family, of the Van Buren County Agricultural Society, for one year, by paying one dollar into its treasury, and may then continue a member by paying fifty cents per annum. Any officer of the society may receive and forward to the treasurer the fee requisite to a membership. By paying ten dollars into the treasury of the society, any person may become a life member, and shall be entitled to a certificate of such membership, signed by the president and secretary, which certificate shall entitle the holder thereof to all the benefits of membership, together with his family.

Art. 10. This constitution shall be altered only by a vote of two-thirds of the members present at the annual meeting of the society.

The following named persons were then duly elected officers of the Society, to serve until the regular annual meeting of the Society, viz:

President—J. R. Monroe.

Vice Presidents—Henry Barnum, M. L. Fitch, A. Herron, D. Morris, J. N. Hinckley, N. S. Marshall, J. K. Pugsley, O. Sisson, P. Haydon, C. M. Morrill, Allen Briggs, Isaac Brown, John Southard, B. A. Olney and W. H. Keeler.

Secretary—Wm. H. Harrison.

Treasurer—Benoni Hall.

The following named gentlemen were chosen to form, with the president and secretary, the executive committee, viz: J. A. Sheldon, E. T. Spencer, E. Barnum, Wm. H. Hulbert, and Robert Nesbitt.

On motion, it was

Resolved, That the proceedings of this meeting be prepared for publication by the secretary, and that the publishers of newspapers in this village be requested to publish the same, together with the constitution, in their respective papers.

On motion, the meeting was adjourned until the first Saturday in August next, at one o'clock P. M., at the Court House.

J. R. MONROE,
President.

E. A. PARK,
Secretary.

Pursuant to notice previously given, the Van Buren County Agricultural Society met at the court house in the village of Paw Paw, on Saturday, the 2d inst., at one o'clock P. M.

Present, J. R. Monroe, the president, several of the vice presidents, the secretary, a majority of the members of the executive committee, and a respectable number of citizens.

Although the meeting was not large, yet it was highly gratifying to see some of our most substantial farmers in the county present, who seemed to take a decided interest in the meeting.

After the object of the meeting was stated by the president in a few brief and pertinent remarks, the Hon. F. C. Annable was called upon, who arose and said:

"He favored this movement, and that the farmers of his town were getting aroused, and began to feel considerable interest in agriculture. He then dwelt on the importance of a knowledge of agricultural chemistry, that if we could obtain an analysis of the soil of this county, the results might be highly beneficial; that we ought to have a fair this fall to exhibit the state of agriculture in this county, and another in five years, to see if we had made any improvement."

The Hon. P. Haydon was then called upon, who arose and said:

"Although he had farmed it pretty extensively in this country, yet he had to confess that he was very ignorant of scientific farming; that he had never attended a fair in his life; he thought that the society might be the means of doing a great deal of good; that the raising of wheat was a hard business under the present system of farming—the amount raised not paying the cost, as the statistics of the State would show, being only an average yield of 8 bushels or thereabout to the acre. He thought that we might do better in sheep husbandry, and might compete successfully with eastern farmers. The cheapness and facilities of transportation of wool, compared with other productions, being greatly in its favor. He also thought that if we should turn our attention to the improving and raising of cattle and horses, we would find it profitable."

The Hon. M. L. Fitch, being called upon and complimented as being "one of our best farmers," was solicited to give his views and experience. He accordingly arose and said, that he did not come for the purpose of making a speech—that he had read the notice in the "*Free Press,*" that several interesting speakers were expected to address the meeting, and therefore he came to hear others. He disclaimed his being "one of the best of farmers," and considered himself far in the back-ground of improvement; he thought the farmer's profession had been spoken of rather disparagingly. He then compared the farmers chances of success in accumulating wealth as being more favorable than other classes, also the probabilities of failure and bankruptcy as being much less than the mercantile class, so often envied. He alluded to the many comforts and conveniences of the farmer, which was never taken into account, and thought that the farmer needed only to be enlightened to make his calling the most desirable of any.

An opportunity was then offered, and nearly all of the citizens present came forward and subscribed their names and paid the amount, (one dollar,) entitling them to become members for one year.

The following amendment to the constitution was adopted:

"The Treasurer of the Society shall, on receipt of any moneys belonging to the Society, whether for fees of membership or otherwise, give a receipt therefor to the person paying the same, which receipt shall be countersigned by the Secretary; and the secretary shall charge the amount of such receipts to the said treasurer, and the secretary shall credit the treasurer with all moneys paid out by him for the society, on presentation of the proper vouchers."

The following resolution was adopted:

Resolved, That the Secretary issue to the proprietors of the Paw Paw *Journal*, and the Paw Paw *Free Press,* life certificates of membership to this Society, properly authenticated, provided they will one or both of them do the printing in their papers gratis.

It was also resolved to hold a fair this (next) Fall. The meeting then adjourned.

The executive committee then met; after consultation, adjourned to meet again on Saturday, the 9th day of August, at 2 o'clock P. M., to make arrangements for the annual meeting and fair.

W. H. HARRISON,
Secretary.

At a regular meeting of the executive committee, held at the Court House on the 9th of August, the following list was agreed upon, the amount of premiums being discretionary with the committee, and depending upon the amount of funds that may be accumulated up to the time of the fair:

HORSES.

For the best, and second best team of draught horses, owned in the county two years or over.

For the best, and second best single horse, owned in the county two years or over.

For the best, and second best colt, three years of age, raised in the county.

CATTLE.

For the first, second and third best team of working oxen, owned in the county two years or over.

STEERS.

For the best, and second best pair of steers, not exceeding three years of age, raised in the county.

BULLS.

For the best, and second best bull three years old or over, owned in the county two years or over.

For the best, and second best bull two years old or over, raised in the county.

For the best, and second best yearling bull, raised in the county.

" " bull calf.

FAT CATTLE.

For the best, and second best fat ox.

" " " cow.

The exhibitor must state the process of fattening—the kind of food and length of time—the age of the animal.

COWS.

For the first, second and third best milch cow. The exhibitor must state the amount of milk and butter made, food used and kind.

HEIFERS.

For the best and second best heifer three years old or under.

For the best and second best heifer two years old or under.

For the best and second best yearling heifers.

For the best and second best calf.

SHEEP.

For the best and second best buck.

For the best three ewes; for the second best.

" " lambs; " " lambs.

SWINE.

For the best and second best boar over one year of age.

For the best and second best one year old or under.

" " " breeding sow.

" " " pen of six pigs five months old or under.

PLOWING MATCH.

With horses—$\frac{1}{4}$ of an acre of land. A premium to the first best and to the second best. Also to the best and second best with oxen.

CORN.

For the best and second best one acre of corn.

WHEAT.

For the best and second best sample of winter wheat, not less than one bushel.

HOUSEHOLD AND MANUFACTURED ARTICLES.

For the best lumber wagon, with box made in the county.

" plow made in the county.

For the best set of two horse harness made in the county.
" pair of coarse boots made in the county.
" " fine " " "
" side sole leather and for the best upper leather made in the county.
" ten yards or over of sheep's gray cloth made in the county.
" pair of woolen hose made in the county.
" and second best ten yards or over of domestic flannel.
" barrel of flour (superfine) made in the county.
" and second best sample of bread in a loaf.
" " " butter in roll not less than five pounds.
" and second best sample of cheese of ten pounds or over.
" " " honey of five pounds or over, in the comb.
" and second best sample of maple sugar, 5 lbs. or over.
" half bushel of fall apples.

GENERAL RULE RELATIVE TO ALL ARTICLES.

No premiums will be given to persons not belonging to the society. Where there is but one exhibitor and the animal or article is not worthy, no premium will be given. Exhibitors will be required to answer all questions put by the examining committees, that relate to the articles offered for premiums, and written statements must accompany them if required. To all articles not mentioned in the foregoing list, discretianory premiums will be awarded to the extent of the societies means, if deemed worthy.

EXAMINING COMMITTEES.

Horses—Josiah Hill, A. S. Downing, Peter Harwick.

Cattle—A. H. Phelps, Josiah Gilman, John A. Ranney.

Sheep and Swine—Eben Smith, Hiram Mather, D. Morris.

Flour—I. W. Willard, John Reynolds, Lyman A. Fitch.

Plows, Plowing, Wheat and Corn—Joshua Bangs, Joseph Woodman, Morgan L. Fitch.

Manufactured articles: Wagons, Plows, Leather, Boots, Harness, &c.—Joseph Bardwell, E. Knowles, E. A. Thompson.

Household Articles—Mrs. Eliza Rice, Mrs. Hiram Mather, Mrs. Jas. Crane, Mrs. Junia Warner, Mrs. Oliver Warner.

Committee to examine Farms and report the best—Calvin Fields' Dr. J. Andrews, Morgan L. Fitch.

Committee of Arrangements to make preparations for the Fair—I. W. Willard, James Crane, Charles Selleck, Hiram Mather, E. J. House.

The executive committee adjourned to meet again on the morning of the fair at 9 o'clock.

J. R. MONROE, *Ch'n*,
E. BARNUM,
E. T. SPENCER,
J. A. SHELDON,
R. NESBITT,
W. H. HURLBUTT,
Executive Committee.

W. H. HARRISON, *Sec'y.*

The first annual meeting and fair of the society was held in the village of Paw Paw, October 15th. The number of people in attendance, the number of cattle and articles displayed, exceeded all expectation. The number of animals and articles of all kinds were of horses, 65; cattle, 66; sheep, 25; swine, 15. Entries of fruit, apples, 18; peaches, 4; quinces, 1. Vegetables, 10; wheat, 10; corn, 9; manufactures, 60; butter, bread, cheese and honey, 10; poultry, hens, 26; turkeys, 2.

The neglect of proper arrangements, and of having things fitted up with order and taste, produced great inconvenience. We hope we shall have things better managed in this respect another year. An address was delivered in the afternoon by the Hon. C. E. Stuart, of Kalamazoo. His remarks and observations were plain, sensible, and well adapted to the occasion, and evinced a thorough knowledge of farming and farm management.

OFFICERS ELECTED AT THE ANNUAL MEETING, FOR '52.

President—Hon. Jay R. Monroe.

VICE PRESIDENTS.

Antwerp—Morgan L. Fitch.

Almena—Henry Barnum.

Arlington—Allen Briggs.

Bloomingdale—Ashbel Herron.

Columbia—Elijah Knowles.

Decatur—Dolphin Morris.

Hamilton—Philotus Hayden.

Hartford—Charles P. Sheldon.

Keeler—J. P. Rosevelt.

Lawrence—Hiram Mather.

Lawrence—N. S. Marshall.

Porter—Orrin Sisson.

Pine Grove—Charles M. Morrill.

South Haven—John Southarg.

Waverly—William Murch.

Treasurer—Benoni Hall.

Secretary—William H. Harrison.

EXECUTIVE COMMITTEE.

J. R. Monroe, Jason A. Sheldon, Eladcit T. Spencer, Robert Nesbitt, Edwin Barnum, Wm. H. Harrison, Wm. H. Hurlbut.

PREMIUMS AWARDED.

HORSES.

Best pair of draught horses, 7 yrs old. A. S. Downing,	$1 00
2d " mares, 5 " John Campbell,	50
Best single horse, not entered, and no competition, A. V. Pantlind.	
Best 3 year old colt, Josiah Hill,	1 00
2d " Jacob Charles,	50
Best 2 year old colt, J. Greenwood,	1 00
2d " Legrand Anderson,	50
Best 1 year old cold, Peter Harwick,	1 00
2d " Henry Barnum,	50

Best nursing colt, Thos. Clark, $1 00
2d " P. Hayden, 50
Best breeding mare, 13 years old, Alexander Sloan, 1 00
2d " " Dolphin Morris, 50
Best stallion, 9 yrs old, Dandy Duroc, Chas. McArthur, 1 00
2d " 5 " C. Tittle, 1 00

Judges on Horses—P. Harwick, A. S. Downing, Josiah Hill—report signed by Peter Harwick and A. S. Downing.

CATTLE.

Best team of working oxen, 6 years old, Orrin Sisson, $1 00
2d " " 5 " R. M. Haynes, 75
3d " " 4 " H. Humphrey, 50
Best cow, 14 years old, S. Godfrey, 1 00
2d " 11 " O. Sisson, 75
3d " 7 " J. Campbell, 50
Best bull, durham and devon, 5 yrs old, Samuel Hoppin, 1 00
2d " $\frac{3}{4}$ durham, 4 yrs old, J. P. Rosevelt. 50
Best 2 years old bull, durham and devon, J. W. Abbot, 1 09
2d " " " Chas. B. Sheldon, 50
Best 1 yr old bull, $\frac{1}{4}$ durham, A. Briggs, 1 00
" bull calf, U. T. Barnes, 1 00
2d " 3 months old, S. Hoppin, 50
Best pair 3 years old steers, Saml. Gilman, 1 00
2d " " W. N. Taylor, 50
Best fat cow, 4 yrs old, O. Sisson, 1 00
2d " " " 50
Best 3 year old heifer, S. W. Abbott, 1 00
2d " C. P. Sheldon, 50
Best 2 " P. Haydon, 1 00
2d " " 50
Best yearling heifer, Orrion Sisson, 1 00
" heifer calf, Henry Colman, 1 00

Judges on Cattle—B. A. Olney, J. A. Ranney, J. Gilman.

SHEEP & SWINE.

Best boar 16 months old, E. Barnum, 1 00
" sow 12 " " 1 00
2d " " A. S. Downing, 50

Best 6 pigs, A. S. Downing, ... $1 00
One fat hog, no competition, Jerry Palmer, ... 50
Best buck, French merino, D. Morris, ... 1 00
2d " ½ " S. N. Gantt, ... 50
Best 3 ewes, H. Barnum, ... 1 00
3d " E. Barnum, ... 50
Best 3 buck lambs, D. Morris, ... 1 00
2d " " E. Barnum, ... 50

Judges on Sheep and Swine—Hiram Mather, D. Morris, Eben Smith.

FRUIT & VEGETABLES.

Best box cultivated apples assorted, P. Haydon, ... 50
2d largest single apple, D. Morris, ... 25
Largest and best basket, natural, D. A. Alexander, ... 50
Best peck, natural, E. T. Spencer, ... 25
Four cling stone peaches, cultivated, N. P. Conger, ... 50
7 large " " A. V. Pantlind, ... 25
Best bunch onions, J. Lyle, ... 50
2d " J. R. Monroe, ... 25
Best 3 pumpkins, B. Van Sickles, ... 50
2d ½ doz. " J. R. Monroe, ... 25
Best squashes, not numbered.
2d best, 2 large, C. H. Mitthofer, ... 25
Best beets, bazonna, no competition, S. H. Blackman, ... 25
" carrots, C. H. Mitthofer, ... 25
" 8 quinces, F. C. Annable, ... 25

Judges—Geo. Smith, S. Godfrey, Wm. K. Butler.

MANUFACTURES.

Best single harness, N. M. Fowler, ... 75
2d " Bird & Ocobock, ... 50
One trunk valise, good workmanship but no competition, Bird & Ocobock, ... 25
One rocking chair, Godfrey, Giles & Co., ... 25
Six sides upper leather, 4 sides harness leather, M. Hannahs, 1 00
One cook stove, 1 plate stove, 1 reaction water wheel and mully saw gearing, Kinne, Hawley & Co., ... 1 00

One bunch shaved shingles, C. H. McArthur, $ 25

Viewing committee—J. Bardwell, E. Knowles, E. A. Thompson.

REPORT OF THE LADIES.

Best cheese, T. Conklin,	1 00
2d " Mrs. A. Warner,	50
Best butter, Mrs. B. Hall,	1 00
2d " John Campbell,	50
Best honey, A. F. Moon,	1 00
2d " B. F. Chadwick,	50
Best rag carpet, J. Smolk, jr.,	50
2d " H. McNeil,	50
3d " Mrs. Mills,	25
Best quilt, Mrs. N. Rice,	50
2d " Mrs. H. Barnum,	25
3d " Mrs. Downing,	25
Two quilts pieced by the Misses Conger, at 7 and 9 years of age, 12½ each,	25
Best fancy work stand, Mrs. A. Warner,	50
2d " " Mrs. Dr. Andrews,	25
Best coverlid, Mrs. J. Crane,	50
2d " Mrs. E. Barnum,	25
Best lamp mat, Mrs. E. Mears,	25
2d " Mrs. A. Warner,	12
Fancy needle work, Mrs. J. H. Simmons,	25
" " Mrs. Arch. Stewart,	25
" " Madames Durkee and Darling,	25
Card basket, Miss I. P. Simmons,	25
Beautiful specimens of silk made in the county, (raw and sewing silk,) H. Dowd,	50
Drawing, scene on the Hudson, Miss H. M. Stewart,	25
Book mark, Miss Godfrey,	12
" Miss J. Butler, 10 years old,	25
Artificial basket and fruit, Mrs. N. Rice,	50

The committee were Mrs. J. Crane, Mrs. A. Warner, Mrs. Mather and Mrs. Downing; report signed Mrs. E. Crane, Ch'n.

REPORT OF THE COMMITTEE ON GRAIN, PLOWS, &C.

The committee on wheat, corn, plows and plowing, beg leave to report that the "Lockman Plow" we think is the best calculated to meet the wants of farmers in all kinds of plowing. It combines the two qualities of sward breaking and fallow plowing in an eminent degree, and would supercede the necessity of purchasing a plow for each kind of plowing, and would award the first premium to this plow.

The "Starbuck plow" would not compete with the "Lockman plow" in turning a heavy sod, in the opinion of the committee, although it might equal the other in all kinds of plowing, we would award the second premium to this plow.

Best plow, Stewart, Mason & Co., ... 50
2d " Kinne, Hawley & Co., ... 25
Best sample of corn, D. Abbott, no premium.
2d " " H. Dowd.
Best acre of corn, 144 bus. ears, U. T. Barns, ... 1 00
2d " 85¾ " H. Mather, ... 50
Best wheat, (Soul's) E. T. Spencer, ... 1 00
2d " (Hutchinson) E. T. Spencer, ... 50

Report not signed—Judges on wheat, plows, &c., F. C. Annable, J. Woodman, M. L. Fitch.

FLOUR.

The committee on flour have examined one barrel, marked "Paw Paw Mills, Extra Superfine flour, A. Sherwood & Co., 196 lbs.," and found it to be of good quality, and feel bound to award the first premium to this barrel, as there is no other presented in competition.

Committee—J. Reynolds, I. W. Willard.

Saturday, Nov. 18th, the executive committee met, and awarded premiums on the following articles:

Specimens of printing, three blank books, and a variety of show bills, cards, &c., S. T. Conway, ... $ 50
One bbl. flour, J. Palmer, ... 50
" piece of sheeps-grey cloth, H. Mather, ... 50
" spinning wheel and 1 reel, J. Davis, ... 50
Pair of turkeys, white, C. Mitthofer, ... 25
Eighteen Poland hens, Godfrey, ... 25

The committee then voted to have the transactions of the society published in pamphlet form, and appointed W. H. Harrison and E.

Barnum a committee to make arrangements with T. S. Conway, for publishing the same—after which, the committee adjourned.

Cash in the Treasury, received from members,..........$124 00

Cash appropriated for premiums,.................... 62 25

With the funds remaining, after paying premiums, and with the amounts raised from the county, together with what we shall raise from members, and persons that may become members, we shall probably be able to offer liberal premiums another year.

STATISTICAL REPORT OF VAN BUREN COUNTY FOR THE YEAR, 1851.

BY W. H. HARRISON.

Van Buren county, though smaller in population than many other counties in the State, is not surpassed by any in the fertility of its soil and its adaptation to all kinds of grains, grasses, fruits and vegetables that flourish in the same latitude. While the advantages of its climate in regard to health, and particularly for the cultivation of fruit, are superior to many other counties in Michigan and Wisconsin. Its nearness to Lake Michigan abates the cold of winter and the heat of summer, and hard frosts are less common than would be otherwise. Extending to Lake Michigan with a convenient site for a harbor at the mouth of Black river, it possesses commercial advantages superior to many of the interior counties in the State, while the Paw Paw and Black rivers, with their numerous branches, afford many excellent situations for mills and other manufacturing establishments.

TIMBER.

Pine exists in considerable quantities in the northern parts of the county and is manufactured into lumber and shingles. Oak of several kinds abounds in all parts of the county, and is made into staves, wagons, &c. Hemlock exists in the northern parts of the county; the bark is used in the manufacture of leather. Whitewood is plenty and made into lumber. Cherry and black walnut exist and are used for furniture. Sugar maple is plenty. Soft maple, white and black ash, hickory, butternut, bass-wood, beech, elm and several other kinds of timber exist in considerable quantities.

Beds of iron ore, marl, and superior clay for the manufacture of brick, exist in several places.

ROADS, &C.

The Michigan Central Railroad runs a distance of twenty miles through the southern part of the county, having depots at Mattewan Paw Paw station and Decatur.

The Paw Paw river is navigable for boats of twenty tons burthen from Paw Paw to St. Joseph, running a distance of twenty miles through the county.

PLANK ROADS.

Charters for plank roads exist from Paw Paw to Paw Paw Station, 5 miles, completed.

From Paw Paw to Lawrence, 9 miles, in course of construction.
" " Allegan, 25 "
" " Schoolcraft, 18 miles.
" Lawrence to Breedsville, 10 "
" " St. Joseph, 28 "
" " Decatur, 10 "
" Breedsville to South Haven, 10 miles.
" " to Kalamazoo, 28 "
" Decatur to St. Joseph, 28 miles.

Besides these, good wagon roads exist throughout the county.

VILLAGES.

Paw Paw, the county seat, is a flourishing village of about 800 inhabitants. It contains a court house, 3 churches, (Methodist, Presbyterian and Congregationalist,) 8 dry goods stores, 1 hardware, 2 drug, 1 book store, 2 flouring and grist mills, 2 saw mills, 2 distilleries, 1 brewery, 2 foundries and machine shops, 4 blacksmith shops, 1 tannery, 1 wool carding and cloth dressing establishment, and 1 printing office, besides several groceries and other shops.

Decatur, Lawrence and Breedville contain thirty or forty houses each.

The other villages are South Haven, Jericho, Mattewan, Paw Paw Station and Pine Grove.

POPULATION.

The principal settlers are mostly from the State of New York; a few from the New England States and other parts. The number of inhabitants were in 1840, 1910; in 1845, 3743; in 1850, 5804.

Table showing the number of acres of Government, Non-Resident and Resident Lands in the several Townships, compiled from the Land Office and Census Statistics, with an estimate for the year 1851.

TOWNS.	Towns South.	Range West.	U. S. lands, 1848.	Occupied Farms, '50. Whole No.	Acres impr'd.	Acres unimproved.	Estimated non-resident lands, 1851.	Description of timber and soil.
Almena, ...	2	13	3080	31	2051	2497	15000	⅓ openings and ⅔ timbered lands.
Antwerp, ...	3	13	400	68	4330	3781	14000	Mostly oak openings.
Arlington, ...	2	15	2280	19	492	1530	18000	Timbered lands.
Bloomingdale . .	1	14	8640	6	206	254	13000	Heavy timbered land; some pine.
Columbia, . . . }	1	15	9080	16	562	1360	28000	{ Heavy timbered land with con-
}	1	16	6400	. . . .	. . . .	. . . .	. . . . .	siderable pine and hemlock.
Decator	4	14	6840	18	866	1722	13000	½ oak openings, ½ timbered.
Hamilton, ...	4	15	3240	38	2077	3049	14000	⅔ oak openings, rest timber.
Hartford, ...	3	16	1600	31	1098	2522	17000	⅓ oak openings, ⅔ timber.
Keeler, ...	4	16	1160	52	2531	4737	14000	⅔ oak openings, rest timber.
La Fayette, ...	3	14	1942	84	2751	4758	13000	½ oak openings, ½ timber.
Lawrence, ...	3	15	None.	41	1130	3498	18000	Mostly heavy timber.
Pine Grove, ...	1	13	5840	6	150	200	16000	Pine, oak openings, heavy timber.
Porter, ...	4	13	2120	52	2046	3917	14000	⅔ oak openings, rest timber.
}	2	16	7520	. . . .	. . . .	. . . .	. . . . .	{
South Haven, }	1	17	320	17	529	1609	33000	Heavy timbered with pine, hem-
}	2	17	12320	. . . .	. . . .	. . . .	. . . . .	lock, sugar maple; good land.
}	2	18	640	. . . .	. . . .	. . . .	. . . . .	}
Waverly, ...	2	14	5640	15	324	908	16000	Heavy timb'ed, some pine, sugar m.
			79,060	494	21,167	36,342	256000	

The whole number of acres of land in the county is probably about 400,000. Of this number there are probably about 25,000 acres of improved land or about one sixteenth of the whole, with 50,000 acres of unimproved land divided into about five hundred farms, owned by residents, making 75,000 acres. The United States lands are compiled from a list from the United States land office at Kalamazoo of 1848. Since that time little has been sold. There are probably about 75,000 acres of United States and school lands unsold; these with the lands owned by resident farms, taken from the 400,000 acres, leaves 250,000 acres owned by non-resident land holders or speculators. These lands are offered for sale at prices varying from 2 to 5 dollars per acre.

STATISTICS OF VAN BUREN COUNTY FOR 1851.

TOWNS.	Census statistics of 1850.		Manuf'ing statistics, 1851.									From the statistics of 1850. Live stock, June 1, 1850.					
	Dwelling houses.	Population.	Saw mills Water.	Saw mills Steam.	Grist mills.	Distilleries.	Asheries.	Wool carding.	Tanneries.	Foundries.	Brick yards.	Horses.	Milch cows	Working oxen.	Other cattle.	Swine.	Sheep.
Almena,	68	420	3					1				51	82	41	117	218	534
Antwerp,	140	614	2								1	105	174	98	247	437	886
Arlington,	50	240										8	49	47	63	167	174
Bloomingdale	28	160										8	25			47	32
Columbia,	48	265	2	1					3		1	24	45	24	63	105	118
Decator,	68	386	1	1		1						30	52	17	85	191	343
Hamilton,	63	370	1									45	92	79	138	348	958
Hartford,	58	296										28	75	46	134	247	418
Keeler,	92	486										59	124	110	157	361	616
La Fayette,	213	1145	2	1	2	2	1	1	1	2	1	72	168	98	162	425	754
Lawrence,	92	510	3		1		1					33	98	70	129	274	615
Pine Grove,	12	62		2								Not	returned.				
Porter,	81	444									1	62	114	74	156	302	560
South Haven,	38	220	1	1								8	48	30	67	111	169
Waverly,	34	186										17	26	16	44	84	83
			15	6	3	3	2	2	4	2	4	550	1,172	750	1502	3317	6260

The manufacture of flour, lumber, whiskey, leather, boots and shoes, wagons, plows, stoves, machinery castings, is carried on to considerable extent.

PRODUCE

During the year ending June 1st, 1850, Compiled from the Census Statistics.

TOWNSHIPS.	Wheat, bushels of.	Corn, bushels of.	Oats, bushels of.	Potatoes, bus'ls of.	Wool, pounds of.	Value of orchard products.	Butter, pounds of.	Cheese, pounds of.	Hay, tons of.	Buckwheat, bushels of.	Clover seed, bushels of.	Rye, bushels of.	Barley, bushels of.	Maple sugar, lbs. of.
Almena,	5870	10790	2087	4330	1318	$123	6312	4370	288	175	....	593	325	880
Antwerp,	13857	29455	2570	10780	2473	578	10011	1490	529	437	1	467	140	3450
Arlington,	1335	3750	1475	1605	389	175	2910	......	93	127	....	....	....	6050
Bloomingdale	865	975	260	625	56	......	1600	......	31	......	....	....	10	880
Columbia,	1146	3240	985	2375	332	225	3500	880	175	128	....	....	....	4786
Decatur,	4653	7600	3675	1620	972	165	3220	300	85	72	....	....	50	830
Hamilton,	5079	11285	4890	4800	2179	285	5745	150	331	657	....	....	10	500
Hartford,	5301	7215	1831	2909	1105	265	4890	400	232	111	....	....	40	12925
Keeler	10359	16195	4869	4176	1569	153	8075	845	373	40	....	36	254	1148
Lawrence,	5028	7815	2620	3370	1565	292	6640	230	314	408	....	....	....	8690
La Fayette	11910	17805	3136	6732	1509	280	16556	800	622	90	....	340	20	6086
Pine Grove,*	....	....	....	....	....	......	......	......	....	......	....	....	....	....
Porter	6984	11420	4070	3768	1725	192	8745	400	301	238	....	....	120	7650
South Haven,	1438	2935	1605	1287	583	208	3150	120	107	......	....	....	....	7180
Waverly,	1290	2300	645	1664	269	......	2350	100	86	......	....	....	....	2820
Total,	75,115	132,780	34718	50,041	15,994	$2,940	83,704	10,085	3567	2483	1	1436	969	63875

*Agricultural Products not reported.

By this, and the preceding table, it appears that Antwerp produces the greatest number of horses, cattle, (except working oxen,) swine, bushels of wheat, corn, potatoes, and fruit, and pounds of wool. Keeler, the greatest number of working oxen and pounds of butter. Hamilton, the greatest number of sheep, bushels of oats, and buck-wheat. Almena, the greatest number of bushels of rye, barley, and pounds of cheese. Hartford, the greatest number of pounds of maple sugar. And Lafayette, the greatest number of tons of hay.

Comparative Table, exhibiting the difference of statistics, between the years 1840 *and* 1850.

Years.	No. of Inhabitants.	No. of horses.	No. cattle.	No. sheep.	No. swine.	Bushels of wheat.	Bushels of corn,	Bushels of oats.	Bus. rye.	Bushels of potatoes.	Bushels of Barley.	Pounds of wool.	Tons hay.
1840	1910	240	2125	538	3422	15640	23587	16176	120	20832	835	900	1342
1850	5804	550	3484	6260	3317	75115	132780	34718	1436	50041	969	15994	3567

It will be seen by this table that our population has gained over two hundred per cent, or more than trebled in ten years. In wool we have gained 1677 per cent; in sheep, 1063 per cent; in wheat, 379 per cent; in corn, 364 per cent; in horses, 129 per cent; in cattle, 64 per cent; in swine, we have lost 3 per cent; in hay, oats and potatoes, we have more than doubled.

APOLOGY FROM THE EXECUTIVE COMMITTEE.

The committee feel bound to offer the following apologies to the society for not making more extensive preparations for the fair: 1st. They were wholly unacquainted with the business, not knowing what preparations were necessary; 2d. They had little or no aid from the citizens of Paw Paw; and 3d. The animals and articles for exhibition exceeded four fold the expectations of your committee.

WM. H. HARRISON,
Sec'y Van Buren Co. Ag. Society.

TRANSACTIONS

OF THE

STATE AGRICULTURAL SOCIETY,

WITH REPORTS OF

COUNTY AGRICULTURAL SOCIETIES,

FOR THE YEAR 1851.

PUBLISHED BY ORDER OF THE LEGISLATURE.

VOL. III.

J. C. HOLMES,
SECRETARY MICHIGAN STATE AGRICULTURAL SOCIETY.

LANSING:
INGALS, HEDGES & CO., PRINTERS TO THE STATE.

1852.

www.ingramcontent.com/pod-product-compliance
Lightning Source LLC
LaVergne TN
LVHW020912110826
845150LV00004B/648

9781425556938